SYSTEM INNOVATION FOR A TROUBLED WORLD

System Innovation for a Troubled World: Applied System Innovation VIII contains roughly one quarter of the total of 250 contributions from 8 different countries that were presented at the IEEE 8th International Conference on Applied System Innovation (ICASI 2022, Sun Moon Lake, Nantou, Taiwan, 21–23 April 2022). This book aims to provide an integrated communication platform for researchers from a wide range of topics including:

- Information Technology
- Communication Science
- Applied Mathematics
- Computer Science
- Advanced Material Science
- Engineering

System Innovation for a Troubled World: Applied System Innovation VIII aims at science and engineering technologists, academics and professionals from the above mentioned areas.

Smart Science, Design and Technology

ISSN: 2640-5504
eISSN: 2640-5512

Book Series Editors

Stephen D. Prior
Faculty of Engineering and Physical Sciences
University of Southampton
Southampton, UK

Siu-Tsen Shen
Department of Multimedia Design
National Formosa University
Taiwan, R.O.C.

VOLUME 5

PROCEEDINGS OF THE IEEE 8TH INTERNATIONAL CONFERENCE ON APPLIED SYSTEM INNOVATION (ICASI 2022), APRIL 21–23, 2022, SUN MOON LAKE, NANTOU, TAIWAN

System Innovation for a Troubled World

Applied System Innovation VIII

Editors

Artde Donald Kin-Tak Lam
Fujian University of Technology, China

Stephen D. Prior
University of Southampton, United Kingdom

Siu-Tsen Shen
National Formosa University, Taiwan

Sheng-Joue Young
National United University, Taiwan

Liang-Wen Ji
National Formosa University, Taiwan

CRC Press
Taylor & Francis Group
Boca Raton London New York Leiden

CRC Press is an imprint of the
Taylor & Francis Group, an **informa** business

A BALKEMA BOOK

CRC Press/Balkema is an imprint of the Taylor & Francis Group, an informa business

Typeset by MPS Limited, Chennai, India

Library of Congress Cataloging-in-Publication Data
A catalog record has been requested for this book

First published 2023
Published by: CRC Press/Balkema
4 Park Square, Milton Park, Abingdon, Oxon, OX14 4RN
e-mail: enquiries@taylorandfrancis.com
www.routledge.com – www.taylorandfrancis.com

ISBN: 978-1-032-45525-9 (hbk)
ISBN: 978-1-032-45527-3 (pbk)
ISBN: 978-1-003-37739-9 (ebk)
DOI: 10.1201/9781003377399

System Innovation for a Troubled World – Tak Lam et al. (Eds)
© 2023 The Editor(s), ISBN: 978-1-032-45525-9

Table of contents

System Innovation for a Troubled World – Tak Lam et al. (Eds)
© 2023 The Editor(s), ISBN: 978-1-032-45525-9

Preface

We have great pleasure in presenting this conference proceeding for technology applications in engineering science and mechanics from the selected articles of the International Conference on Applied System Innovation (ICASI 2022), organized by the International (Taiwanese) Institute of Knowledge Innovation and the IEEE, held in Sun Moon Lake, Nantou, Taiwan April 21–23, 2022.

The ICASI 2022 conference was a forum that brought together users, manufacturers, designers, and researchers involved in the structures or structural components manufactured using smart science. The forum provided an opportunity for exchange of the research and insights from scientists and scholars thereby promoting research, development and use of computational science and materials. The conference theme for ICASI 2022 was "System Innovation for a Troubled World" and tried to explore the important role of innovation in the development of the technology applications, including articles dealing with design, research, and development studies, experimental investigations, theoretical analysis and fabrication techniques relevant to the application of technology in various assemblies, ranging from individual to components to complete structure were presented at the conference. The major themes on technology included Material Science & Engineering, Communication Science & Engineering, Computer Science & Engineering, Electrical & Electronic Engineering, Mechanical & Automation Engineering, Architecture Engineering, IOT Technology, and Innovation Design. About 150 participants, representing 8 countries came together for the 2022 conference and made it a highly successful event. We would like to thank all those who directly or indirectly contributed to the organization of the conference.

Selected articles presented at the ICASI 2022 conference will be published as a series of special issues in various journals. In this conference proceeding we have some selected articles from various themes. A committee consisting of experts from leading academic institutions, laboratories, and industrial research centres was formed to shortlist and review the articles. The articles in this conference proceedings have been peer reviewed to maintain the high standard. We are extremely happy to bring out this conference proceeding and dedicate it to all those who have made their best efforts to contribute to this publication.

Professor Siu-Tsen Shen & Dr Stephen D. Prior

System Innovation for a Troubled World – Tak Lam et al. (Eds)
© 2023 The Editor(s), ISBN: 978-1-032-45525-9

Editorial Board

System Innovation for a Troubled World – Tak Lam et al. (Eds)
© 2023 The Author(s), ISBN: 978-1-032-45525-9

Using embedding learning to improve the accuracy of binary sentiment analysis on few-shot text data

Xian-Zhong Lin*, Wei-Chun Wang & Kuen-Fang Jea
Department of Computer Science and Engineering, National Chung-Hsing University, Taichung, Taiwan

ABSTRACT: Sentiment analysis of text data is an important task in natural language processing. Taking the product review system as an example, proper use of sentiment analysis can reveal the opinions of consumers, and thus adjustments can be made to products. However, most current sentiment analysis methods assume that there is a large amount of training data available for the classification model. The classification accuracy is not ideal when little data are available for analysis. Aiming at sentiment analysis of text data with few data samples, in this paper, we propose an embedded learning model, which combines convolutional neural network and bidirectional long short-term memory learning models to improve the relevance of data by dimensionality reduction and thus strengthen the classification accuracy of few-shot learning. In addition, we segment the training data and use them to train the proposed model in batches to avoid the overfitting problem that is often observed in few-shot learning models. According to the experimental results shown in this study, the accuracy of our model is about 11% on average, which is higher than that of the traditional classification methods.

1 INTRODUCTION

In recent years, with the vigorous development of science and technology, the amount of data on the Internet has grown exponentially. Research in the field of data science is also booming. One of the most important tasks in this field is text classification, which is a classic problem in natural language processing. The goal of text classification is to classify textual data such as sentences, paragraphs, or documents by attaching tags. Text classification is widely used, including spam detection, sentiment analysis, and news classification (S. Minaee et al. 2021). Sentiment analysis of the text is to classify articles by analyzing the emotions exhibited by people in the text, such as complex emotions (e.g., happiness, anger, sadness, or simple positive) and negative emotions, for binary classification. This research mainly focuses on the binary sentiment analysis problem.

Previously, most of the related studies on sentiment analysis was based on having a large amount of data as the starting point for the training of classification models. However, sentiment analysis is not the only case where data volumes are overwhelming. For example, when a new product or service is presented, the ability to understand consumers' likes and dislikes by analyzing its few reviews at the beginning of its life cycle and dealing with it is often the key to its success. Consequently, traditional sentiment analysis methods will face the problem of insufficient comment data, resulting in poor training results, reduced accuracy, and misjudgment. This paper proposes a training model that can improve the accuracy of sentiment analysis problems with a small amount of data.

Previously, most of the related studies on improving the accuracy of few-shot classification models used prior knowledge (i.e., background knowledge in related fields) to strengthen the classification effect of the model. According to the survey of Y. Wang et al., the three research directions for the utilization of prior knowledge are as follows. The first pone is for data, such as obtaining references from other related fields to make up for the lack of original data. The second one is to adjust the training model to reduce the range of finding the best classification method when the classification model is trained, to improve the classification accuracy with few samples. The third one is to improve the algorithm so that the classification model can more efficiently find the correct classification method. For example, the training experience in related fields is used to make the classification model start training from a place closer to the optimal classification method during training, which further improves the classification accuracy of the few-sample situation. Among the three, this study chooses the second direction to adjust the training model, because there remains sufficient research space and variability in adjusting the model.

To address the problem of binary sentiment analysis of text data when samples are few, this paper proposes an embedded learning model with dimensionality reduction as the main axis. Using a training model combining convolutional neural network (CNN)

*Corresponding Author: jameslin@cs.nchu.edu.tw

DOI 10.1201/9781003377399-1

(Y. Lecun et al. 1998) and bidirectional long short-term memory (BiLSTM) (M. Schuster et al. 1997) to reduce the dimensionality of text data makes the correlation between data more clear and makes it easier to learn classification results with high accuracy in the context of a small number of samples. In addition, to further improve the classification accuracy and avoid the overfitting problem, the training model refers to the concept of meta-learning to further cut the training data so that the training situation is closer to the testing situation. This study uses review data from the IMDb movie review webpage to conduct experiments to verify the effect of the proposed model. The experimental results show that the proposed embedded learning model has higher accuracy than other traditional classification models that require a large amount of data when the number of sample data is small. Compared with classification models that are also similar to neural networks, the overfitting problem is greatly reduced.

2 RELATED WORK

There are four ways to adjust the training model to improve the effectiveness of few-shot learning: multitask learning, embedded learning, learning with external memory, and generative modeling. Embedded learning uses an embedding function to embed the training and test datasets into lower dimensions, respectively. This makes similar data closer to each other and dissimilar data easier to distinguish, thus constructing a smaller hypothesis space to improve the accuracy of few-shot learning. The training the embedding function is the focus of embedded learning.

M. Fink et al. 2004 originally proposed the embedded learning method that only used the kernel to train the embedded function. Most recent embedding function training uses more complex neural-like networks. O. Vinyals et al. used meta-learning to train different embedding functions for the training and test datasets, respectively. Subsequently, H. Altae-Tran et al. further improved and applied it in the field of drug testing. P. Bachman et al. added the mechanism of active learning to the method of (O. Vinyals et al. 2016) to enhance the training effect. J. Choi et al. expanded the embedding function of (O. Vinyals et al. 2016) from embedding a single data to embedding a set at a time. J. Snell et al. proposed another special type of embedded learning. They calculated all the training data after embedding to obtain a prototype, and the test dataset was compared with the prototype for classification. M. Ren et al. added semi-supervised learning to the former method to exploit unlabeled data. The advantage of this type of method is that the use of embedding technology can effectively improve the correlation between the data, making more effective use of the dataset with few samples. However, the embedding process will lead to data loss. If the training result of the embedding function is poor, the classification accuracy of the subsequent classification model will easily be reduced.

3 PROPOSED METHOD

Sentiment analysis is a quite popular project in text classification work. Its main purpose is to analyze the opinions expressed by people in textual data and extract the polarity or viewpoint to classify them (S. Minaee et al. 2021). Textual data refer to text-based data, such as product reviews, movie reviews, and tweets. Binary sentiment analysis, which is the focus of this study, is a sentiment analysis task that categorizes textual data as positive or negative according to opinion polarity. The problem to be solved in this study is as follows: How to improve the accuracy of binary sentiment analysis in small-sample text data? Here, the accuracy is defined as

$$\text{Accuracy} = (TP + TN)/\text{T}.$$

Here, T is the total number of test data; TP and TN are the true positive and true negative parameters in the confusion matrix, respectively; and $TP + TN$ indicates the correct number of data for the model classification. This study focuses on binary sentiment analysis in the case of a few samples and defines the few samples as follows: the data in the training dataset are less than 500, and those in the test dataset are less than 100.

To exhibit a good learning effect even with a small amount of data, this study uses embedded learning to make up for the deficiency of the insufficient amount of data. The embedded learning technology divides the original high-dimensional dataset into training and test sets and conducts training and testing, respectively. After finding the best embedding function, the data generated by the embedding function are then embedded into a lower-dimensional dataset for similarity comparison. The role of the embedding function here is to reduce the dimension of the data. Embedded learning technology has three important functions: embedding the training set into a lower-dimensional embedding function g, embedding the test set into a lower-dimensional embedding function f, and the similarity function s that compares the similarity of the data after embedding. The architecture of embedded learning is shown in Figure 1.

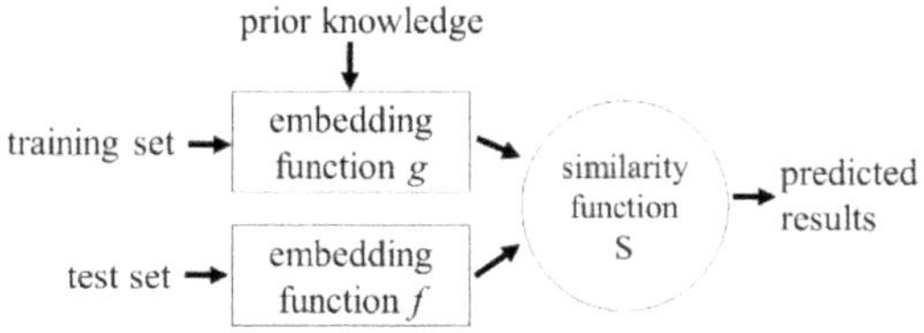

Figure 1. Embedded learning architecture (Y. Wang et al. 2020).

For the embedding function of the training and test sets, this study adopts the same training method. First, the skip-gram model is used to convert the text data into a form of vectors. It is then trained by a model combining CNN and BiLSTM. The similarity function used in this study adopts the cosine similarity function to compare the training and test sets after embedding and predicts the classification of the test set. Finally, the accuracy can be calculated based on the predicted results to evaluate the experimental results of this study. In addition, this study hopes to make the training situation match the testing situation as much as possible to improve the accuracy of the model. Therefore, the meta-learning model is used for reference, and all the training data will not be placed into the training at once. Instead, the training process is divided into multiple episodes, and each episode will only use a small amount of data randomly selected from all training data as the training set during training.

The deep learning model of this study is used to express the classification prediction method as shown by the formula:

$$\hat{y} = \sum_{i=1}^{k} a(\hat{x}, x_i)y_i$$

where $\hat{y}$ represents the prediction result, $\hat{x}$ is the test data to be classified, k represents the number of training samples selected for each episode, x_i and y_i are a set of samples in the training set-category pairing, and a $(\hat{x}, x_i)$ is the attention mechanism in the model. Using the attention mechanism, the model can identify xi that is more important (i.e., closer) to $\hat{x}$ and improve the classification accuracy.

To effectively train text data, the Word2Vec model should be used to convert the words in the text into low-dimensional vectors. This process is also known as embedding, which fits well with the model that this study hopes to improve the effect of few-shot learning through dimensionality reduction. The skip-gram used in this study is most widely used in the Word2Vec model. The feature of the skip-gram model is to generate the corresponding word vector according to the input word and predict the probability of other words appearing near the word. To improve the efficiency of the skip-gram model, this study uses a negative sampling technique during training. Different from general neural network training, many operations must be performed to update all neuron weights for each training sample. Negative sampling only picks a few words that rarely appear near the selected word (i.e., negative cases) to update the weights. Hence, the computational load can be reduced, and the training efficiency can be greatly improved.

After converting the training set into vectors, the next step in this research method is to feed these vectors into the CNN model in the form of matrices for training. The training purpose of the CNN model is to learn to recognize the features in the space. When using CNN in the classification of text data, CNN can still identify the keywords in the text (such as "I like" emotional fragments), thereby improving the accuracy of classification. The composition architecture of CNN can be roughly divided into three layers: convolution, pooling, and fully connected layers. After multiple convolutional and pooling layers, the features are extracted, and then the classification is judged through the fully connected layer. However, in this study, the extracted features do not pass through the fully connected layer, but are input into the BiLSTM model as sequence data for encoding and further embedded in low dimensions. In this study, BiLSTM is combined with the CNN model to perform deep learning of the embedding function and train an embedding function that can obtain higher classification accuracy. The embedded function architecture is shown in Figure 2.

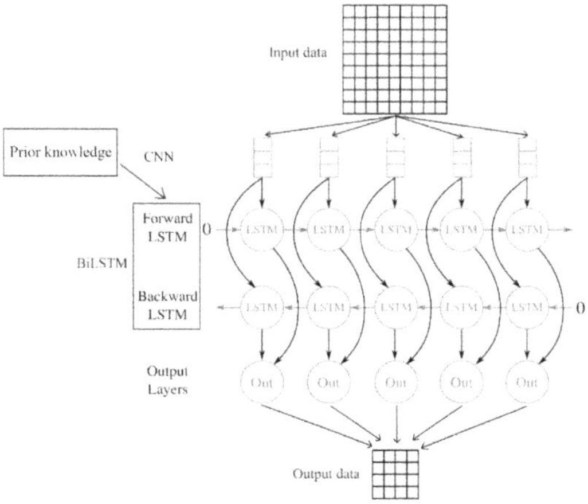

Figure 2. Embedding function architecture.

4 EXPERIMENT RESULTS

The experimental data used in this study are the dataset of the IMDb online movie database established by A. Maas et al. This dataset categorizes each movie's reviews into positive and negative categories; each category contains 1000 text files. The name of the file represents the number and source HTML address of the movie review. The content of the file is the original film review. The polarities of movie reviews are classified based on the ratings of the original materials. On the original 0–10 rating, reviews with a score greater than or equal to 7 are classified as positive, whereas reviews less than or equal to 4 are classified as negative reviews. Comments with intermediate scores are discarded. Although the IMDb dataset is not a few-sample dataset, this study uses the IMDb dataset to only sample it and simulate the few-sample training dataset. The rich samples of the IMDb dataset itself are used as a test dataset to verify the classification accuracy of the model after training so that the accuracy obtained after testing is higher. During the experiment in this study, 250 movie reviews were randomly selected from positive and negative movie reviews as the training dataset, and 50 reviews were used as the test dataset.

The experimental results are exhibited in Figure 3, which show that, in the case of few samples, the classification accuracy of this study is the highest among all models. Compared with MLP, LSTM and

BiLSTM, which are also neural networks, they were outperformed by about 7%, 11% and 7%, respectively. Compared with other simpler machine learning classification models, naive Bayes and SVM, the accuracy of this study is 16% and 9% higher, respectively.

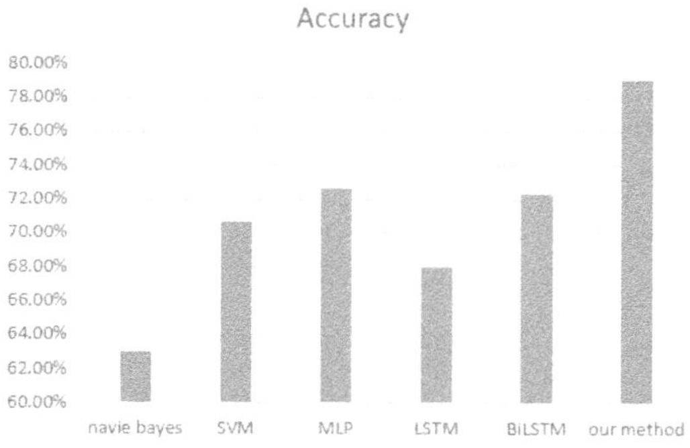

Figure 3. Accuracy comparison.

Figure 4 shows the average training accuracy and test accuracy of the method in this study in 20 epochs. It can be observed that, compared with other deep learning methods, the method in this study not only outperforms the accuracy rate, but also avoids the overfitting problem.

In this study, the embedded learning model combining CNN and BiLSTM can effectively utilize the dataset with a small number of samples and obtain a high accuracy rate. The training method in this study is to divide the training dataset for training in stages in each epoch so that the training situation matches the test situation as much as possible. Hence, the training model can be prevented from over-chasing the accuracy of a specific training dataset, thereby reducing the overfitting problem.

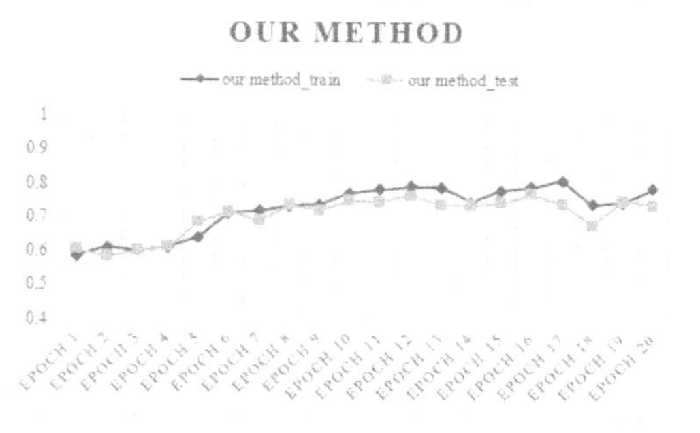

Figure 4. Average training accuracy and test accuracy of our method.

However, to achieve high accuracy, the price paid by the proposed method is that the model training time is the longest among all methods. Simple Bayesian classifiers, SVMs, and MLPs can all be trained within 30 s. The more complex LSTM takes about 2 min. However, the proposed method takes 505 s, which is nearly five times that of LSTM.

5 CONCLUSION

In this study, aiming at binary sentiment analysis of few-sample text data, an embedded learning method based on the concept of dimensionality reduction was proposed to improve the classification accuracy and alleviate the overfitting problem. This method combines CNN and BiLSTM models to embed text data into lower dimensions for comparison and classification prediction and borrows the meta-learning architecture to make the training situation closer to the test situation. Experimental results show that the method in this study outperforms other traditional classification models in terms of accuracy. Compared with the same neural network deep learning model, the overfitting problem is greatly reduced, and data can be used effectively in situations where the sample size is small. This method makes the use of sentiment analysis work more broad and not limited to situations where a large amount of data exist.

There are two future directions in this research. One is to adjust for the training of the embedding function. To utilize as limited data as possible, the current training model takes longer to train than traditional neural networks. As the size of the dataset increases, it may face the dilemma of the training time being too long and the accuracy improvement being not significant enough. Therefore, this research hopes to find a training model that simplifies the embedding function without unduly sacrificing accuracy. Another future research direction is to challenge more diverse sentiment classification tasks. Human emotions are not limited to positive and negative; in fact, they also contain more diverse categories of emotions, such as happiness, anger, and sadness. Classifying multiple emotions is more challenging than binary classification.

REFERENCES

H. Altae-Tran, B. Ramsundar, A. S. Pappu, and V. Pande, "Low Data Drug Discovery with One-Shot Learning," *ACS Central Science*, vol. 3, no. 4, pp. 283–293, 2017.

P. Bachman, A. Sordoni, and A. Trischler, "*Learning algorithms for active learning*," Proceedings of the 34th International Conference on Machine Learning vol. 70, pp. 301–310, 2017.

J. Choi, J. Krishnamurthy, A. Kembhavi, and A. Farhadi, "*Structured Set Matching Networks for One-Shot Part Labeling*," Proceedings of the IEEE/CVF Conference on Computer Vision and Pattern Recognition, pp. 3627–3636, 2018.

M. Fink, "*Object Classification from a Single Example Utilizing Class Relevance Metrics*," Proceedings of the 17th International Conference on Neural Information Processing Systems, pp. 449–456, 2004.

Y. Lecun, L. Bottou, Y. Bengio, and P. Haffner, "*Gradient-based Learning Applied to Document Recognition*," Proceedings of the IEEE, vol. 86, no. 11, pp. 2278–2324, 1998.

A. L. Maas, R. E. Daly, P. T. Pham, D. Huang, A. Y. Ng, and C. Potts, "*Learning Word Vectors for Sentiment Analysis*," Proceedings of the 49th Annual Meeting of

the Association for Computational Linguistics: Human Language Technologies, vol. 1, pp. 142–150, 2011.
S. Minaee, N. Kalchbrenner, E. Cambria, N. Nikzad, M. Chenaghlu, and J. Gao, "Deep Learning–based Text Classification: A Comprehensive Review," *ACM Comput. Surv.*, vol. 54, no. 3, 2021.
M. Ren et al., "*Meta-Learning for Semi-Supervised Few-Shot Classification*," presented at the 6th International Conference on Learning Representations, 2018.
J. Snell, K. Swersky, and Richard, "*Prototypical Networks for Few-shot Learning*," Proceedings of the Annual Conference on Neural Information Processing Systems, pp. 4077–4087, 2017.
M. Schuster and K. Paliwal, "*Bidirectional Recurrent Neural Networks*," IEEE Transactions on Signal Processing, vol. 45, pp. 2673–2681, 1997.
O. Vinyals, C. Blundell, T. Lillicrap, K. Kavukcuoglu, and D. Wierstra, "*Matching Networks for One Shot Learning*," Proceedings of the 30th International Conference on Neural Information Processing Systems, pp. 3630–3638, 2016.
Y. Wang, Q. Yao, J. T. Kwok, and L. M. Ni, "Generalizing from a Few Examples: A Survey on Few-shot Learning," *ACM Comput. Surv.*, vol. 53, no. 3, pp. 1–34, 2020.

System Innovation for a Troubled World – Tak Lam et al. (Eds)
© 2023 The Author(s), ISBN: 978-1-032-45525-9

Classifying and recognizing face types of thin-shell models for volume decomposition in CAE analysis

Pradiktio Putrayudanto & Jiing-Yih Lai*
Department of Mechanical Engineering, National Central University, Taiwan, China

Yao-Chen Tsai, Ming-Hsuan Wang & Chia-Hsiang Hsu
CoreTech System (Moldex3D) Co., Ltd., Hsinchu, Taiwan, China

ABSTRACT: Volume decomposition aims to split a computer-aided design (CAD) model into sub-volumes that can be meshed with existing hexahedral mesh generation algorithms. The possibility of making automatic decomposition has been investigated, but the challenge remains whenever a CAD model has many extrusion features. Therefore, we propose an approach for separating extrusion features from the main body of a thin-shell model by assigning different face types. In addition, a novel method for classifying thin-shell models into regular and irregular wall types is introduced. This study mainly focuses on the type of irregular wall and divides the face-type recognition algorithm into the following four steps: separation of inner and outer regions, classification of regular and irregular-wall types, further classification of irregular-wall types, and assigning face types on each sub-type of irregular walls. As a result, a set of attributes that are usable for decomposition derivation is available for each face on the model. To verify the feasibility of the proposed method, several realistic thin-shell models are employed to demonstrate the face types recognized.

In addition, the application of the face types obtained in volume decomposition is demonstrated.

1 INTRODUCTION

Thin-shell models are mostly produced in the segment of plastic manufacturing. Models with thin-shell have a wide range of geometrical characteristics. Adding extrusion features that are useful for reinforcement and rigidity addition, such as ribs or columns, escalates the complexity of a thin-shell model. Consequently, analysis using finite element for thin-shell model encounters many challenges, especially on volume decomposition process.

On the decomposition process, the ability of the user to analyze and divide the model into sub-volume becomes essential. The model is then converted into a solid mesh after the subdivision of volumes are complete. In order to get an accurate simulation result, hexagonal (hex) mesh are preferable comparing to tetragonal (tet) mesh. Several algorithms have already been investigated to generate hex meshes, such as sweeping and mapping. A good hex mesh can be generated with a sweep algorithm if the geometry and topology criteria of a sub-volume meets the requirement. Due to the highly complex geometry of thin-shell models, finding sub-volumes that meet the division criteria becomes complicated. Several attempts of decomposition-based approaches to create all-hex meshes have been initiated (Lu et al. 1999; Sheffer A 2000; Tautges et al. 1997). According to these studies, the capabilities of the algorithm still have low level of automation (Tautges TJ 2001). Hence, the simplification of the decomposition process is a necessity, since it can lead to extending various thin-shell models that could be decomposed automatically.

A new framework for identifying thin-shell models and assigning the face types to the surface is presented in this study. A thin-shell model is classified according to the topology and geometry criteria of the face types, called wall face types. For the regular-wall thin-shell model, the surfaces of wall face type are adjacent to each other and the normal vector ($\boldsymbol{N}_i$) is relatively perpendicular to the reference axis (as a default we use positive-z). Nonetheless, the irregular-wall thin-shell model does not meet both or one of the criteria. Models that have already been recognized as an irregular-wall thin-shell models will later be classified according to their sub-type criteria. The subtypes of the irregular-wall thin-shell model can be separated into three categories: "Partial Open", "Freeform", and "Simple". Face Identifier is introduced as well on the next section as a marker to identify the sub-type of irregular-wall model. Finally, every face type on inner and outer regions will be assigned according to sub-type of the irregular-wall model.

*Corresponding Author: jylai@ncu.edu.tw

 DOI 10.1201/9781003377399-2

2 STEP ALGORITHM OVERVIEW

Prior to recognizing the model types and determining the face types, several calculations need to be performed, which are called preliminary functions. These functions consist of: calculating edge and face attributes, determining hole and fillet features, and recognizing rib feature (Lai et al. 2018). Fillet feature, such as edge blended faces (EBF) and vertex blended faces (VBF), might be suppressed during the simulation process due to its low influential to the result, yet this feature always exists and has an impact on model classification. Figure 1 illustrates the example of regular and irregular-wall thin-shell model after preliminary functions. Here yellow color represents hole feature, while red and green color represent EBF and VBF, respectively.

Generally, for each regular and irregular-wall model, the face attribute on a part can be divided into: inner and outer regions. Inner region is a region where the faces can only be perceived from +Z (top of the model) of the bounding box, while outer region is visible to any side (+X, -X, +Y, -Y, -Z) of the bounding box, except +Z. Inner region can be assigned with four face types: transition, ridge, wall, bottom and protrusion. Similarly, outer region has the following four face types: flange, outer wall, outer bottom and outer protrusion. The transition and flange face types represent the boundary between inner and outer regions. Due to the differences in topology and geometry conditions for each subtype of irregular-wall, the list of face types is determined by those subtypes.

(a) (b)

Figure 1. Results of preliminary functions for: (a) regular, and (b) irregular wall thin-shell CAD models.

2.1 *Separation of Inner and Outer Regions*

Initially, all faces on a thin-shell model are divided into three categories based on their normal vector-to-reference axis angle (θ_f), and some specific topological requirements. The categories from this initial step are listed as follows: candidate outer region face (c_{or}), candidate inner region face (c_{ir}) and candidate wall face (c_w). The last category is used for any face which cannot be determined yet as c_{or} or c_{ir}.

In this step, faces that are assigned previously as a hole feature is not counted. All the other faces in the model are calculated. Assume a face f_i is considered. Assign f_i as c_{or} if $\theta_f > 170°$ and no intersection between a line (L_F) and any other face of the model, where L_F is created from the centroid of f_i parallel to N_i. Otherwise, f_i is not added into any category. Then the calculation is repeated for all faces except the c_{or}. A face will be recognized as c_{ir} if θ_f meets the following conditions: $0° \leq \theta_f < 90°$ and the neighbor face of f_i is not c_{or}. If the criteria of θ_f is satisfied but it is neighboring to c_{or}, then assign f_i as c_{or}. Otherwise, collect those faces for further calculation. At the end, all faces in the model will be categorized as c_{or}, c_{ir} and c_w.

After faces are already assigned with those categories, those faces are grouped according to category types and adjacency relationship. All c_{ir} faces that are adjacent to each other are regarded as $\boldsymbol{G}_{ci}$, while the others type using the same concept will be $\boldsymbol{G}_{co}$ for c_{or} and $\boldsymbol{G}_{cw}$ for c_w. $\boldsymbol{G}_{ci}$ is checked by identifying any intersections (x) among the members by generate L_F on each $\boldsymbol{G}_{ci}$ face. If $x < 2$, then convert all of faces in $\boldsymbol{G}_{ci}$ to be $\boldsymbol{G}_{cw}$.

To clearly separate inner and outer regions, a boundary as a set of edges is employed. These edges are called transition edges (e_{tr}). Calculation of transition edges is started by considering any edge (e_{ci}) of the faces in $\boldsymbol{G}_{ci}$ with the following conditions: (a) if convex e_{ci} is an edge of c_{or} or c_w, then e_{ci} is regarded as a candidate transition edge, (b) if face on $\boldsymbol{G}_{ci}$ is a fillet and the e_{ci} is adjacent to c_w, then e_{ci} is considered as a candidate transition edge. Figure 2 illustrates transition edges e_{tr} on both conditions, where e_{tr} is colored by green.

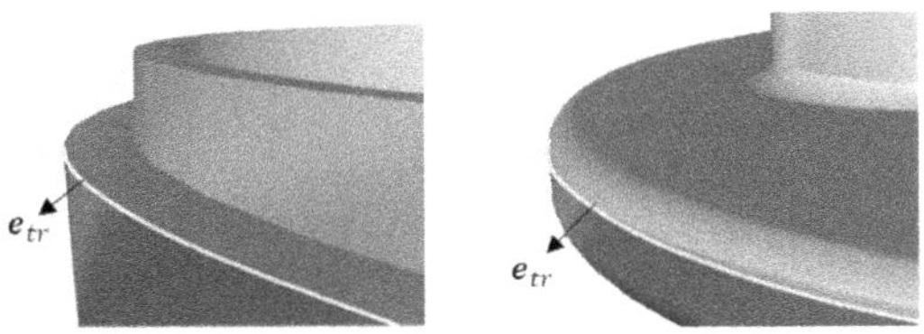

(a) (b)

Figure 2. Assigning curve as e_{tr} on: (a) $\boldsymbol{G}_{ci}$ with convex edge, and (b) $\boldsymbol{G}_{ci}$ as a fillet face.

Subsequently, the candidate transition edges are used as a separator to create groups of candidate inner faces ($\boldsymbol{G}_f$). This can yield multiple of $\boldsymbol{G}_f$. To determine the real $\boldsymbol{G}_f$, do the calculations below: find each face in $\boldsymbol{G}_f$ with $\theta_f < 3°$ and generate L_F. If an intersection is encountered, regard all faces in $\boldsymbol{G}_f$ as outer region faces. Otherwise, convert all faces in $\boldsymbol{G}_f$ as inner region faces. Figure 3 illustrates that multiple $\boldsymbol{G}_f$ are generated on a thin-shell model and each group of $\boldsymbol{G}_f$ is represented by different colors. $\boldsymbol{G}_{f1}$, $\boldsymbol{G}_{f3}$ and $\boldsymbol{G}_{f3}$ will be converted as outer region faces. Figure 4 illustrates the result of separation between inner and outer region faces. Light blue color represents inner region faces, and grey color represents outer region faces.

2.2 *Classification of Regular and Irregular-Wall Types*

For the algorithm to classify thin-shell models, transition face types (f_{tr}) must first be recognized. Our

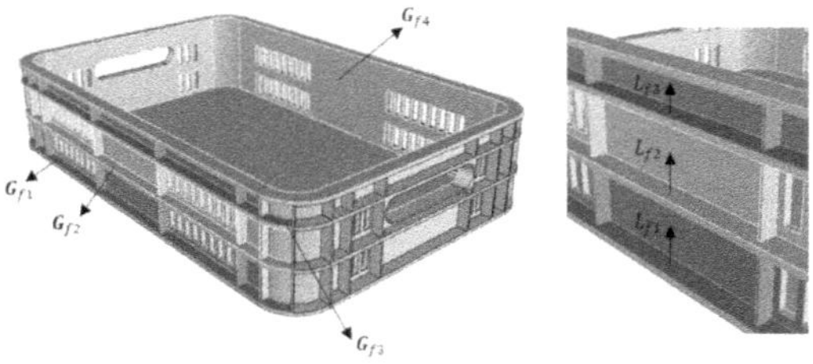

Figure 3. Multiple G_f created and the illustration to calculate the intersection for each G_f.

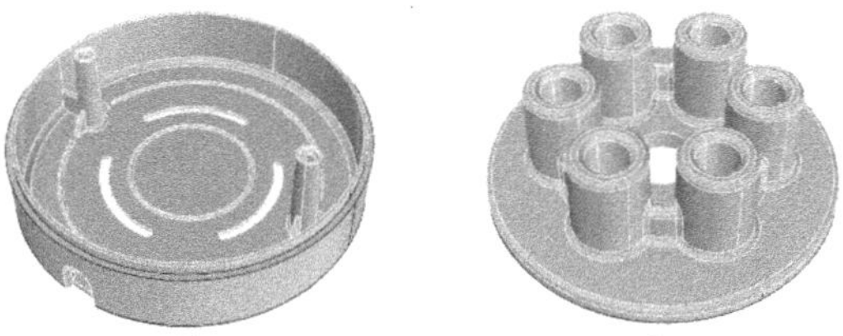

(a) (b)

Figure 4. Results of separation between inner and outer region for: (a) regular, and (b) irregular wall thin-shell CAD models.

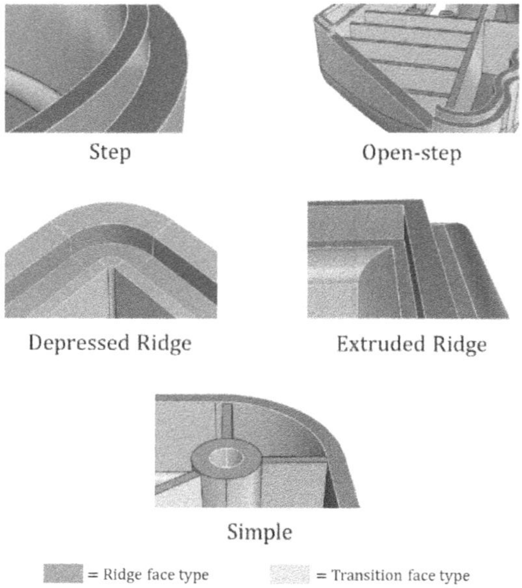

Figure 5. Several types of transition face that might occurs: step, open-step, depressed ridge, extruded ridge and simple.

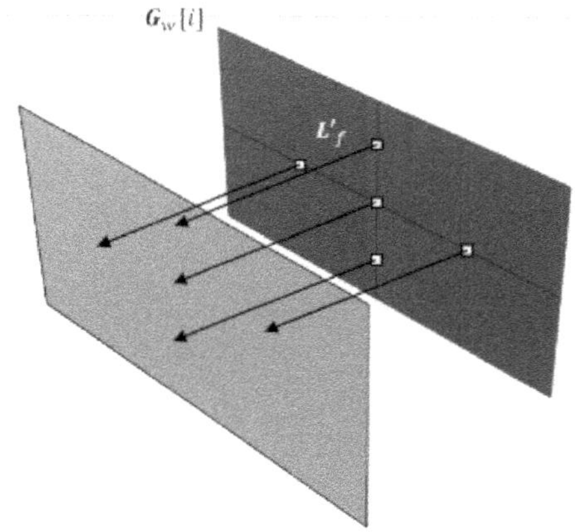

Figure 6. Illustration of seeding points on isocurve and generate L'_f to check the intersection with any other face.

algorithm is able to recognize five different types of transition faces, including step, open-step, depressed ridge, extruded ridge type and simple. "Step" transition type is characterized by a complete ladder-shaped groove along perimeter of the model, while "open-step" is similar to "step" type, yet there is a region created without the groove. The depressed and extruded ridge types are shown by a closed depressed or extruded profile along the perimeter of the model, respectively. Figure 5 illustrates the geometry of five types of transition faces.

After f_{tr} has been recognized, the next step is classifying the model based on the topology and geometry criteria. Only a set of candidate inner wall faces ($\boldsymbol{G}_w$) will be computed in this classification step. Computing $\boldsymbol{G}_w$ using geometry criteria is described as follows:

1. For each f_{tr}, find its neighboring face located on inner region and connected by convex edge. Collect those faces in $\boldsymbol{G}_w$.
2. For each face in $\boldsymbol{G}_w$, generate a set of testing points (p_t) along the iso-parametric curve. The number of p_t is 3 and the points are separated equidistant on each isocurve.
3. Using p_t, generate a line (L'_f) starting from p_t anti-parallel to the face normal vector. Compute the nearest intersection between L'_f and any other face on the model. Figure 6 depicts how the intersection between any face and L'_f is occurred.
4. If the nearest face is regarded as a hole feature, outer face or transition face, remain the face in G_w. Otherwise, delete face data from G_w.

Set of faces in $\boldsymbol{G}_w$ will then be analyzed by using topology criteria. A thin-shell model will be assigned as a regular-wall case if faces in $\boldsymbol{G}_w$ are closely connected. Connection is only established if and only if the numbers of faces in $\boldsymbol{G}_w$ is more than or equal to 2. Some particular conditions might occur and faces in $\boldsymbol{G}_w$ may not be connected directly. If a face in $\boldsymbol{G}_w$ is adjacent concavely to f_{tr} or extrusion feature, then it will be counted as a connection between faces in $\boldsymbol{G}_w$. Figure 7 (a) illustrates the connection between faces in $\boldsymbol{G}_w$, where 4 faces in G_w exist and all faces form a closed connection. Figure 7 (b) depicts the results after classifying the model types for irregular-wall type.

2.3 *Classification of Sub-types Irregular Wall*

In the next step, all irregular-wall models will be categorized into three sub-types based on their face

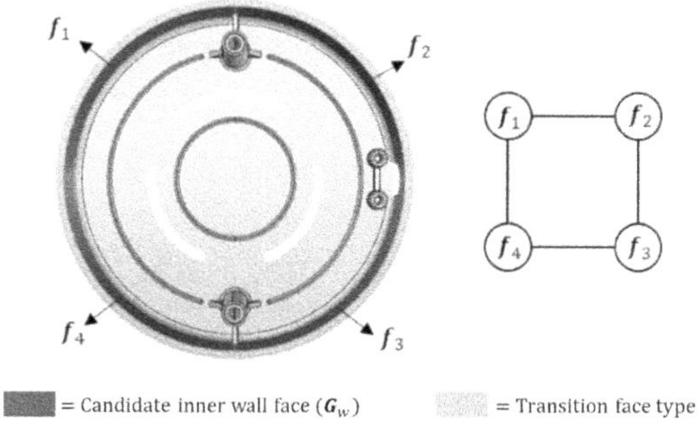

(a)

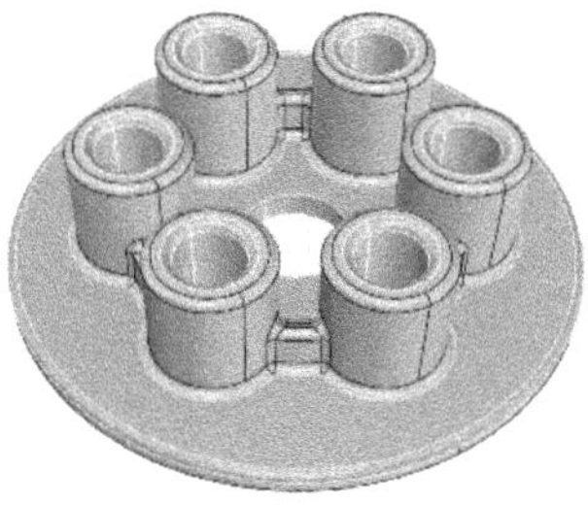

(b)

Figure 7. (a) Illustration of the connection among G_w on regular wall model and (b) irregular wall thin-shell CAD models result after classification of regular and irregular-wall model.

identifier (f_{dt}). This face is used as a marker to represent each sub-type of the irregular-wall models. The extrusion feature must first be recognized before the face identifier can be calculated. This extrusion feature can be located either on the inner or outer region. Therefore, the faces on the irregular-wall model will only consist of the face types except protrusion face types.

The classification begins with seeking the face identifier for the first sub-type (partial open) of irregular wall. First. calculating each face on the model (except hole feature), the requirements of the f_{dt} of this sub-type should meet the following criteria: (1) located on inner region without any inner face types, (2) adjacent to transition faces concavely, and (3) geometrically nearly vertical ($(90° - \alpha) < 90° < (90° + \alpha)$), where α is an allowable tolerance. In this work, $\alpha = 3°$. This sub-type has similar list of face types on inner and outer regions for regular-wall models. The only difference is that no closed connection exists among the wall faces within the inner region. The identification can be extended to the second subtype if no face meets the above conditions. Figure 8 illustrates the example of f_{dt} for partial-open sub-type for an irregular-wall model.

The second sub-type is called "freeform" type. This subtype has specific characteristic which is a complex surface without rigid radius and adjacent to f_{tr}. Listed below are the types of faces associated according to this subtype: transition/flange, bottom and protrusion for either inner or outer region. To detect the f_{dt} for

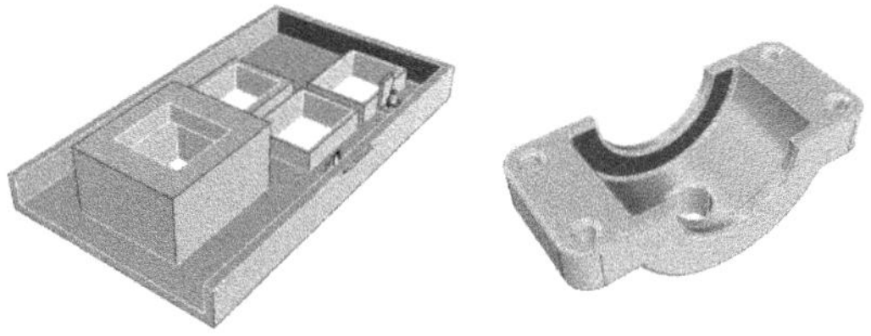

Figure 8. Illustration of f_{dt} represented by blue color on partial-open sub-type of irregular-wall model.

this subtype, calculate any face that meets the following conditions: (1) located on inner region, and (2) not detected as a fillet or hole feature. Then calculate the isocurve and detect the radius at $t = 0$, $t = 0.5$ and $t = 1$. If inconsistency of radius at one of the isocurves is found on the face, and the face is neighboring to f_{tr} as well, then the model will be recognized as a freeform irregular-wall model. If no face is matched, then assign the model to be the third sub-type of the irregular-wall model. The last sub-type, namely simple, only has two face types for both inner and outer regions, which are bottom and protrusion face types.

2.4 *Face types insertion for each sub-types of irregular wall*

Inserting face types on irregular-wall model needs to follow the sub-types of irregular-wall, since each sub-type has different result from the previous step. For partial open sub-type, some transition faces might need to convert into another face type. Therefore, it starts from f_{dt} and detects whether if there exists a transition face concavely connected to f_{dt}. Let N_1 be the normal vector of a surface and N_z be the unit vector of positive-z axis. Those f_{dt} and the adjacent transition faces will be converted as inner wall face if $G_1 \cdot N_z < \cos\alpha$. Otherwise, convert those faces as inner bottom face. The second sub-type of irregular-wall face type consists of transition and bottom faces, yet some transition faces might need to be recalculated. To correct this error, convert any transition face without e_{tr} into bottom face. Otherwise, remain the face types. For the simple sub-type, list of the face type is similar with freeform sub-type. The difference is that some of transition face type might be recognized on outer region. The algorithm will assign bottom on a face that geometrically horizontal either located on inner or outer region based on the following equation:

$$\theta < \alpha$$

$$\theta = \cos^{-1}\left(|N_i \cdot N_z| / N_i N_z\right)$$

where N_z is the normal vector of the reference axis.

3 RESULTS & DISCUSSION

This section describes the result of the classification and recognition of the face types on thin-shell models. In general, the result consists of two main parts: separation between inner and outer regions, and insertion face types on inner and outer regions. Two cases of regular-wall model and six cases of irregular-wall model are provided in this section to verify the feasibility. The algorithm is written using C++ using openNURBS functions and implemented on Rhinoceros 5.0 CAD platform. The input data is a B-Rep model.

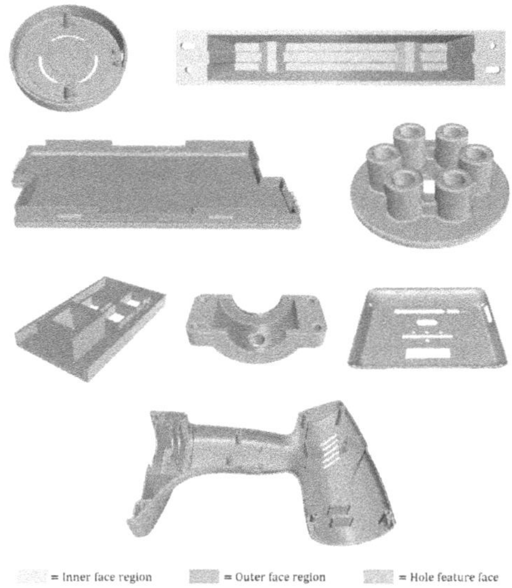

Figure 9. The results of the separation between inner and outer region on several test CAD thin-shell models.

On Figure 9 illustrates the result of separation between inner and outer region. The light blue color represents the inner region, while outer and hole are illustrated by grey and yellow color respectively. All of the test case of thin-shell model is successfully recognized by the inner and outer region. Figure 10 depicts inner face types recognition results for both regular and irregular-wall models. The irregular-wall model is divided into three subtypes: "partial-open", "freeform" and "simple", and shown on the figure as well, respectively. Fillet faces which connect to extrusion feature are successfully recognized as a protrusion face type. Figure 11 shows the picture of the outer region face types recognition. Extrusion feature located on outer region can also be recognized with our algorithm.

4 CONCLUSION

A novel procedure to ease the decomposing model process of thin-shell models by classifying and recognizing face type attributes was proposed in this study. Main-shell body and the complex extrusion feature can successfully be separated on thin-shell cases. Face type attributes can also be inserted to every face located on inner and outer region. Furthermore, a previous study

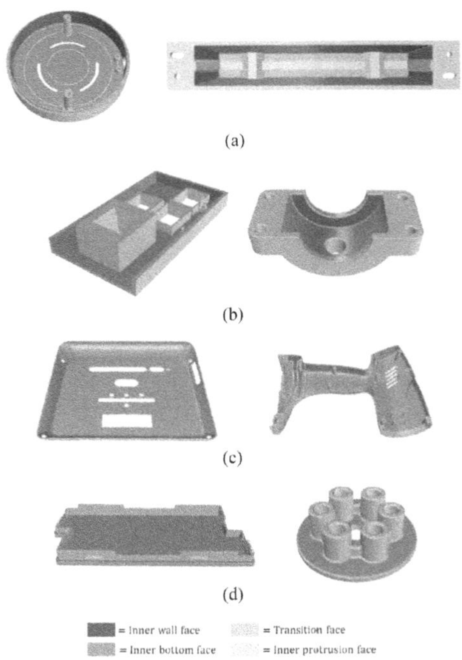

Figure 10. The results of the inner face types recognition of: (a) regular model, (b) "partial-open" sub-type, (c) "freeform" sub-type, and (d) "simple" sub-type of irregular-wall model.

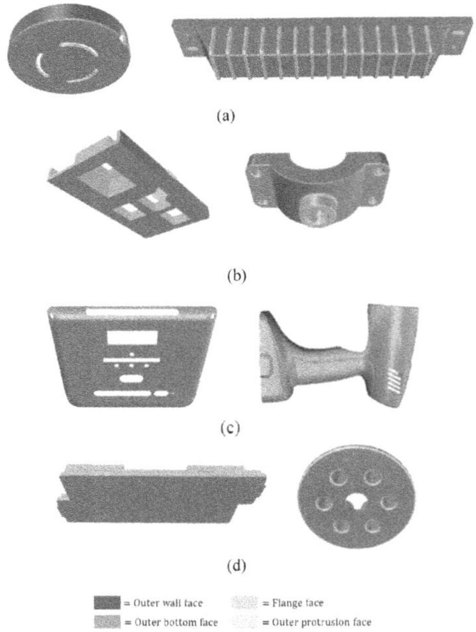

Figure 11. The results of the outer face types recognition of: (a) regular model, (b) "partial-open" sub-type, (c) "freeform" sub-type, and (d) "simple" sub-type of irregular-wall model.

(Lai et al. 2021) is proposed as an attempt to decompose the thin-shell model into simple sub-volume that can be meshed using sweep algorithm. In order to determine the contour of a sub-volume, it is necessary to match the contours of faces on the inner and outer regions. The recognition of face types in this study can enhances and extends the capability of previous research, to deal with more complex thin-shell models.

REFERENCES

Lai, J.-Y., Wang, M.-H., Song, P.-P., Hsu, C.-H., & Tsai, Y.-C. (2021). Hybrid mesh generation for the thin shell of thin-shell plastic parts for mold flow analyis. *Engineering with Computers*.

Lai, J.-Y., Wu, J.-W., Song, P.-P., Chou, T.-Y., Tsai, Y.-C., & Hsu, C.-H. (2018). Automatic recognition and decomposition of rib features in thin-shell parts for mold flow analysis. *Engineering with Computers*, 801–820.

Lu, Y., Gadh, R., & Tautges, T. J. (1999). Feature decomposition for hexahedral meshing. *ASME Design Automation Conference* (pp. 12–15). Las Vegas, Nevada: ASME.

Sheffer A, E. M. (2000). Hexahedral meshing of non-linear volumess using voronoi faces and edges. *International Jouenal for Numerical Methods in Engineering*, 329–351.

Tautges TJ. (2001). The generation of hexahedral meshes for assembly geometry:s survey and progress. *International Journal For Numerical Methods in Engineering*, 2617–2642.

Tautges, T., Liu, S.-S., Lu, Y., Kraftcheck, J., & Gadh, R. (1997). Feature recognittion application in mesh generation. *Trends in Unstructured Mesh Generation*.

System Innovation for a Troubled World – Tak Lam et al. (Eds)
© 2023 The Author(s), ISBN: 978-1-032-45525-9

A unique festival shaped by geographical environment and historical events: Thanksgiving ritual in Zhanghu Village

Shyh-Huei Hwang
College of Design, National Yunlin University of Science & Technology, Yunlin, Taiwan

Chiu-Mei Lai*
Graduate School of Design, National Yunlin University of Science & Technology, Yunlin, Taiwan

ABSTRACT: Zhanghu Village is located in the Gukeng Mountains of Yunlin County, with faults lying across the mountain, steep mountain walls, dense forests with native camphor trees, ancient trees covering the sky, and the dangerous landscape, which was regarded by the ancestors as an obstacle to reclamation. During the Qing Dynasty, there was a hideout for bandits. During the Japanese occupation period, this was the main battlefield of the "Yunlin Incident". After Taiwan returned under the Chinese occupation, it was the place where the civilian army hid and was annihilated in the 228 Incident "Battle of Zhanghu". As a result, the Gukeng Mountains became a place of large number of lonely ghosts. In this study, through interviews and actual participation in the Thanksgiving ritual in Zhanghu Village, combined with the relevant geographic and historical context of colonial wars and historical events, this research concludes that the Thanksgiving ritual in Zhanghu Village includes "worship in front of the door (拜門口)", "reward soldiers and generals (賞兵犒將)", "Pudu (普度)", "Worship Tiangong (拜天公)", "gods Pilgrimage (遶境)", and "Jiang Ye Temple Rewarding Soldiers (將爺廟賞兵)". Unlike Thanksgiving rituals in other places, which only focus on "worshiping gods", their ceremony is also unique. The emphasis is on appeasing invisible ghosts and rewarding the five battalions.

Keywords: Zhanghu Village, Thanksgiving ritual, Ritual design, Gukeng Mountains.

1 INTRODUCTION

In the modern age of globalization, the pace of life is evolving rapidly. Although there are still urban and rural areas in which people live, life patterns are influenced by each other, and the original unique living culture in various places is likely to be no longer valued due to factors such as complexity and time-consuming activities. In 2003, UNESCO's "Convention for the Protection of Intangible Cultural Heritage" listed ceremonies and festivals as "intangible cultural heritage"[1].

This research explores the ritual culture of Zhanghu Village, Gukeng Township. After participating in the Thanksgiving Festival in Zhanghu Village in 2021, it's hard to understand why each part of the Thanksgiving ritual seems to focus on rewarding generals of the five battalions[2] and worshipping wandering ghosts. In addition to the annual small-scale Pudu, there is also an expanded Pudu once every 12 years, which inspired this study to explore the considerations when creating the Thanksgiving ritual in Zhanghu Village.

2 METHODS

2.1 *Field interviews*

For the collection of the subject content data, this research starts with interviews. Since April 2021, several in-depth interviews with local elders, ritual workers, community association personnel, village chiefs, and enthusiastic residents have been conducted. The audio recordings of interviews were transcribed, and classified by decoding; analyzed information including the temples of Zhanghu Village and their beliefs, and gradually sorted out the details of the beliefs and rituals of Zhanghu Village, and also visited all the temples in Zhanghu Village.

*Corresponding Author: a0912190626@gmail.com

[1] Bureau of Cultural Assets, Convention for the Safeguarding of Intangible Cultural Heritage (article source: UNESCO website) https://twh.boch.gov.tw/taiwan/learn_detail.aspx?id=32

[2] The five battalions refer to the places where intangible soldiers and generals are stationed, located in the east, west, north and south of the village. They are used to protect villages from ghosts. The five battalions often exist in the form of Jiang Ye Temple.

 DOI 10.1201/9781003377399-3

2.2 *Participate in observation*

After many interviews, the researcher clearly understands the details of Thanksgiving rituals in Zhanghu Village and then participate in the Thanksgiving activity in 2021 to experience and verify the details provided by the residents.

2.3 *Review relevant historical documents*

In order to understand the background of the current Thanksgiving ritual in Zhanghu Village, this study refers to research reports related to the Zhanghu area to verify the considerations when developing the Thanksgiving ritual.

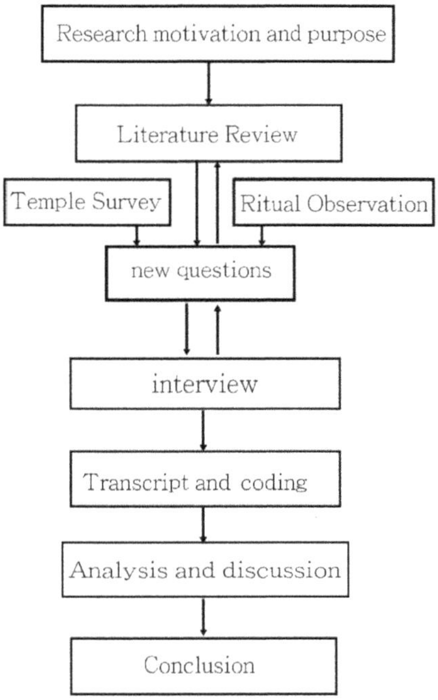

Figure 1. Research flow chart.

3 ANALYSIS

3.1 *The geographical environment and the background of the colonial wars in Zhanghu Village*

3.1.1 *The rugged geographical environment of the Gukeng Mountains has become a natural cover for wars in history*

Zhanghu Village is located in the mountainous area of Gukeng Township, Yunlin County. There are five mountain villages in Gukeng Township, including Yongguang, Zhanghu, Huashan, Guilin, and Caoling. There are faults in the mountainous area, the mountain walls are steep, the forest is densely covered with native camphor trees, the ancient trees cover the sky, and the terrain is dangerous and complex. It was the main battlefield of the "Yunlin Incident" that lasted six years during the Japanese occupation era. It was also an important base and cover for the anti-Japanese guerrillas (Zhang Suying 2021), and the "Battle of Zhanghu" in the 228 Incident after Taiwan's return to China. It is where the civilian army retreated to hide but was annihilated[3].

3.1.2 *A large number of lonely ghosts stacked up by historical trauma events*

Taiwan, which was regarded by the Qing dynasty government as a "land of undeveloped", was a treasure island coveted by foreign powers. The history accumulated by colonization not only impacted the culture of the ancestors but also suffered from the confrontation between people and the colonial government. There were also many traumatic incidents. From the Treaty of Shimonoseki that ceded Taiwan to Japan after the Sino-Japanese War in 1895, to the return of Taiwan to China after the end of World War II in 1945, Taiwan has always been regarded as a negotiating target for victory or defeat in wars. Although the Taiwanese citizens once tried to resist colonial rule, the forces of the civilian army ultimately found it difficult to compete with the military forces of the government. In addition to the bloody defeat, they bravely tried to make a comeback in the mountains.

The Gukeng Mountains are naturally shielded by the "precipitous terrain," "steep mountain walls," and "ancient trees covering the sky," so they will not be absent from historical traumatic events. There are two major traumatic incidents related to the Gukeng Mountains. One is "the massacre incident" and "the Zhanghu incident," in the Yunlin Incident that lasted for six years from 1895 to 1902. The second one is the "Battle of Zhanghu" in the 228 Incident in 1947. In particular, the Yunlin incident was the deadliest casualty. In Zhang Sufen's article, "Archives, Documents, and Historical Memory: The Yunlin Incident in Gukeng," there are descriptions of the casualties of both sides' large and small war operations:

> "The thrilling offensive and defensive battle that took place in the Gukeng area Mountains more than 100 years ago[4] ", *"When the Japanese sent troops to scout Dapingding, they were surrounded and attacked by the civilian army. Lieutenant Nakamura, who led the army, was injured and committed suicide, more than half of his subordinates were killed* [5]*. When the Japanese attacked from the foot of Dapingding Mountain, they were killed by the civilian army, and the casualties were very heavy*[6] ", *"After the*

---

[3] 228 Incident Battle of Zhanghu Lake, written by Xu Fenglin, Yunlin National Cultural Memory Bank, http://data.yunlin-memory.cool/data/Event/3007

[4] Zhang Sufen, "Archives, Documents and Historical Memory: The Yunlin Incident in Gukeng", "Taiwan Literature", Vol. 72, No. 1, January 2021, p. 135

[5] Ibid, p. 140

[6] Ibid, p. 140

Japanese army entered Xiaopingding, they were ambushed, and the Japanese lieutenant Nakamura Masaki was captured alive and died[7]", "After the Japanese army was defeated and retreated at Dapingding, they did not dare to go deep without reinforcements, so they burned the village along Dapingding all the way to Gukeng[8]", "26 Japanese policemen including Commander Sasaki Hidenosuke were annihilated in the mountains[9]", "In order to retaliate for the assassination of Sasaki Hidenosuke and others, the Japanese launched a massacre operation in Caoling. The residents died or fled, and the villagers were left empty[10]", "The civilian army attacked the camp of the Zhanghu garrison and killed six people including the captain and second lieutenant Fujiwara Masajuro. The Japanese called it the "Zhanghu Incident[11]". (Zhang Suying 2021).

Although only a few historical records show the death toll of the Japanese army, the residents interviewed also described what they heard from older generations: "If there is nowhere to hide, life will disappear" and "Kill people when they see them" (2022.03.19, interviewee: SQT), "In the Japanese era, many people were killed in the field behind the school" (2022.03.19, interviewee: ZHM), it is not difficult to imagine how many souls died in various wars in the Gukeng Mountains.

3.2 *Taiwan's folk beliefs have regional characteristics*

The early Han people immigrated to Taiwan and cultivated in a harsh environment with a shortage of medical resources. Death was the biggest threat. In order to pray for survival, whether ghosts or gods, as long as they are effective, they all had beliefs that met the needs of the people. In other words, most of the folk beliefs of worshiping gods and ghosts originated from the purpose of praying for the family's health and safety, a good harvest, success, etc. Folk beliefs are generated under the conditions of this very free and selective belief and evolve according to the environment. They have different characteristics in different places (Liu Huanyue 2001, pp. 118~122).

Zhanghu Village is located in a mountainous area with dangerous scenery. Regardless of the safety doubts caused by the environment or the invisible fear caused by a large number of dead souls in historical traumatic events. In order to pray for peace in life and peace of mind, the beliefs may be according to various factors unique to the place, and the ceremonies with local characteristics have been developed.

3.3 *The ceremony of Thanksgiving ritual in Zhanghu Village is divided into six major items*

3.3.1 *To appease the ghosts and hoping they don't make trouble during the ceremony, follow the rule: worship in front of the door*

Before the beginning of the Thanksgiving ceremony in Zhanghu Village, there was a seemingly informal ceremony "worship in front of the door". It was a very carefully designed ceremony for the successful completion of the ceremony. The purpose was to eliminate possible interference variables in the ceremony first. The interviewer said, *"There are some ghosts near the place where we want to hold the ceremony. We must feed these ghosts before the ceremony"* and told them *"Here, we have to worship the Great God, stop making trouble."* (2021.10. 16, interviewee: ZHH 08_05). Therefore, the intention of "worship in front of the door" is to worship the ghosts around the ceremony site to eat before the ceremony begins. From the item of "worship in front of the door", it can be seen that the residents believe that the ghosts are usually around their homes, and they are not all kind. When necessary, they need to be appeased to avoid trouble.

3.3.2 *Thanks to the soldiers and generals of the Jiang Ye Temples for guarding Zhuangtou to prevent the invasion of ghosts: Reward soldiers and reward generals*

The existence of the five battalions is to protect the village. According to Ji Yuqian's field investigation, the dates of the temple-related "reward soldiers and generals" rituals are mostly fixed on the first and fifteenth days of each month, and the rituals are different between temples (Ji Yuqian 2019).

"Reward soldiers and generals" in Zhanghu Village is one of several rituals of the Thanksgiving Festival, rather than being independent of the festival. Although the target of "reward soldiers and generals" is the generals of the five battalions the residents will still prepare "108 bowls of meat wine" to invite the heavenly soldiers and heavenly generals, that is, "36 Tiangangs (天罡) and 72 Dishas (地煞)", to come down to earth to participate in and assist the Taoist priests in exorcism[12].

This shows that the residents have an invisible fear of ghosts. This fear aggravates the dependence on the soldiers of the five battalions in the village. Therefore, the "reward soldiers and generals" are also included in the Thanksgiving ceremony.

3.3.3 *Comfort nearby ghosts: Pudu*

In the process of reclamation, the early ancestors inevitably suffered casualties due to unacceptable soil and water or fighting for territory. Especially the wandering bachelors who came to Taiwan alone to reclaim

[7] Ibid, p. 141

[8] Ibid, p. 141

[9] Ibid, p. 160

[10] Ibid, p. 162

[11] Ibid, p. 164

[12] Li Jiayun (2006). Research on the architectural art of Taoist temples in Taiwan-Taking the reconstruction design of the Temple of God in Xuantian "Beixuan Palace" as an example. Thesis. Pingtung City, Pingtung University of Education. Page 227.

the land and had no family and became lonely ghosts after death. People used Pudu rituals to comfort them in order to live without the interference or harm of the ghosts (Liu Huanyue 2001, p146).

It can be seen from historical events that the Gukeng Mountains have a history of a large number of deaths and injuries. The villagers of Zhanghu have used the Thanksgiving ritual to save the souls of the dead year after year to appease the nearby ghosts. What is special is that in Taiwan, it is said that July is the month when the gate of hell opens, and in 1952, in order to correct the worship atmosphere, the government stipulated that the 15th day of the seventh lunar month was the date of the unified Pudu. However, the Pudu in Zhanghu Village was combined with Thanksgiving rituals held, which makes it special.

3.3.4 *Thanks to the emperor for leading the gods to give peace: Worship Tiangong*

The most important object of thanksgiving in Thanksgiving rituals is Tiangong. Tiangong refers to the Jade Emperor (玉皇大帝) who rules the gods. Because of his noble status, the sooner he worships, the more sincere he is. The time chosen is usually in the early morning or at the beginning of the day[13]. In 2021, Thanksgiving rituals of Zhanghu Village were held at 5:00 in the morning in accordance with the birthday of the Censer Keeper. The gods participating in the "worship Tiangong" include the gods who are usually placed in Zhongying[14] (中營): Guanyin (觀音佛祖) and Third Mazu (三媽), Syuanwu (玄天上帝), Shancai (善財), Longnv (龍女), and Fude (福德正神). All private gods are welcome to participate. Residents participating in the worship are represented by luzhu (censer keeper, 爐主) and toujia (local chiefs, 頭家). Residents can participate if they want, but it may be too early. There are not many residents who participate in the worship of Tiangong.

3.3.5 *Patrol and keep the place safe: gods Pilgrimage*

The god's Pilgrimage is an important activity in which the gods walk into the countryside and drive people's feelings towards the gods in the festival activities. The purpose of circumnavigating the border is slightly different depending on the temples. For example, to visit every corner of the village, to protect the local safety, to be free from disasters at all times, and not to be disturbed by external forces such as evil spirits and epidemic diseases (Yang Yu-lin 2015); or protect people and handle cases on behalf of God, as long as the residents have any problems that cannot be solved, they can stop the sedan chair at any time during the Pilgrimage, and ask God to deal with them (Lin Yi-jun 2020).

The Pilgrimage in Zhanghu Village was arranged on the second day of the activity. It was a continuation of the entire Thanksgiving ritual activity. The gods patrolled the territory and cared for residents and temples one by one. The detour team was led by the black command flag with the "Tai Chi diagram (太極圖)" and the incantations of "long life and wealth" and "the great general personally slays the east, west, north and south demons". On the bamboo pole, the weeping of the bamboo leaves symbolizes the purification of the route. It can be seen that in addition to the patrol of the gods, the Pilgrimage of Zhanghu Village also has the effect of driving away the ghosts.

3.3.6 *Once again, pray the generals to guard the border of village: Jiang Ye Temple rewards soldiers*

Jiang Ye Temple rewards soldiers is another ritual that is not in the official Thanksgiving one, but the residents take the initiative to execute the project.

After the festival, the five battalion flags that had taken back the altar due to the festival were set up again, and the residents would carry offerings to the Jiang Ye Temple near their homes to reward the soldiers. This once again shows the residents' belief in and dependence on Jiang ye Temple, and the meaning behind it is to ask the generals to guard the border of the village and prevent evil spirits from entering.

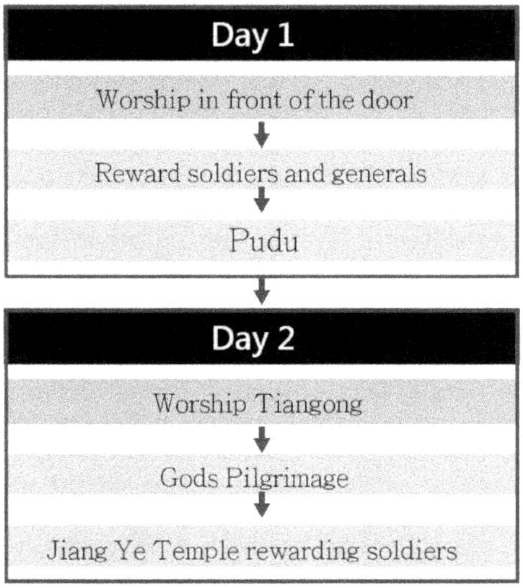

Figure 2. The 2021 Zhanghu Thanksgiving ritual process.

4 DISCUSSION

4.1 *The proportion of Thanksgiving rituals in Zhanghu Village is more related to invisible ghosts*

As far as literature is concerned, Thanksgiving rituals in general temples or communities focus on the worship of Tiangong. However, Thanksgiving rituals in Zhanghu Village' have a wider scope.

[13] Li Xiu'e (2015), "Illustration of Taiwanese Folk Festivals", Taichung City: Morning Star Publishing. page 94

[14] Zhongying is The middle battalion which is the central command of the five battalions

Figure 3. Photos of Thanksgiving ceremonies in Zhanghu Village in 2021.

In this study, the researcher actually participated in the observation of Thanksgiving rituals in Zhanghu Village on December 24 and 25, 2021, and found that: (1) The whole and complete ritual process is not just to worship the Lord of Heaven, it starts from worship in front of door on the first day, then rewards soldiers and generals, and it focuses on Pudu in the afternoon. On the second day, they worship Tiangong and then Pilgrimage around the whole settlement, which includes all households and temples. After the end, the residents go to Jiang Ye Temple to reward soldiers. (2) From the perspective of the guardian deities of the village, Fude Temple and Jiang Ye Temple are two kinds of temples that belong to Zhanghu Village. Although both are the same as the guardian of the village and the object of the villager Thanksgiving rituals, FudeTemple was not listed as one of Thanksgiving ritual projects. In addition to being invited to the altar to participate in the Thanksgiving Ceremony, there is no ritual designed to highlight Fude's importance in the Thanksgiving Ceremony like in Jiang Ye Temple. (3) From the point of view of the full participation of the residents, the rituals that residents participated in focus on these five parts: "Worship in front of the door", "Reward soldiers and generals", "pudu", "welcome to the Pilgrimage", and "Jiang Ye Temple rewards soldiers".

Therefore, from the content of the ceremony, it is obvious that Zhanghu Thanksgiving not only thanked God for a safe and fruitful harvest but also thanked Jiang Ye for being safe from invisible ghosts. To sum up, the etiquette is very original, focusing on thanking the five battalions that guard the village to avoid the invasion of ghosts. And it also focuses on appeasing the ghosts that live in peace with the villagers.

4.2 *Thanksgiving rituals in Zhanghu Village have the thinking of peaceful coexistence among gods, ghosts, and people*

In the ritual of Thanksgiving, although the ritual related to the gods is the worship Tiangong at night, at the beginning of the whole ritual, the deity will be seated at the altar and wait, and the residents will also be there for the Pudu ritual. The staff prepared meat and vegetarian meals for the gods. Therefore, during the ceremony, gods, ghosts, and people coexist peacefully in the same place. It can be seen that the design of this ceremony has the idea of the peaceful coexistence of gods, ghosts, and people.

4.3 *Zhanghu Village's Thanksgiving Ritual Ceremony is a collection of all different rituals in one*

Although in the early agricultural society, Thanksgiving mostly thanked God for a bumper harvest, in Zhanghu Village: (1) The residents of Zhanghu Village one hundred years ago had seen various incidents of death from wars in the mountains. There was more sense of impermanence in life; (2) From the perspective of historical events, there were many people who died unjustly in the Gukeng Mountains and became wandering ghosts. A large number of lonely and wild ghosts in the mountains entered the village would interfere with the lives of the villagers. Therefore, villagers attached great importance to the function of the Jiang Ye Temple and were grateful. (3) The government at that time strictly stipulated that the Thanksgiving festival should be held once a year. And now the villagers can no longer test the original intention of this ritual design. This study is based on the aforementioned environmental factors, historical events and death factors, and legal regulations, and then speculates that the ancients incorporated various ceremonies into the Thanksgiving ceremonies due to their needs, so as to the needs and regulations are taken into account.

5 CONCLUSION

(1) Influenced by historical events, the Thanksgiving rituals in Zhanghu Village focus on gratitude to the soldiers and generals of the Jiang Ye Temple and the appeasement of the souls of the dead.
(2) The Thanksgiving rituals in Zhanghu Village were limited by the festival saving regulations a hundred years ago, and many common folk rituals such as "worship in front of the door", "reward soldiers and generals", "Pudu", "Worship Tiangong", "gods Pilgrimage", "Jiang Ye Temple Rewarding

Soldiers" were combined into the Thanksgiving Festival.

(3) The Thanksgiving rituals in Zhanghu Village is a very unique festival due to factors such as environment, history, and regulations.

REFERENCES

Convention for the Protection of Intangible Cultural Heritage (article source: UNESCO website), 2021.03.01 from the "*CulturalAssets Bureau*": https://twh.boch.gov.tw/taiwan/learn_detail.aspx?id=32

Ji Yuqian (2019). *A Study of the Belief in Xuantian God in Zhushan*, Nantou. Thesis. National Chiayi University, Chiayi County.

Liu Xuanwei (2011). *Tzu Chi July Guxiang and Zhongyuan Purdue: A Dialogue between Modernity and Tradition*. Thesis. Fu Jen Catholic University, New Taipei City.

Li Jiayun (2006). *Research on the architectural art of Taoist temples in Taiwan — Taking the reconstruction design of the Temple of God in Xuantian "Beixuan Palace" as an example*. Thesis. Pingtung University of Education, Pingtung City. Page 227.

Liu Huanyue (2001). "*Taiwanese People's Worship and Ritual*". Taipei City: Changmin Culture.

Li Xiu'e (2015). "*Illustration of Taiwanese Folk Festivals*". Taichung City: Morningstar Publishing.

Yijun Lin (2020). *Research on Pingtung King's Peace Ceremony - Taking Donggang, Xiaoliuqiu and Nanzhou as examples in Pingtung County. PhD thesis*. National Kaohsiung Normal University, Kaohsiung City.

Yang Yulin (2015). *Research on the ritual ceremony of the Prince of Danshui Sufu—Focusing on the annual celebration and the ceremony of sending the king boat*. Thesis. Truth University, New Taipei City.

Xu Fuquan, editor-in-chief (2005), "*Taiwan Folk Sacrificial Etiquette*", Hsinchu City: National Hsinchu Society Education Center.

Zhang Sufen, "*Archives, Documents and Historical Memory: The Yunlin Incident in Gukeng*", "Taiwan Literature", Vol. 72, No. 1, January 2021, pp. 131–176.

228 Incident Battle of Zhanghu Lake, written by Xu Fenglin, 2021.03.01 from "*Yunlin National Cultural Memory Bank*": http://data.yunlin-memory.cool/data/Event/3007

System Innovation for a Troubled World – Tak Lam et al. (Eds)

Uplink throughput optimization for multiple gateway systems in LoRaWAN class A

Jeng-Ji Huang* & Jhen-Wei Chen
Department of Electrical Engineering, National Taiwan Normal University, Taipei, Taiwan

ABSTRACT: LoRaWAN (long range wide area network) is widely considered as a promising wireless communication technology that can meet the requirements for thousands or millions of IoT (internet of things) end devices to be able to connected to the Internet. In this paper, a scheme is proposed for end devices to select a spreading factor for uplink transmissions in LoRaWANs. In our proposed scheme, connectivity between end devices and gateways is first constructed, which depends on the received signal strength. As an end device may be able to be connected to multiple gateways at the same time, the reliability of uplink transmissions improves. Nonetheless, the traffic load seen at each gateway increases, thereby leading to more packet collisions during uplink transmissions. In order to deal with this tradeoff situation, a systematic probability-based approach is employed in the selection of a spreading factor for each end device, and a water-filling algorithm is utilized to balance traffic loads between spreading factors, from which the proabability distribution for the spreading factor selection of each end deivce can then be determined. Numerical results show that our proposed scheme significantly outperforms ADR (adaptive data rate), which is a scheme recommended by the standard.

1 INTRODUCTION

We consider a typical LoRaWAN system, where end devices (EDs) are in association with a network server through one or several gateways (GWs) (D. Garlisi et al. 2021). Uplink packets are assumed to be generated from ED $i \in \mathcal{E}$ according to a Poisson process with mean rate λ_i, where $\mathcal{E}$ is the set of EDs. Furthermore, packet lengths are assumed to be independent, exponentially distributed with mean L.

A LoRaWAN system can be either homogeneous or heterogeneous. The former occurs when SFs are all selectable by each ED, while the latter occurs when selectable SFs may be different between EDs. Define $c_{i,g,x}$ as the connectivity of an ED i to a GW $g \in \mathcal{G}$ using SF $x \in \mathcal{X}$, where $\mathcal{G}$ is the set of GWs and $\mathcal{X}$ the set of SFs. That is, if the sensitivity requirement η_x of SF x can be met between ED i and GW g, $c_{i,g,x} = 1$; otherwise, $c_{i,g,x} = 0$. For example, suppose the transmit power of ED i is P_t. Since the path loss (in dB) for ED i whose location is d away from GW g can be modelled as (F. Cuomo et al. 2018)

$$\mathcal{L}(d) = \mathcal{L}'(d_0) + 10\gamma log\left(\frac{d}{d_0}\right) + \chi$$

where d_0 is a reference distance equal to 40 m, $\mathcal{L}'(d_0) = 127.41$ dB, $\gamma = 2.08$, and $\chi \sim \mathcal{N}(0, \sigma^2)$ is normal distributed, the receive power P_r can therefore be obtained as $P_r = P_t - \mathcal{L}(d)$. Obviously, raising transmit power can lead to better connectivity, at the expense of a shorter battery life. In this paper, power control is not taken into account, and it is assumed that each ED has negotiated in advance with the network server for transmit power settings.

Denote as R_x the bit rate of SF x, which can be expressed as $\frac{x \cdot B}{2^x}$ (Semtech 2015), where B represents the bandwidth. The transmission time for an uplink packet of length l can be obtained as $t_x = \frac{l}{R_x}$. Taking into account the duty cycle (DC) restriction imposed on a frequency band (LoRa Alliance Technical Committee 2017), denoted as δ, the service time of an uplink packet transmission can be expressed as $s_x = \frac{t_x}{\delta}$. Since packet lengths are exponentially distributed, it can readily be shown that s_x is exponentially distributed with mean equal to $\frac{L}{R_x\delta}$. The mean service rate of SF x is therefore $\mu_x = \frac{R_x\delta}{L}$.

In this paper, we consider maximizing the overall system throughput $T = \sum_{i \in \mathcal{E}} T_i$, where T_i is the throughput of ED i.

2 THE PROPOSED SCHEME

In our proposed scheme, the SF x for uplink transmissions is selected randomly by ED i based on the probability mass function (PMF) of X_i, $P_{X_i}(\cdot)$, which is determined in order to maximize T. The traffic generated from ED i and allocated to SF x can be expressed as $\lambda_i \cdot P_{X_i}(x)$. The traffic load, $\rho_{g,x}$, of SF x as seen at GW g can then be expressed as

$$\rho_{g,x} = \frac{\sum_{i \in \mathcal{E}} \lambda_i \cdot P_{X_i}(x) \cdot c_{i,g,x}}{\mu_x}. \quad (1)$$

*Corresponding Author: hjj2005@ntnu.edu.tw

DOI 10.1201/9781003377399-4

In the following, we focus on finding $P_{X_i}(\cdot), \forall i \in \mathcal{E}$.

Define the packet delivery ratio (PDR), $q_{i,x}$, of ED i using SF x as the ratio of the number of packets successfully received at one (or more than one) GW to the total number of packets sent by ED i using SF x. Since ALOHA is employed, the PDR of using SF x at GW g equals $e^{-2 \cdot \rho_{g,x}}$. Let $G_{i,x} = \{g | g \in \mathcal{G}, c_{i,g,x} = 1\}$ be the set of GWs that are reachable by ED i using SF x. Therefore, the PDR of ED i using SF x can be written as

$$q_{i,x} = 1 - \prod_{g \in G_{i,x}} \left(1 - e^{-2 \cdot \rho_{g,x}}\right). \tag{2}$$

The overall system throughput T can then be rewritten as

$$T = \sum_{i \in \mathcal{E}} T_i = \sum_{i \in \mathcal{E}} \sum_{x \in \mathcal{X}} \lambda_i \cdot P_{X_i}(x) \cdot q_{i,x}. \tag{3}$$

In the following, a theorem for maximizing T is proven.

Theorem 1: T is maximized if $q_{i,x} + P_{X_i}(x) \frac{\partial q_{i,x}}{\partial P_{X_i}(x)} = q_{i,x'} + P_{X_i}(x') \frac{\partial q_{i,x'}}{\partial P_{X_i}(x')}, \forall i, x \neq x'$.

Proof: The optimization problem can be formulated as

$$\begin{aligned} \max & \sum_{i \in \mathcal{E}} \sum_{x \in \mathcal{X}} \lambda_i \cdot P_{X_i}(x) \cdot q_{i,x} \\ s.t. & \sum_{x \in \mathcal{X}} P_{X_i}(x) = 1, \forall i \end{aligned} \tag{4}$$

The Lagrangian equation is therefore

$$\begin{aligned} \mathcal{L}(\boldsymbol{p}, \boldsymbol{\alpha}) = & \sum_{i \in \mathcal{E}} \sum_{x \in \mathcal{X}} \lambda_i \cdot P_{X_i}(x) \cdot q_{i,x} \\ & + \sum_{i \in \mathcal{E}} \alpha_i \left(\sum_{x \in \mathcal{X}} P_{X_i}(x) - 1 \right). \end{aligned} \tag{5}$$

where $\boldsymbol{p}$ is the $|\mathcal{E}| \times |\mathcal{X}|$ matrix formed by $P_{X_i}(\cdot)$ and $\boldsymbol{\alpha}$ is the $|\mathcal{E}| \times 1$ vector formed by α_i. Then, we have at optimum

$$\begin{aligned} \frac{\partial \mathcal{L}(\boldsymbol{p}, \boldsymbol{\alpha})}{\partial P_{X_i}(x)} = & \lambda_i q_{i,x} + \lambda_i P_{X_i}(x) \frac{\partial q_{i,x}}{\partial P_{X_i}(x)} \\ & + \alpha_i = 0 \quad \forall i, x. \end{aligned} \tag{6}$$

It is clear that at optimum

$$\begin{aligned} & \lambda_i q_{i,x} + \lambda_i P_{X_i}(x) \frac{\partial q_{i,x}}{\partial P_{X_i}(x)} \\ & = \lambda_i q_{i,x'} + \lambda_i P_{X_i}(x') \frac{\partial q_{i,x'}}{\partial P_{X_i}(x')}, \quad \forall i, x \neq x', \end{aligned} \tag{7}$$

which completes the proof.

In order to make **Theorem 1** a reality in a multiple gateway system, EDs are first classified into groups $G_1, \cdots, G_{|\mathcal{X}|}$ according to the number of SFs selectable by them for uplink transmissions; for instance, G_1 refers to the group of EDs that only SF 12 can be selected, while $G_{|\mathcal{X}|}$ refers to those that all SFs can be selected. Nonetheless, $P_{X_i}(\cdot)$ is separately determined, due mainly to the distinct connectivity to GWs for each ED.

First of all, the packet failure rate (PFR) as seen by ED i using SF x can be obtained from Eq. (2) as

$$(1 - q_{i,x}) = \prod_{g \in G_{i,x}} \left(1 - e^{-2 \cdot \rho_{g,x}}\right). \tag{8}$$

The condition in **Theorem 1** can thus be rewritten as

$$\begin{aligned} (1 - q_{i,x}) + P_{X_i}(x) \frac{\partial (1 - q_{i,x})}{\partial P_{X_i}(x)} = & (1 - q_{i,x'}) \\ & + P_{X_i}(x') \frac{\partial (1 - q_{i,x'})}{\partial P_{X_i}(x')}, \quad \forall i, x \neq x'. \end{aligned} \tag{9}$$

Note that the second term in Eq. (9), i.e., $P_{X_i}(x) \cdot \frac{\partial(1-q_{i,x})}{\partial P_{X_i}(x)}$, can be viewed as an adjustment to the first term, i.e., $(1 - q_{i,x})$ or PFR, so as to let the PFR become more balanced between xs for ED i. Note that the balance of PFR between xs is only necessary for those that are selectable by ED i, since they are the ones that contribute to T_i, as defined in Eq. (3). Since

$$\begin{aligned} \frac{\partial(1 - q_{i,x})}{\partial \rho_{g,x}} = & 2\rho_{g,x} e^{-2\rho_{g,x}} \cdot \\ & \prod_{g' \neq g, g' \in G_{i,x}} \left(1 - e^{-2 \cdot \rho_{g',x}}\right) > 0, \end{aligned} \tag{10}$$

it is clear that $(1 - q_{i,x})$ grows if $\rho_{g,x}$ is increased. As a result, it can be obtained that

$$\begin{aligned} \frac{\partial(1 - q_{i,x})}{\partial P_{X_i}(x)} & = \sum_{g \in G_{i,x}} \frac{\partial(1 - q_{i,x})}{\partial \rho_{g,x}} \cdot \frac{\partial \rho_{g,x}}{\partial P_{X_i}(x)} \\ & = \sum_{g \in G_{i,x}} \left[2\rho_{g,x} e^{-2\rho_{g,x}} \cdot \prod_{g' \neq g, g' \in G_{i,x}} \left(1 - e^{-2 \cdot \rho_{g',x}}\right) \right] \end{aligned} \tag{11}$$

As can respectively be seen in Eqs. (8) and (11), $\rho_{g,x}$ plays a critical role in determining $(1 - q_{i,x})$ and $\frac{\partial(1-q_{i,x})}{\partial P_{X_i}(x)}$; they are positive if $\rho_{g,x} > 0$ and are 0 if $\rho_{g,x} = 0, \forall g \in G_{i,x}$.

The balance in Eq. (9) can then be achieved using the proposed algorithm based on a water-filling algorithm (P. He et al. 2013). The height and the width of a stair in a water tank equal respectively $(1 - q_{i,x})$ and $1/\frac{\partial(1-q_{i,x})}{\partial P_{X_i}(x)}$, and the water is $\sum_{x \in \mathcal{X}} P_{X_i}(x) = 1$ in our proposed algorithm, which are illustrated as follows.

The proposed algorithm mimics pouring water of amount 1 into each of $|\mathcal{E}|$ tanks. However, instead of determining the allocation of λ_i to SFs altogether at a time for ED i, it is done in $(L - 1)$ rounds where

$L = \frac{1}{\in}$ and $\in$ stands for a very small proportion. In each round, an amount equal to 1 of water is poured into a tank containing k stairs for each ED $i \in G_k$, and $\hat{P}_{X_i}(x_1), \cdots, \hat{P}_{X_i}(x_k)$, which represents the allocation of $\in \cdot \lambda_i$ to SFs $x_1, \cdots, x_k$ in a round, is determined such that $(1 - q_{i,x}) + \hat{P}_{X_i}(x) \frac{\partial(1-q_{i,x})}{\partial P_{X_i}(x)}$ can be as balanced between $x_1, \cdots, x_k$ as possible.

The reason why there are so many rounds is stairs in tanks might constantly alter in heights and widths as water is poured in, due to that any change of $P_{X_i}(x)$ may lead $\rho_{g,x}$ to vary, which in turn causes $(1 - q_{i,x})$ and $1/\frac{\partial(1-q_{i,x})}{\partial P_{X_i}(x)}$ to fluctuate. However, by letting $\in$ very small, e.g., 0.001, the change of the heights and widths of the stairs would be so slight that its impact on the balance in Eq. (9) can more easily be readjusted in the following rounds.

3 RESULTS AND DISCUSSION

In this section, numerical results are given and compared between our proposed scheme, ADR (LoRa Alliance Technical Committee 2017), and Uniform, a uniformly random selection scheme (R. B. Sorensen et al. 2017). The carrier frequency is assumed to be 860 MHz, the bandwidth is 125 KHz.

The system used in the experiment consists of 500 EDs and 5 GWs, all of them uniformly distributed in circular areas, the former and the latter of which are of radius 150 and 50 (m), respectively. An example is shown in Figure 1.

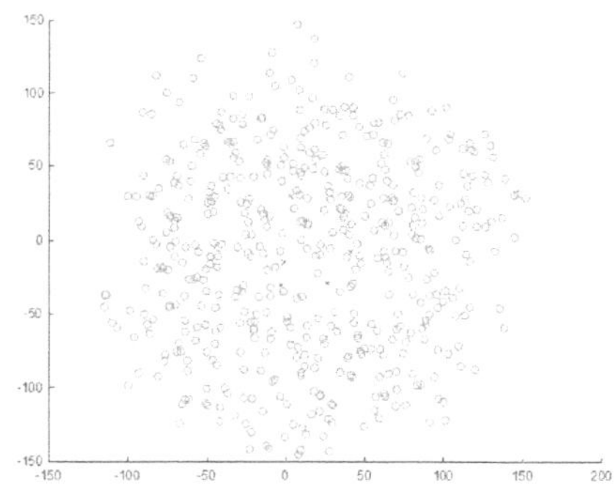

Figure 1. The uniform distribution of 500 EDs (red circles) and 5 GWs (blue crosses) in a circular service area of radius 150 m.

Two kinds of traffic are considered: either uniform or bipolar. When traffic is uniform, the arrival rate of a packet is uniformly distributed between (0,1) around the mean. When traffic is bipolar, only two different rates r_1, r_2 are possible with $r_1 = 400 \cdot r_2$.

As can be seen in both Figures 2 and 3, the performance in terms of data extraction rate (DER) under the proposed scheme is much better than any of the other two schemes. The reason is that our proposed scheme, based on Theorem 1, tries to keep a balance of traffic load between SFs, while the other two don't have this

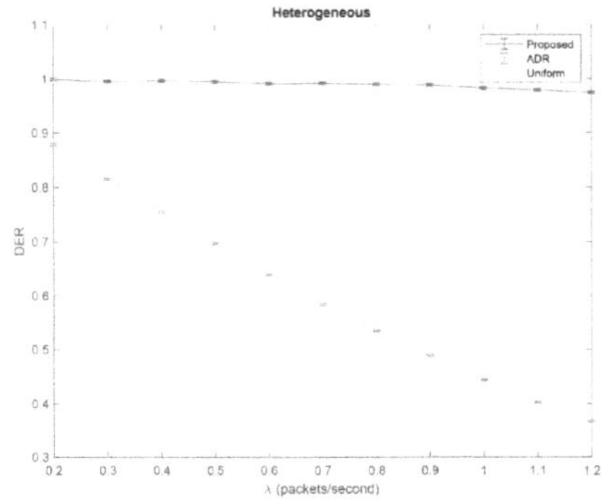

Figure 2. A comparison of data extraction rate (DER) between the proposed scheme, ADR and Uniform under uniform traffic.

in mind. Compare further between Figures 2 and 3, both the proposed scheme and Uniform have the same trend, no matter whether the traffic is uniform or bipolar. However, ADR has a worse DER under bipolar traffic than under uniform traffic.

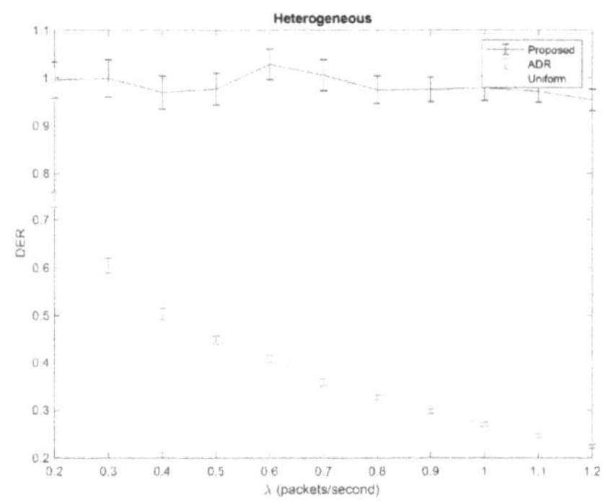

Figure 3. A comparison of DER between the proposed scheme, ADR and Uniform under bipolar traffic.

4 CONCLUSION

In the paper, a scheme is proposed to balance traffic load between spreading factors (SFs). Simulation results demonstrate that our proposed scheme is able to provide a much better data extraction rate (DER) than the other two schemes in the literature.

REFERENCES

F. Cuomo, M. Campo, E. Bassetti, L. Cartella, F. Sole and G. Bianchi, 2018. *IEEE Med-Hoc-Net.*

D. Garlisi, I. Tinnirello, G. Bianchi and F. Cuomo, 2021. *IEEE Trans. Wirel. Commun.* 20(3) 2019–2033.

P. He, L. Zhao, S. Zhou and Z. Niu, 2013. *IEEE Trans. Wirel. Commun.* 12(7) 3637–3647.

LoRa Alliance Technical Committee, 2017. *LoRa Alliance.*

R. B. Sorensen, D. M. Kim, J. J. Nielsen and P. Popovski, 2017. *IEEE Wireless Communication Letters* 6(5) 566–569.

Semtech, 2015. http://www.semtech.com/.

System Innovation for a Troubled World – Tak Lam et al. (Eds)
© 2023 The Author(s), ISBN: 978-1-032-45525-9

Interesting bookkeeping application-based on constructing financial new knowledge

Shou-Cih Chen, Li-Feng Lin, Shih-Han Huang*, Jia-Yi Shih, Chia-Fu Chang
Department of Management, Air Force Institute of Technology, Kaohsiung, Taiwan

ABSTRACT: Since most teenagers have a weak concept of financial management and lack the habit of continuous bookkeeping, they cannot achieve the effect of controlling their cash flow. According to the Technology Acceptance Model (TAM), developers can use external factors to influence users' attitudes and intentions. There are many kinds of Bookkeeping Application programs on the market, but most of them would not continuously motivate users to use them. If users have interesting game functions, they often regard the programs as games which have losene their meaning because the games are too distracting. Our research mainly designs a set of application programs that combine interesting bookkeeping, financial news, and stock market information. By combining the bookkeeping application program and interesting games, it attracts users and develops bookkeeping habits. It is different from the bookkeeping application program on the market. One of our functions to connect financial news sites, allowing users to learn about investment, financial news, and stock market trends directly through this app, and to cultivate users' sensitivity to the financial market environment.

1 INTRODUCTION

Since we were born, everyone's life is inseparable from money, which highlights the importance of rational planning for money. Most of the parents provided food, clothing, shelter, and transportation to their children. When growing up without worrying about food and clothing, it is easier for people to have no concept of the source and management of money, especially for a teenager who has just grown up. Therefore, our study uses the relevant literature on the level of financial literacy among teenagers. To explore it, Lusardi, A. et al. pointed out that the personal characteristics of young people are important factors that affect the level of financial literacy. (Lusardi A. et al. 2009) Adolescents with high average annual household income have relatively higher financial literacy levels. In addition, the initiative immediacy, and source of updating financial information from young people also have a significant positive impact on improving financial literacy (Doyle 1992). From the three sources of financial information: newspapers, magazines, the Internet, and TV news, they would obtain the highest financial knowledge. Finally, the level of financial literacy has a significant positive impact on adolescents' financial behaviors (habits of keeping accounts, habit of budgeting for income and expenditure, and behavior of saving).

When smartphones are very common, the bookkeeping applications are very diverse, but users are still easily interrupted in use because of the lack of motivation. Davis (1989) proposed that the Technology Acceptance Model (TAM) is mainly explore the influence of external factors on users' beliefs, attitudes, and intentions. Therefore, it is widely used in explaining IT acceptance behavior or predicting the influencing factors of IT use. We hope to use it and design this bookkeeping program which users are willing to use it continuously. Besides, there is no application that combines bookkeeping functions and financial news at present, so the financial management concept education function that other accounting programs have is weaker.

According to the bookkeeping applications on the market, we found that bookkeeping applications should focus on the simple and efficient recording of consumption behavior, and how to improve the basic input function and effectively enhance the bookkeeping motivation is one of the key points of our design. The bookkeeping applications should adop flexibly classify, which could use simple icons to mark classification items, input intuitively consumption data, and intuitive design to operation. The position of the component should not be biased or small and prevent the component from being not able to be seen, especially important function keys need to pay attention to this point. Lastly, it is necessary to continuously strengthen the function of accounting development which aimed to avoid the interesting presentation of distracting users and losing the meaning of accounting.

Therefore, this research focuses on the shortcomings of the bookkeeping program and developes the improvement of the accounting program, in order to better meet the needs of users and allow them to continuously use. App Inventor 2 is one of the few

*Corresponding Author: huangsh0416@gmail.com

DOI 10.1201/9781003377399-5

platforms that provide free services and has the most complete functions, and its drag-and-drop building block design makes the design process easier to use. To maximize the utilization of all functions, this research chooses this platform for development. In order to avoid the game being too distracting, the game with Bookkeeping Application programs will be designed to allow users to continuously log in and record to obtain rewards, so as to achieve the purpose of bookkeeping. In the part of the knowledge game, we choose to use the scoring method of answering questions, so that users can gradually accumulate correct financial knowledge through small tests. We ensure that young people more willing to actively learn financial-related knowledge.

2 CONCEPTUAL FRAMEWORK AND RESEARCH METHODS

We use App Inventor 2 to design this accounting application of bookkeeping application program. In terms of application framework, we design accounting, search, edit, financial knowledge, financial games, other functions for use, and in the financial knowledge, and established a web database. By linking to various well-known financial websites, such as Yahoo stock market, Commonwealth magazine, etc., it is easier for users to access and learn the changes and knowledge of the relevant financial environment, and also achieve the purpose of de-signing application programs. The architecture is shown in Figure 1.

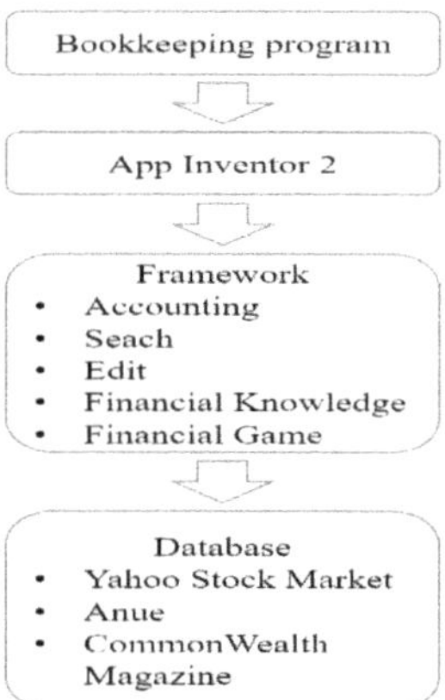

Figure 1. System architecture.

In the part of the code, the functions of the financial knowledge page and the financial game are different from those accounting programs in the market. In the following, we will aim to the operation of these two parts of the code to explain and introduce in more depth.

First, it is in the code section of financial knowledge. When the button is clicked, the web browser of the APP with the set URL will be opened and linked to the web page, as shown in Figure 2.

In addition, the user should be required to presses the selected random question and answers. If the answer is equal to the option (correct answer) in the second column of the list, the following things will be executed for judging right or wrong in the financial game. Then, the step will set to false in order to avoid this question become the next question again. It will show the words "Great, you are right", and the number of correct answers is increased by 1. As shown in Figure 3.

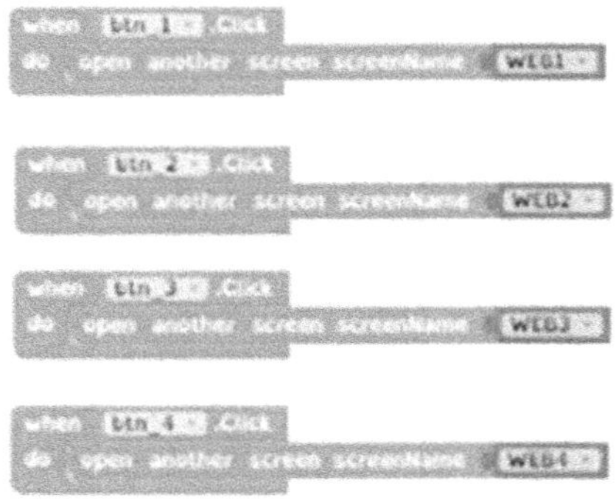

Figure 2. Code for financial knowledge webpage.

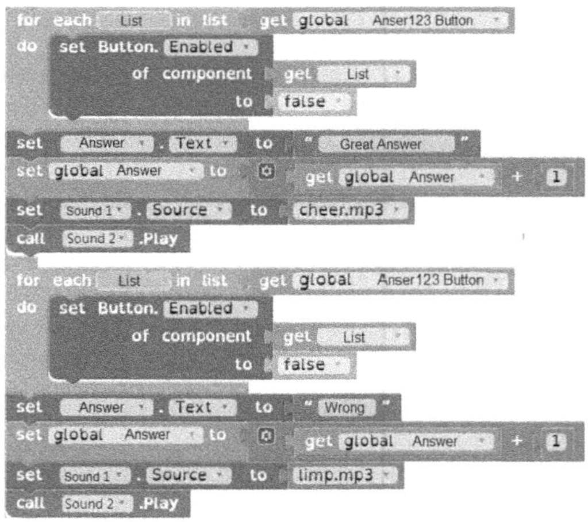

Figure 3. Code for financial game.

3 EXPERIMENTAL RESULTS AND DISSCUSION

In order to verify the difference between this accounting application and the existing accounting applications in the market, when testing the functions, the research team noticed that before setting up the whole system, some user tests are required to know which functions are designed correctly, and find errors, and then seek new solutions.

In the user test, since this study uses adolescents as the main user group, 120 adolescents between the ages of 18 and 25 are selected to conduct a simple usage test. The test time for each adolescent is about 3 to 5 minutes, and user feedback is obtained from it. After the test, about 80% of the young people have significantly improved their willingness to learn financial knowledge, and because the interface is easy to familiarize with, it also increases the willingness of users to continue to use it. Another 20% of users are less willing to learn financial knowledge.

In the testing process, it was found that the intuitive operation interface of the accounting program allows users to easily get started. It also has a bonus effect on interesting games, making users interested in testing their financial knowledge and using the financial management website. Links allow users to click on today's headlines to view relevant financial knowledge and stock trends, and the use effect is very good.

In order to enable users to operate this application more intuitively, the icons used in this research are symbols that conform to the usage habits of ordinary users, such as the calendar symbol for the query date, the upper and lower symbols for income and expenditure, and the financial knowledge symbol. The game handle symbols used in the game are all designed in line with the usage habits of ordinary users, but each function also uses a clear interface, so that users can easily understand how to operate, such as the billing interface, which is a simple click Date, enter expenses or income, and finally enter the amount.

As soon as you enter the application, you can clearly know the interface design of this program, so that users can intuitively click on the functions of the accounting program, the calendar symbol is the function of inputting the date, and the first upward and downward pattern is the input of income and payout function, the curve pattern represents the function of financial news, and the game handle symbol used in the financial knowledge game, as shown in Figure 4.

Users can intuitively add expenses or income, first select the date, enter the date, then fill in the amount of income or expenditure and select the type of money, for example, income is divided into salary, bonus, investment, etc., expenses It is divided into food, mobile phone, transportation, leisure and entertainment, etc. After inputting, it can be stored, as shown in Figure 5.

The user can click on the date and choose to increase or decrease the previous billing amount to modify it. After the modification is completed, save it and it will be updated to new data, as shown in Figure 6.

Links to websites of financial news and international stock markets, so that users can directly learn about investment and financial news and stock market trends through this application. The link name will display the news of which website, so that users can clearly click to browse the websites they are interested in, as shown in Figure 7.

In this study, we want to understand the difference with the popular accounting applications on the market. Therefore, the treasure account accounting program is compared with the accounting city which is the first and second in the App store, CWMoney and

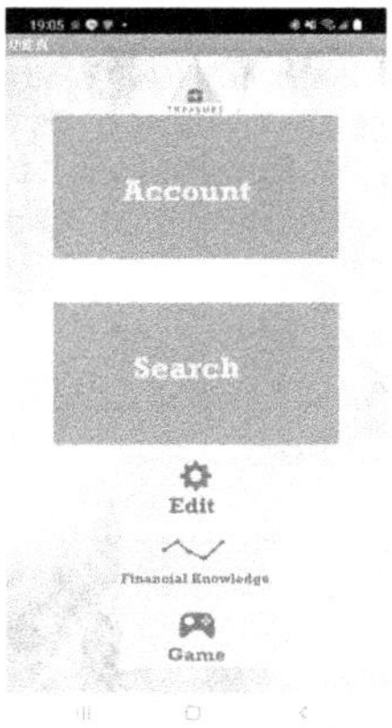

Figure 4. Main interface.

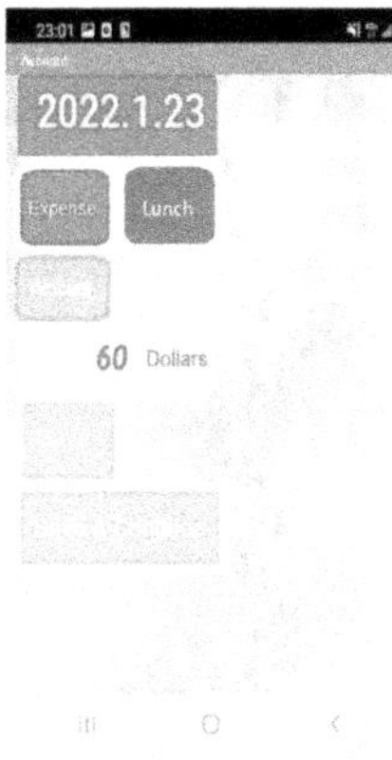

Figure 5. Account.

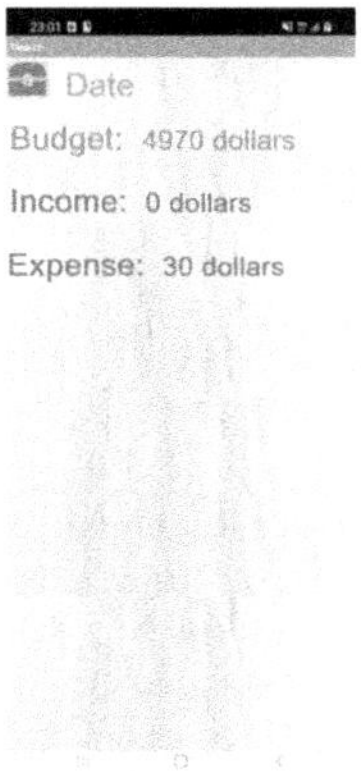

Figure 6. Search.

Figure 7. Financial news.

Monny, and has interesting games. (A. Ferreira 2014) The compared functional items are: Basic accounting functions, graphical options etc. Tick means the app has this function, and cross means the app does not have this function, as shown in the Table 1.

The comparison result shows that all accounting programs have the basic functions, except for financial information, only Treasure Account has this function, allowing users to inquire or read the financial information and headline news of the day through this program. As for the function of interesting games, this accounting program is slightly better, because although the game in Accounting City is very cute and lively and interesting, it attracts users' attention, but it is easy to cause distraction and interfere with the use of the interface, resulting in confusion for users, for example, because they want to play The game is messed with accounting or the game board is too rich, so that the user has to study and explore how to modify the accounting data, and cannot operate intuitively. The following will explain the differences in the use of each function one by one.

Since it is found that none of the accounting programs on the market have the function of integrating financial information, so that users cannot get the latest financial information through the accounting program, so this research has created a function that combines financial information, so that users can use the treasure account accounting application, to learn and acquire financial knowledge and information.

Table 1.

	Treasure account	CWMoney	Huobao Rabbit	Accounting city
Basic accounting functions	V	V	V	V
Graphical options	V	V	V	V
Modify historical data	V	V	V	V
Query historical data	V	V	V	V
Budget	V	V	V	V
Financial Information	V	X	X	X
Fun game	V	X	X	V

After comparing the advantages and disadvantages of the functions of various accounting programs, the interface design of the accounting programs in this study is consistent with the design on the market. In addition, a new financial management function that has not been designed in the accounting programs on the market has been created. This research has added this function to Treasure Account The accounting program allows users to directly view the financial information of related websites, and the fun game design is simple and interesting, and it will not take away the original intention of the accounting program.

4 CONCLUSION

This research mainly combines the links of common mini games and new knowledge about financial management, as well as the daily required bookkeeping function, using games combined with bookkeeping to achieve fun and increase users' willingness to use, in order to verify the external factors (interesting) on users' interest. The relationship between beliefs, attitudes and intentions, and design a set of accounting programs that conform to users' cognition and achieve accounting habits.

In the past, although there were many kinds of billing programs, most of them focused on the billing function, and ignored the direction of increasing users' willingness to use. Therefore, many users often gave up halfway on the long road of billing, so that they could not really achieve. For the purpose of controlling the flow of money, the accounting program of this topic takes the user's willingness to use as the starting point, and uses interesting games to attract users to develop the habit of accounting, which can be closer to the user's usage habit, and links to websites related to financial news Links that allow users to increase their financial knowledge through this app. It also provides future related programmers who can use this accounting program as an example to make a more complete and user-friendly accounting program, not just an accounting tool, but also a way to help users improve their accounting habits and establish financial management concepts. application.

The accounting program of this study will continue to strengthen the development of the accounting function. And it attracts users to start billing in a variety of interesting ways, and for those who want to continue to maintain their billing habit, it provides an efficient and simple billing function interface, so that users can maintain their billing habit. The application of this research will aim to operate more intuitively and continuously improve the user experience to better meet the needs of users.

In the future, we expect to use other software to design this accounting program, so that the overall interface of the game can be more optimized, and the accounting function can be more complete. As the application of this research currently only provides domestic related financial news and lacks the trend of the international economic environment, it is hoped that more diversified financial related news will be added in the future.

This research uses App Inventor 2 to develop accounting applications. The interface optimization may not be as complete as the listed accounting applications described in the previous chapters. However, the main purpose of this study is to combine the functions of new financial knowledge, and the user interface will be continuously optimized in the future. It will make the screen monotonous and bright, in order to attract more users to use this billing application.

REFERENCES

Doyle, K. O, 1992. Toward a Psychology of Money. American Behavioral Scientist, 35(6), 708–724.

Davis, F. D, 1989 Perceived Usefulness, Perceived Ease of Use, and User Acceptance of Information Technology. MIS Quarterly, 13 (3), pp. 319–340.

A. Ferreira, 2014 Personal finances mobile application for seniors. Faculdade de Engenharia da Universidade do Porto.

Gurney, K, 1988. Your Money Personality: What It Is and How You Can Profit from It. New York: Doubleday.

Lusardi, A., Mitchell, O. S. and Curto, V.,2009. Financial Literacy Among the Young. Michigan Retirement Research Center, University of Michigan, No. 2008-191.

Zeinab Alizadeh Barmi and Amir Hossein Ebrahimi, 2011 Automated testing of non-functional requirements based on behavioural scripts. Chalmers University of Technology. Department of Computer Science and Engineering.

System Innovation for a Troubled World – Tak Lam et al. (Eds)

Improving the throughput performance of a multichannel MAC protocol in RSU-assisted vehicular ad hoc networks

Jeng-Ji Huang* & Sheng-Hsiang Lin
Department of Electrical Engineering, National Taiwan Normal University, Taipei, Taiwan

David Shiung
Department of Electronics Engineering, National Changhua University of Education, Changhua, Taiwan

ABSTRACT: Vehicular Ad Hoc Networks (VANETs) have widely been considered as a promising wireless communication technology to offer both vehicle safety and infotainment. Much research attention has been devoted to designing a multichannel MAC protocol in VANETs, with the aim of maximizing data throughput while ensuring collision-free deliveries of safety messages. For instance, in the coordinated multichannel medium access control (C-MAC) protocol, a time division multiple access (TDMA) approach is employed for the latter, and a balance of the rates is stricken for the former between successful requests for data transmissions and data transmissions that can be performed. However, C-MAC may suffer from channel underutilization as a contention-based mechanism is used to make reservations for data transmissions as well as to identify new vehicle arrivals at the coverage of a road side units (RSU). Two designs are thus proposed in this paper to improve channel utilization. Not only is TDMA better leveraged under both designs, but local information is utilized in one of the designs, which is regularly gathered at a vehicle by receiving beacons from neighboring vehicles. Simulations show either design can substantially outperform C-MAC in terms of throughput. Compare further between the two designs, though the one using local information can achieve a higher throughput than another retaining a vehicle identification process when vehicle intensity is low on a road, the reverse is true when it is high. A switching rule is then proposed and its accuracy is confirmed from numerical results.

1 INTRODUCTION

Vehicular ad hoc networks (VANETs) are an emerging wireless communications technology that plays a crucial role in the realization of an intelligent transportation system (ITS) (M. N. Ahangar et al. 2021). For example, they can be helpful in collecting relevant information about vehicles regarding automated driving systems. Numerous research efforts have been devoted to various technical issues that pertain to potential applications in VANETs, e.g., clustering of vehicles (W. Ahsan et al. 2020), routing (C. Tripp-Barba et al. 2019), real-time path planning (M. Wang et al. 2015), and connectivity (S. Jobaer et al. 2020), etc. Roadside units (RSUs) can greatly help to enhance the performance of a VANET in various aspects, e.g., facilitating communications between vehicles and network infrastructure (H. Ghafoor et al. 2019), cooperating with nearby vehicles for safety message forwarding (G. Xiao et al. 2021), or, as Internet access points, playing the role of base stations in cellular networks (S. Kim 2019).

Several designs include an RSU to assist in channel allocations to vehicles located within its coverage in order to achieve better channel utilization. For instance, a collision-free time division multiple access (TDMA) MAC protocol is proposed in (W. Guo et al. 2012) that the number of TDMA slots can be adapted to vehicle intensity at an RSU. Though transmissions can be guaranteed to be collision-free, a waste of channel bandwidth may occur when a vehicle has no data to transmit during the time slots allocated to it. A scheduling algorithm is proposed to schedule data transmissions on SCHs during SCHIs at an RSU in (K. Liu et al. 2016); however, no detail is elaborated about how reservations are made for data transmissions. In (Y. Kim et al. 2016), a coordinated multichannel medium access control (MAC) protocol, called C-MAC, is proposed for VANETs. In C-MAC, the boundary within an SI is dynamically adjusted in such a way that the rate of reservations made for data transmissions equals the rate of data transmissions that can be performed. Though safety messages can be guaranteed to be delivered in a collision-free manner as TDMA is employed, C-MAC has the drawback that a contention-based mechanism is used to make reservations for data transmissions as well as to identify new vehicle arrivals at the coverage of an RSU.

Thus, two designs are proposed in this paper to improve the channel utilization in RSU-assisted VANETs. In one of the proposed designs, the local

*Corresponding Author: hjj2005@ntnu.edu.tw

DOI 10.1201/9781003377399-6

information about neighboring vehicles, which is gathered regularly at a vehicle by receiving beacons from other vehicles, is utilized. An RSU can therefore be notified of new vehicle arrivals by those that are located within the coverage of the RSU and have already been identified. In addition, TDMA is better exploited under the proposed designs; as a result, not only notifications of new vehicle arrivals but reservations for data transmissions can be made in a collision-free manner by a vehicle.

2 SYSTEM MODEL

Without loss of generality, a bidirectional E-W road or highway is considered, along which RSUs are deployed. For simplicity, it is assumed that a vehicle and an RSU share a common transmission range, which is denoted as R (meters), and there may or may not be an overlap between the communication ranges of two neighboring RSUs. Note that only when vehicle traffic intensity is relatively high will an RSU be activated. It is because, in general, a contention-based mechanism works well under low vehicle traffic intensity, and assistance from an RSU is needed when vehicle traffic intensity is high.

The speed of a vehicle, denoted as V (meters per second), follows an independent Gaussian (μ_V, σ_V^2) distribution, where μ_V, σ_V^2 are the mean and variance of V. Let λ (vehicles per meter) be the traffic intensity in either direction of the road, and n be the number of vehicles within the coverage of an RSU in an SI. It should be clear that n is a random variable, which correlates with λ, V, and R. On the other hand, it is assumed that each vehicle is equipped with only one wireless transceiver. Let N_{SCH} be the number of SCHs available for vehicles located within the coverage of an RSU.

An SI, whose length is denoted as T_{SI}, is subdivided into a CCHI and an SCHI, as depicted in Figure 1, and the former typically contains a safety message phase (SMP) and a channel reservation phase (CRP). The SMP is basically made up of TDMA slots, each of which is allocated to a vehicle that is located within the coverage of an RSU and has been identified by the RSU. Safety messages are sent by vehicles in the time slots allocated to them. Note that safety messages sent by a vehicle serves to inform the RSU that the vehicle is still located within its coverage so the same time slot can be allocated to the vehicle in the next SI. The delivery of safety messages can therefore be collision-free. Denote by T_{SMP} the length of the SMP, by T_{SS} the length of an SMP slot time, and by m the number of vehicles that have been identified. As will be shown in Sec. 3, a vehicle can always be identified shortly after it moves into the coverage of an RSU, suggesting that $m \approx n$. Thus, $T_{SMP} = m \cdot T_{SS} \approx n \cdot T_{SS}$.

3 THE PROPOSED DESIGNS

As mentioned previously, using IEEE 802.11p to make reservations for data transmissions is one of main drawbacks in C-MAC. It is because a contention-based mechanism may cause a waste of bandwidth as there inevitably exist idle or collision slots.

In order to tackle this problem, it is proposed in both Ω_1 and Ω_2 that for vehicles that are located within the coverage of an RSU and have been identified by it, whenever they intend to make reservations for data transmissions, requests are simply piggybacked in their own safety messages, which are sent in their respective SMP time slots. Likewise, after the reception of a request by the receiving vehicle, a response is replied via piggybacking in its safety messages, which are sent in its own SMP time slot. Thus, as long as the sending and receiving vehicles have both been identified by the RSU, reservations for data transmissions can be made in a collision-free manner. After a request and the corresponding response have been collected, the transmission can then be scheduled, at the RSU, and the schedule is included in the LI packet that is broadcasted at the beginning of the next SI.

By removing CRP, either Ω_1 or Ω_2 enjoys a longer SCHI than C-MAC. Denote the length of SCHI in either of our designs by T'_{SCH}. It is clear that $T'_{SCH} = T_{CRP} + T_{SCH} = (1 + \alpha) \cdot T_{SCH}$. Namely, either Ω_1 or Ω_2 has $(1 + \alpha)$ times more bandwidth for data transmissions than C-MAC.

Though CRP is removed, Ω_1 possesses a vehicle identification process the same as C-MAC. Thus, Ω_1 and C-MAC share a common number of SCHs, i.e., N_{SCH}. The throughput of Ω_1, denoted as G_{Ω_1}, can thus be expressed as $G_{\Omega_1} = \frac{D \cdot N_{SCH} \cdot \frac{T'_{SCH}}{T_{trans}}}{T_{SI}} = (1 + \alpha) \cdot G_{C-MAC}$.

In addition to CRP, the use of a vehicle identification process may cause a waste of bandwidth. It is because the entire SCHI on CCH is dedicated to the process, even if it can quickly be completed when there are only a small number of new vehicle arrivals in an SI. Thus, we propose that the process be removed and that the SCHI on CCH be used only for data transmission in Ω_2. The reason why it can be removed is that an RSU can instead be informed of new vehicle arrivals at the RSU's coverage by utilizing the local information of neighboring vehicles that is gathered and updated regularly by a vehicle from the receptions of beacon signals broadcasted periodically by nearby vehicles.

Assume the information about neighboring vehicles is available at each vehicle, a vehicle that has already been identified can thus report its neighbors to the RSU via piggybacking in safety messages. After the information is received by the RSU, a SMP time slot can then be allocated to each arriving vehicle. Denote by N''_{SCH} the number of SCHs in Ω_2. Clearly, $N''_{SCH} = N_{SCH} + 1$, assuming that the SCHI on CCH can be used for data transmissions. In the following, we illustrate how to estimate the number of new vehicle arrivals in an SI, such that the number of SMP time slots to be allocated can be determined.

New eastbound vehicle arrivals at the coverage of an RSU must emerge from the west communication range of the RSU, and depart from its east communication range. Let n_e be the number of eastbound vehicles located between an RSU and its west communication range at a particular instant of time, e.g., the end of an SI; clearly, $n_e \approx n/4$. For those n_e vehicles, their neighbors which will soon become new arrivals should be located between the RSU's west communication range and the location that is to its west and R away of it. Obviously, the number of such eastbound vehicles is approximately n_e. Therefore, the total number of SMP time slots that will be allocated to those soon-to-be new vehicle arrivals in an SI can be estimated to be $2 \cdot n_e \approx n/2$ in Ω_2.

Denote the length of an SMP as T''_{SMP} in Ω_2. Clearly, $T''_{SMP} = (1 + \beta) \cdot T_{SMP}$, where $\beta = 0.5$. As before, the throughput of Ω_2, denoted as G_{Ω_2}, can thus be expressed as $G_{\Omega_2} = \frac{D \cdot N''_{SCH} \cdot \frac{T''_{SCH}}{T_{trans}}}{T_{SI}}$ where $T''_{SCH} = T_{SI} - T''_{SMP}$. Note that, since T_{SMP} depends on n, G_{Ω_2} is a random variable in an SI.

4 NUMERICAL RESULTS

In this section, results obtained from simulations written in MATLAB are shown and discussed. Relevant system parameters include $R = 150$ m, $\mu_V = 80$ km/hr, $\sigma_V = \sqrt{\frac{\mu_V}{5}}$ km/hr (or 22.2 m/s), $\lambda = 0.1 \sim 0.3$ (vehicles/m), the number of lanes in either direction is 5. Locations of vehicles are initially Poisson distributed with intensity λ on the road. After a simulation begins, they move with a speed V. Each simulation takes 100 seconds, or 1000 T_{SI}s. Note that $\frac{T_{SI}}{T_{SS}} = 200$.

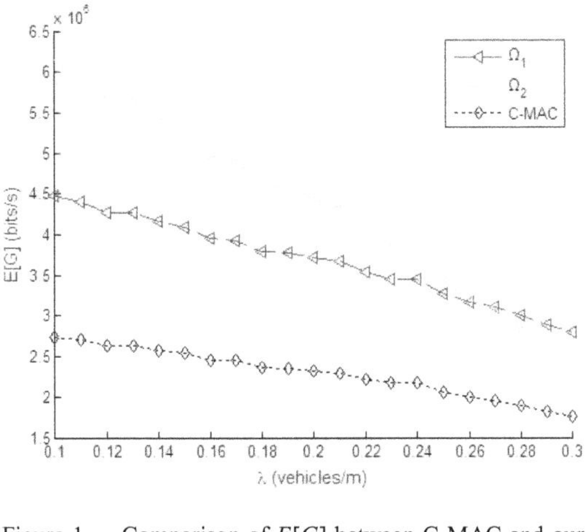

Figure 1. Comparison of $E[G]$ between C-MAC and our proposed two designs.

An average throughput, denoted as $E[G]$, is thus calculated from 1000 samples, and it is compared in Figure 1. As shown, either of our designs outperforms C-MAC under any λ. While, as mentioned above, the gain over C-MAC is 61% on average in Ω_1, it is between 20% and 120%, or 70% on average in Ω_2. Note that $E[G]$ declines as λ increases in either C-MAC, Ω_1, or Ω_2. It is because n, i.e., the number of vehicles within the coverage of an RSU, tends to raise with a growing λ, leading T_{SCH} to shrink, as discussed earlier.

The switching rule can be written as $80 \underset{\Omega_1}{\overset{\Omega_2}{\gtrless}} n$. As discussed previously, though n may fluctuate from SI to SI, the duration to check whether a switch is needed is typically far longer than T_{SI}. Therefore, n can be replaced with $\bar{n}$, which equals $\lambda \cdot R$. The rule can thus be rewritten as $0.267 \underset{\Omega_1}{\overset{\Omega_2}{\gtrless}} \lambda$. As can be seen in Figure 1, the threshold, i.e., 0.267, is very near the value of λ where $E[G]$s of Ω_1 and Ω_2 are equal. This confirms that our proposed switching rule is quite accurate.

5 CONCLUSION

In this paper, two designs are proposed in RSU-assisted VANETs to improve the throughput of C-MAC by leveraging the orthogonality of TDMA and by utilizing the local information about neighboring vehicles gathered regularly at a vehicle. Numerical results demonstrate that either of the proposed designs can significantly provide a higher throughput than C-MAC, with a gain of at least 60% on average. In addition, a rule for switching between the proposed two designs is numerically derived, and results show that the proposed rule is very accurate.

REFERENCES

M. N. Ahangar, Q. Z. Ahmed, F. A. Khan, and M. Hafeez, 2021. *Sensors* 21 706.

W. Ahsan, W, M. F. Khan, F. Aadil, M. Maqsood, S. Ashraf, Y. Nam, and S. Rho, 2020. *Electronics* 9 394.

W. Guo, L. Huang, L. Chen, H. Xu, and J. Xie, 2012. *International Conference on Wireless Communications and Signal Processing* 1-6.

H. Ghafoor and I. Koo, 2019. *Wirel. Netw.* 25 1711–1729.

S. Jobaer, Y. Zhang, M. A. Iqbal Hussain, and F. Ahmed, 2020. *Electronics* 9 1499.

K. Liu, J.-Y. Ng, V. C. S. Lee, S. H. Son, and I. Stojmenovic, 2016. *IEEE/ACM Trans Netw* 24(3) 1759-1773.

Y. Kim, M. Lee, and T.-J. Lee, 2016. *IEEE Trans. Veh. Technol.* 65(8) 6508-6517.

S. Kim, 2019. *Wirel. Netw.* 25 1815-1827.

C. Tripp-Barba, A. Zaldivar-Colado, L. Urquiza-Aguiar, and J. Aguilar-Calderon, 2019. *Electronics* 8 1177.

M. Wang, H. Shan, R. Zhang, X. Shen, and F. Bai, 2015. *IEEE Trans. Veh. Technol.* 64(5) 1664-1678.

G. Xiao, H. Zhang, N. Sun, and Y. Zhang, 2021. *Wirel. Netw.* 27 1335–1351.

System Innovation for a Troubled World – Tak Lam et al. (Eds)
© 2023 The Author(s), ISBN: 978-1-032-45525-9

Form performance of a rotating spiral cone under the apparent movement: Examining different styles of discontinuous lines

Guang-Dah Chen*
Department of Visual Communication Design, National Taiwan University of Arts, Banqiao, New Taipei City, Taiwan

Hsiwen Fan
Department of Design, Graduate School of Science and Engineering, Chiba University, Chiba, Japan

ABSTRACT: Many studies have shown that using visual effects induced by different motion perceptions can be a way to explore new shapes. In this study, the apparent movement of motion perception was taken as the subject. It was found that previous studies mainly focused on the placement of continuous lines on the form and explored the formative changes of these lines, hence, such studies lacked the exploration of discontinuous line formations. Furthermore, this study also aimed to enhance its reliability by using an eye tracker to improve the past observation method, which merely employs human eyes. This study invited 20 subjects to view a rotating spiral cone under the apparent movement while wearing an eye tracker and recorded their saccade amplitudes of eye movement messages of different patterns of discontinuous lines on the cone. The results indicated that, for the saccade amplitude effects of eye movement message, the circle was the geometric shape that showed the least effect. Meanwhile, triangles are larger than squares and circles.

1 INTRODUCTION

As one of the important theories in motion perception, many studies (Chen et al. 2008; Chang & Chen 2015; Chen et al. 2015a, 2015b) have applied the perception of apparent movement in shape-making development. However, these studies were all based on the continuous lines of the form, lacking the discussion of nonlinear and discontinuous forms. Thus, to fill this gap, this study attempts to probe how the discontinuous lines formed by different styles behave on the rotating spiral cone under the apparent movement.

In particular, this study recorded the scan path by an eye tracker to investigate the difference between the perceptual course of the apparent movement and the shape change. The scan path can be measured by recording its changes and the distribution of the kinesthetic response when the subject is viewing a dynamic shape to understand the response of perceptual eye movement message. In this way, such results can practically contribute to the treatment of human visual behaviors and provide a theoretical basis for visual perception and the development of shapes in the related field of shape making about motion perception.

Based on the concept of velocity threshold in psychophysics, this study conducted experiments through the adjustment method. Under the rotational motion, different patterns of discontinuous line changes were measured from the change in the visual trajectory of eye movements and the visual trajectory and eye movement message of eye movements.

2 EXPERIMENTAL

In this study, the independent variables include the lines formed by the three discontinuous line graphs and the absolute upper and lower thresholds of the speed of the seeming motion. On the other hand, the eye-movement message responses were the dependent variables. The study observed whether different patterns of discontinuous line graphical changes affect the distribution and state of the visual trajectory of the apparent movement's perceptual response from the comparison between the reference value of the speed threshold and the change of the eye movement's visual trajectory. It also compared whether there was a difference in terms of line of sight trajectory and regional area distribution. Further, the study investigated whether there is an interrelationship and causality between the seemingly kinetic movements and their eye movement message responses from the intersection of the speed threshold's reference value and the change of the visual trajectory. Moreover, this study is a within-subjects design, of which the three discontinuous line graphs were randomly ordered to avoid sequencing effects.

There were a total of 20 subjects sampled, including 10 male and 10 female, aged between 23 and 35 years, with normal visual acuity.

*Corresponding Author: gd196478@yahoo.com.tw

 DOI 10.1201/9781003377399-7

Table 1. Experimental Samples (drawn for this research).

Graphics on the surface	Square	Circle	Regular triangle
Graphics on the surface Experimental Samples	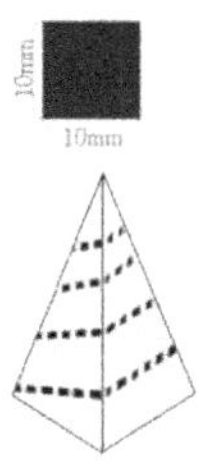	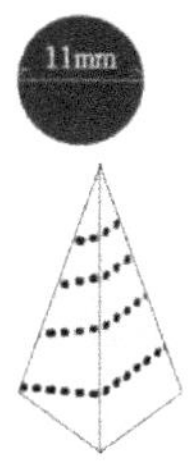	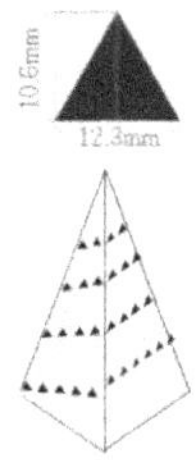

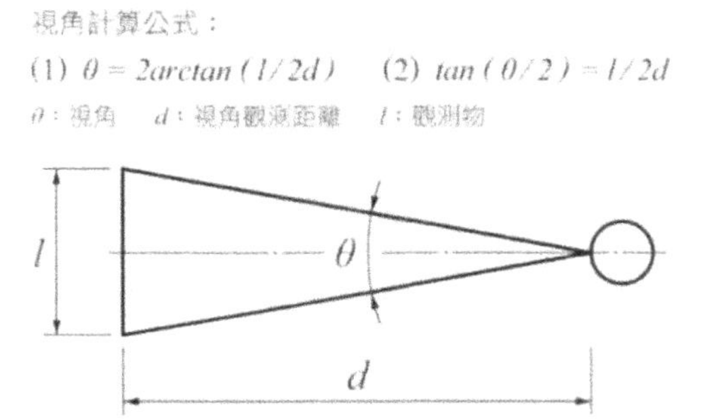

Figure 1. Formula diagram of the angle of view calculator (Oyama et al. 2000).

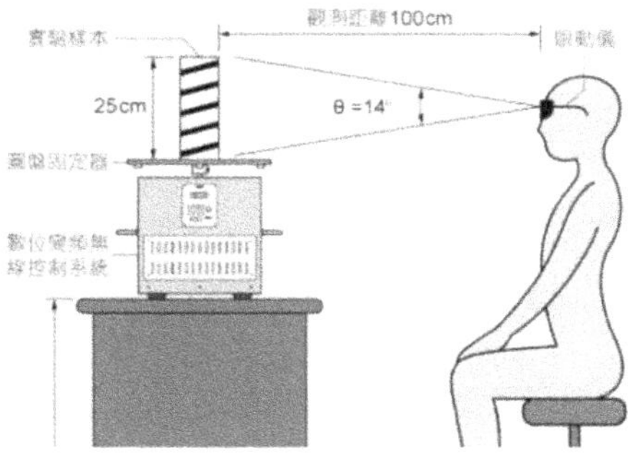

Figure 2. Schematic diagram of the experimental environment.

Based on Chen et al. (2015a) and Chen et al. (2015b) a triangular cone was chosen, which is the most suitable shape to represent the seemingly dynamic motion in this study. People are prone to the illusion of size (or area) in the volume of shape making, which is closely related to the morphological and characteristic elements of the shape. In order to avoid the influence of such illusions on the perceptual response to the dynamic illusion of motion in the experimental samples, the body areas of the three experimental samples (Table 1) were set as follows. (1) With the formula and rules of the mathematical, graphical method, known external circles were used to find the circumscribed polygon. (2) The visual angle was calculated by the formula, "$\theta = 2\arctan$ (l/2d)" (Figure 1), to obtain the best viewing angle of 14 (14.25) degrees at the observation distance of 1 M. (3) Based on the best viewing angle of 14 degrees, the height of the cone was calculated as 25 cm by the replacement formula, "tan ($\theta/2$) = l/2d". (4) According to Yiu (2010), the perceptual preference for the length-to-width ratio of the cone was found to be 2.1:1; thus, the width of the column was calculated as 12 cm. (5) The angle and line widths of the continuous line pattern on the cone surface were set to 10-mm line width and 15-degree angle of the continuous parallel spiral straight line, as suggested by Chen et al. (2008), which has the most dynamic illusion effect (the widest speed threshold). In other words, such a line graph maintains its dynamic illusion effect at different speeds. (6) Solid samples were modeled by a 3D software named SolidWorks and produced by a 3D printing technology, avoiding the tooling effect due to manual production.

Before the experiment, the subjects were instructed to use the digital inverter wireless control system and remote control. They were also instructed to master how to view and discern the form and perceptual effect of spiral discontinuous lines on the cone's surface and the apparent movement during rotation. Subsequently, the subjects were given three practice sessions to familiarize themselves with the experimental procedures. Then, the experiment commenced with giving instructions to the subjects. In the case of uncertainties, the mechanism of the experiment would be explained to the subjects until they understood. The subject sat 1 m in front of the sample at the height of 14 degrees in the experimental environment (Figure 2). Meanwhile, the instructor started the digital inverter wireless control system with experimental samples to gradually increase the speed from 0 to 1300 rpm. The examinees used the remote control buttons to make repeated speed adjustments to confirm the subjective perception of the form and perceptual effect of the apparent movement. In addition to recording the absolute thresholds of upper and lower speed, the instructor recorded the number of gazes, the amplitude of saccades, and the change and distribution of the visual trajectory of the subjects' perceptual response to the apparent movement. In particular, each recording was made in 10-second increments until the experiment

Table 2. Statistical table of the seemingly kinesthetic motion perception and eye movement message experimental saccade amplitude values for three discontinuous line figures (Unit: pixel).

Geometric column		Eyemovement message	Min	Max	Mean	SD
Circle	Upper absolute threshold	Saccade amplitude	4.00	126.41	57.93	33.65
	Lower absolute threshold		3.11	116.54	62.53	34.78
Regular triangle	Upper absolute threshold	Saccade amplitude	1.38	118.67	66.65	30.26
	Lower absolute threshold		6.25	161.88	64.04	34.93
Square	Upper absolute threshold	Saccade amplitude	2.00	131.54	58.57	40.22
	Lower absolute threshold		6.19	130.61	67.92	35.65

Note: UAT = upper absolute threshold, LAT = lower absolute threshold.

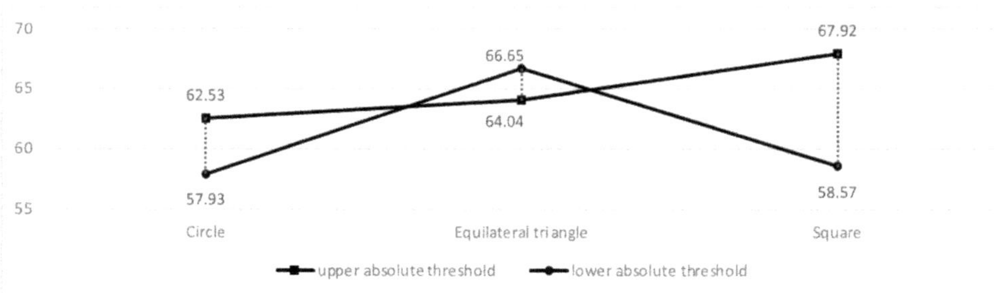

Figure 3. Trend plots of the average saccade amplitude of the apparent movement perception and eye movement messages for three discontinuous line graphs (Unit: pixel).

Table 3. A checklist of sample T for the amplitude of saccades for three types of discontinuous line graphical triangle cones with apparent movement messages, with gender independent.

	Cone samples	Male average	Female average	*p*
Circle	Upper absolute threshold	70.08	54.99	0.33
	Lower absolute threshold	73.08	42.78	0.06
Regular triangle	Upper absolute threshold	68.38	59.70	0.55
	Lower absolute threshold	70.28	63.02	0.66
Square	Upper absolute threshold	78.12	57.72	0.28
	Lower absolute threshold	65.33	51.80	0.42

was completed. The total time of this study reached about 15 minutes.

3 RESULTS

Table 2 and Figure 3 present the experiment results. As the discontinuous line patterns of geometric triangular cones changed, the amplitude of the perceptual eye movement message saccades also changed. This finding shows that when such patterns change, the corresponding amplitude will change accordingly. Further, the eye movement messages in all triangular cone discontinuity graphs showed the following: the average lower absolute threshold saccade amplitude in the circular body was 62.53 pixel, and its average upper absolute threshold saccade amplitude in the circular body was 57.93 pixel; the average absolute threshold saccade amplitude under the triangle was 64.04 pixel, and the average absolute threshold saccade amplitude above it was 66.65 pixel; the average absolute threshold saccade amplitude under the square was 67.92 pixel, and the average absolute threshold saccade amplitude above it was 58.57 pixel. As a result, the circle was the geometric cone with the smallest saccade amplitude for non-continuous line graphical eye movement messages in apparent movement. Finally, the results of the independent sample t-tests for gender showed no significant differences between men and women in the perception of the three discontinuous line graphs of the trigeminal cone and in the response and discrimination of the amplitude of eye movement messages (upper and lower absolute thresholds) (Table 3).

4 CONCLUSION

This study investigated the effect of the rotating spiral cone on the apparent movement and the difference between the discontinuous lines of the three different shapes. According to the experimental results concerning the cone dynamic illusion and eye-movement message, the circular discontinuous line has the smallest effect on the amplitude of eye-movement saccades during the apparent movement, which is the most obvious shape to manifest the illusion effect. Moreover, triangles are larger than squares and circles in the saccade amplitude effects.

ACKNOWLEDGEMENT

This study was financially supported by the Research Grant MOST 108-2410-H-224-016- and MOST 109-2410-H-224-007- from Taiwan's Ministry of Science and Technology.

REFERENCES

Chang, C. C. and Chen, G. D. (2015). A Pilot Study on Form Expressions of Induced Movement and Reversible Figure Shapes. *Journal of Design Research*, 11, 7–12.

Chen, G. D., Chang, C. C., and Lin, P. C. (2008). A Study on the Induced Motion of the Rotative Speed and the Width of Line. *Issue of Basic Design and Art*, 16, 19–22.

Chen, G. D., Lin, C. W., and Fan, H. (2015a). Motion Perception on Column of Rotational Dynamic Illusion in Kinetic Art. *Journal of Design*, 20(3). 1–19.

Chen, G. D., Lin, C. W., and Fan, H. (2015b). The Study of Motion Perception on The Rotational Motion Illusion of Cone. *Journal of Science and Technology*, 24(1), 85–101.

Liu, Y. H. (2010). *The research of preference for ratio of geometry shapes* (Unpublished master's thesis). National Yunlin University of Science and Technology, Yunlin, Taiwan.

Oyama, T., Imai, S., & Wake, T. (2000). *New. Sensory perception psychology handbook.* Tokyo: SeishinShobo Ltd.

System Innovation for a Troubled World – Tak Lam et al. (Eds)
© 2023 The Author(s), ISBN: 978-1-032-45525-9

A Markov-process-based delay analysis for uplink transmissions in LoRaWAN class A

Jeng-Ji Huang* & Han-Hui Chiu
Department of Electrical Engineering, National Taiwan Normal University, Taipei, Taiwan

David Shiung
Department of Electronics Engineering, National Changhua University of Education, Changhua, Taiwan

ABSTRACT: LoRaWAN (long range wide area network) is an emerging wireless communication network that can provide low power transmissions and wide area coverage for IoT (internet of things) end devices. Since it operates in an ISM (industrial scientific medical) band, there is a regulation of duty cycle, e.g., 1%, for each sub-band in uplink transmissions. In addition, there may be several spreading factors available for an end device to select for uplink transmissions, and the service rate of using one spreading factor is different than that of using another. In literature, for simplicity, a uniform service rate is assumed, so that the uplink transmissions are modeled as a M/D/c queueing system. Though an approximated mean queuing delay can be obtained from the simplified model, the accuracy is far from satisfactory. In this paper, a Markov process is first utilized to reflect different service rates between spreading factors, and the uplink transmissions of an end device can then be modeled as a M/G/c queueing system. Numerical results show that more accurate mean queueing delay can be obtained under our proposed analysis.

1 INTRODUCTION

We consider a typical LoRaWAN system, where end devices (EDs) are in association with a network server through one or several gateways (GWs) (D. Garlisi et al. 2021). Uplink packets are assumed to be generated from ED $i \in \mathcal{E}$ according to a Poisson process with mean rate λ_i, where $\mathcal{E}$ is the set of EDs. Furthermore, packet lengths are assumed to be independent, exponentially distributed with mean L.

Denote as R_x the bit rate of SF x, which can be expressed as $\frac{x \cdot B}{2^x}$ (Semtech 2015), where B represents the bandwidth. The transmission time for an uplink packet of length l can be obtained as $t_x = \frac{l}{R_x}$. Taking into account the duty cycle (DC) restriction imposed on a frequency band (LoRa Alliance Technical Committee 2017), denoted as δ, the service time of an uplink packet transmission can be expressed as $s_x = \frac{t_x}{\delta}$ (R. B. Sorensen et al. 2017). Since packet lengths are exponentially distributed, it can readily be shown that s_x is exponentially distributed with mean equal to $\frac{L}{R_x \delta}$. The mean service rate of SF x is therefore $\mu_x = \frac{R_x \delta}{L}$. Let X_i be the SF selected by ED i for uplink transmissions. It is assumed that the probability mass function (PMF) of X_i, $P_{X_i}(\cdot)$, is known.

Let Y_i be the frequency channel (FC) selected by ED i for uplink transmissions. In this paper, the probability mass function (PMF) of Y_i, i.e., $P_{Y_i}(y)$ is determined, where $y = 1, \cdots, N_Y$ and N_Y is the number of frequency channels. The mean service time for a packet transmission by an ED i on frequency channel y using SF x can be obtained as $\mathrm{E}[s_{x,y}] = \frac{1}{\mu_{x,y}} = \frac{L}{R_x \delta}$. Let s_y be the conditional service time, given $Y_i = y$, its conditional mean can then be obtained as

$$\mathrm{E}[s_y] = \mathrm{E}_x[\mathrm{E}[s_{x,y}]] = \frac{L}{\delta} \sum_x \frac{P_{X_i}(x)}{R_x}. \tag{1}$$

Similarly, the conditional second moment of s_y, given $Y_i = y$, can be expressed as

$$\mathrm{E}[s_y^2] = \frac{L^2}{\delta^2} \sum_x \frac{P_{X_i}(x)}{R_x^2}. \tag{2}$$

A state-dependent frequency channel selection policy is considered and $P_{Y_i}(\cdot)$ is solved by utilizing an Embedded Markov Chain (EMC) (D. Gross et al. 1985). In our proposed analysis, we take advantage of the uniformity of N_Y servers when all servers of the M/G/c queue are busy. That is, only states when the queue is empty (or queue length $n = 0$) and those when the queue is nonempty ($n > 0$) need to be considered, without having to record the exact value of a nonzero n. Thus, our proposed analysis can effectively save the state space, as will later be discussed.

Denote as $m_{i,k}$ the state of the kth server at ED i and as $\boldsymbol{m}_i$ a vector of $m_{i,k}$. The value of $m_{i,k}$ either equals

*Corresponding Author: hjj2005@ntnu.edu.tw

DOI 10.1201/9781003377399-8

the SF x_j, i.e., $m_{i,k}=x>0$, being currently used on the kth frequency channel or $m_{i,k}=0$ if the frequency channel is idle. Define the indicator function as

$$\mathcal{J}(x)=\begin{cases}1 & x\neq 0,\\ 0 & otherwise.\end{cases} \tag{3}$$

Let $\mathcal{J}(\boldsymbol{v})$ be a binary vector of length equal to that of $\boldsymbol{v}$, with the kth entry equals $\mathcal{J}(v_k)$, where v_k is the kth entry of $\boldsymbol{v}$.

In EMC, a packet arrival (or generated) at ED i may cause a state transition from $\boldsymbol{m}_i$ to $\boldsymbol{m}_i'$, where $\mathcal{J}(\boldsymbol{m}_i'-\boldsymbol{m}_i)=\boldsymbol{e}_k, \forall\, k=1,\cdots,N_Y$, and $\boldsymbol{e}_k$ represents the $N_Y\times 1$ unit vector with all elements equal to zero except the kth element equal to 1; similarly, a departure can lead to a state transition from $\boldsymbol{m}_i$ to $\boldsymbol{m}_i'$, where $\mathcal{J}(\boldsymbol{m}_i-\boldsymbol{m}_i')=\boldsymbol{e}_k, \forall k=1,\cdots,N_Y$.

As mentioned earlier, to save state space, we focus on discussing only transitions between states $\boldsymbol{m}_i$ containing at least one zero, i.e., $\boldsymbol{1}_{N_Y}^{\mathrm{T}}\mathcal{J}(\boldsymbol{m}_i)<N_Y$\$, and those without any zeros, i.e., $\boldsymbol{1}_{N_Y}^{\mathrm{T}}\mathcal{J}(\boldsymbol{m}_i)=N_Y$, where $\boldsymbol{1}_{N_Y}$ is a vector with all N_Y elements equal to 1 and $\boldsymbol{v}^T$ represents the transpose of a column vector $\boldsymbol{v}$. It is because, when all servers are busy, the system can be considered as a M/G/1 queue, where the single server is made up of N_Y frequency channels. Therefore, the mean queue length and the mean queueing delay can be found from P-K formula (D. Gross et al. 1985) as

$$L_y=\rho_y+\frac{\rho_y^2+\lambda_y^2\sigma_y^2}{2(1-\rho_y)} \tag{4}$$

and

$$\mathrm{E}\left[D_y\right]=\frac{L_y}{\lambda_y} \tag{5}$$

respectively, where $\lambda_y=\lambda_i\cdot P_{Y_i}(y)$, $\sigma_y^2=\mathrm{E}\left[s_y^2\right]-\left(\mathrm{E}\left[s_y\right]\right)^2$ and $\rho_y=\lambda_y\cdot\mathrm{E}\left[s_y\right]$.

The state space required in our analysis is therefore $(N_X+1)^{N_Y}+N_X^{N_Y}$, where $(N_X+1)^{N_Y}$ refers to the total number of states containing at least one zero and $N_X^{N_Y}$ to the total number of states without any zeros. For simplicity, denote as $\mathcal{M}_0$ the set of the former states and $\mathcal{M}_1$ that of the latter states.

2 NUMERICAL ANALYSIS

For ED i, the overall system throughput is the maximum if $\rho_y=\rho_{y'}, \forall y\neq y'$. Given all FCs are available, the event of which is denoted Ω, the conditional PMF of Y_i can be obtained as

$$P_{Y_i|\Omega}(y)=\frac{\mu_{x,y}}{\sum_y\mu_{x,y}}=\frac{1}{N_Y}. \tag{6}$$

Consider $\boldsymbol{m}_i\in\mathcal{M}_0$. Either an arrival or a departure leads to a state transition from $\boldsymbol{m}_i$ to $\boldsymbol{m}_i'$. Since the arrival rate is λ_i and the departure rate is the sum of the service rates of all busy servers, i.e., $\sum_{k=1}^{N_Y}\mu_{m_{i,k}y_k}\mathcal{J}(m_{i,k})$, where $\mu_{m_{i,k}y}=0$ if $m_{i,k}=0$, the rate of a state transition is therefore

$$r_{\boldsymbol{m}_i}=\lambda_i+\sum_{k=1}^{N_Y}\mu_{m_{i,k}y_k}\mathcal{J}(m_{i,k}). \tag{7}$$

Suppose an arrival occurs when $\boldsymbol{m}_i\in\mathcal{M}_0$, since only those idle frequency channels, i.e., $m_{i,k}=0$, are selectable, a conditional PMF of Y_i can be obtained from Eq. (6) as

$$P_{Y_i|\boldsymbol{m}_i\in\mathcal{M}_0}(y)=\begin{cases}\frac{P_{Y_i|\Omega}(y)}{\sum_{k=1}^{N_Y}P_{Y_i|\Omega}(y)\cdot(1-\mathcal{J}(m_{i,k}))} & y=y_k,\forall k\in\mathcal{K}_0\\ 0 & otherwise.\end{cases}, \tag{8}$$

where $\mathcal{K}_0=\left\{k|m_{i,k}=0\right\}$ is the set consisting of the indices of all idle servers. Note that $\left(1-\mathcal{J}(m_{i,k})\right)=1$ if $m_{i,k}=0$; $\left(1-\mathcal{J}(m_{i,k})\right)=0$ otherwise. Thus, the transition probability from $\boldsymbol{m}_i\in\mathcal{M}_0$ to $\boldsymbol{m}_i'$ can be expressed as

$$p_{\boldsymbol{m}_i,\boldsymbol{m}_i'}=Pr\left\{\boldsymbol{m}_i'|\boldsymbol{m}_i\right\}=\frac{\lambda_i}{r_{\boldsymbol{m}_i}}\cdot P_{X_i}(x_j)\cdot P_{Y_i|m_i\in\mathcal{M}_0}(y_k),$$
$$\forall\mathcal{J}(\boldsymbol{m}_i'-\boldsymbol{m}_i)=\boldsymbol{e}_k, m_{i,k}'=x_j. \tag{9}$$

Similarly, suppose a departure occurs when $\boldsymbol{m}_i\in\mathcal{M}_0$, where $\boldsymbol{1}_{N_Y}^{\mathrm{T}}\mathcal{J}(\boldsymbol{m}_i)>\boldsymbol{0}$. Note that no departure will occur when all servers are idle; thus, $\boldsymbol{m}_i=\boldsymbol{0}_{N_Y}$ is excluded, where $\boldsymbol{0}_{N_Y}$ is a $N_Y\times 1$ column vector with all elements equal to 0. Thus, the transition probability from $\boldsymbol{m}_i\in\mathcal{M}_0$ to $\boldsymbol{m}_i'$, where $\boldsymbol{1}_{N_Y}^{\mathrm{T}}\mathcal{J}(\boldsymbol{m}_i)>\boldsymbol{0}$, can be expressed as

$$p_{\boldsymbol{m}_i,\boldsymbol{m}_i'}=\frac{\mu_{x_jy_k}}{r_{\boldsymbol{m}_i}},\forall\mathcal{J}(\boldsymbol{m}_i-\boldsymbol{m}_i')=\boldsymbol{e}_k, m_{i,k}'=x_j. \tag{10}$$

Next, consider $\boldsymbol{m}_i\in\mathcal{M}_1$. Since an arrival event is dealt with in the M/G/1 part, as mentioned earlier, no state transition will take place in our proposed analysis. The rate of a state transtion is therefore the rate of a departure event, which can be written as

$$r_{\boldsymbol{m}_i}=\sum_{k=1}^{N_Y}\mu_{m_{i,k}y_k}\mathcal{J}(m_{i,k}). \tag{11}$$

Note that when $\boldsymbol{m}_i\in\mathcal{M}_0$ the queue is definitely empty, i.e., $n=0$; however, when $\boldsymbol{m}_i\in\mathcal{M}_1$ the queue length can be either zero or greater than zero. That is, for $\boldsymbol{m}_i\in\mathcal{M}_1$, there is a probability $q_0=1-\rho_i$ that $n=0$, according to M/G/1 (D. Gross et al. 1985), where $\rho_i=\frac{\lambda_i}{\mu_i}$, and μ_i is the mean service rate of the M/G/1 queue and can be calculated by

$$\mu_i=\sum_{\boldsymbol{m}_i\in\mathcal{M}_1}\mu_{m_{i,k}y_k}\cdot\pi_{\boldsymbol{m}_i}, \tag{12}$$

and $\pi_{\boldsymbol{m}_i}$ is the steady state probability of $\boldsymbol{m}_i$.

Since the steady probability $\pi_{\boldsymbol{m}_i}$ is unsolved as yet, q_0 will be determined iteratively, with initially equal to 1. From our experiments, q_0 will converge quickly after only a few iterations.

Thus, the transition probability from $\boldsymbol{m}_i \in \mathcal{M}_1$ to $\boldsymbol{m}'_i \in \mathcal{M}_0$, which can happen only if a departure occurs and $n = 0$, can be expressed as

$$p_{\boldsymbol{m}_i,\boldsymbol{m}'_i} = \frac{\mu_{x_j, y_k}}{r_{\boldsymbol{m}_i}} \cdot q_0, \quad \forall \mathcal{J}(\boldsymbol{m}_i - \boldsymbol{m}'_i) = \boldsymbol{e}_k, \boldsymbol{m}'_{i,k} = x_j. \tag{13}$$

where $k = 1, \cdots, N_Y$ and $j = 1, \cdots, N_X$.

On the other hand, a departure may also occur when $\boldsymbol{m}_i \in \mathcal{M}_1$ and $n > 0$. A head-of-line packet will immediately be served upon the departure on the same frequency channel but a different SF may be used. Thus, the transition probability from $\boldsymbol{m}_i \in \mathcal{M}_1$ to $\boldsymbol{m}'_i \in \mathcal{M}_1$ can be expressed as

$$p_{\boldsymbol{m}_i,\boldsymbol{m}'_i} = \frac{\mu_{x_j, y_k}}{r_{\boldsymbol{m}_i}} \cdot (1 - q_0), \quad \forall \left| \mathcal{J}(\boldsymbol{m}_i - \boldsymbol{m}'_i) \right| = \boldsymbol{e}_k, m'_{i,k} = x_j. \tag{14}$$

and

$$p_{\boldsymbol{m}_i,\boldsymbol{m}'_i} \leftarrow p_{\boldsymbol{m}_i,\boldsymbol{m}'_i} + \frac{\mu_{x_j, y_k}}{r_{\boldsymbol{m}_i}} \cdot (1 - q_0), \quad \forall \boldsymbol{m}_i = \boldsymbol{m}'_i, \tag{15}$$

where $k = 1, \cdots, N_Y$ and $j = 1, \cdots, N_X$. Note initially that $p_{i,j} = 0, \forall i, j$ in the transition matrix $\boldsymbol{P}$. Note also in Eq. (14) that $\boldsymbol{m}'_i$ share all common entries with $\boldsymbol{m}_i$ except the kth one, with either $m'_{i,k} > m_{i,k}$, $m'_{i,k} = m_{i,k}$, or $m'_{i,k} < m_{i,k}$. For those with $m'_{i,k} \neq \boldsymbol{m}_{i,k}$, it means a different SF is selected for the newly served packet, while for those with $\boldsymbol{m}_i = \boldsymbol{m}'_i$, the same SF is selected again as the one that just departs. An example of the state transitions for a system with $N_X = N_Y = 2$ is shown in Figure 1.

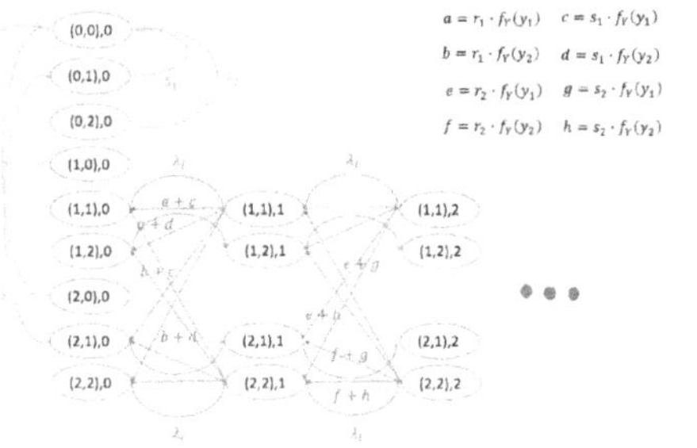

Figure 1. An example of state transition for a system with $N_X = N_Y = 2$.

3 RESULTS AND DISCUSSION

In this section, numerical results are presented in order to confirm the accuracy of our proposed analysis. The selectable SFs are assumed to be $x = 7, \cdots, 12$, and three sub-bands, i.e., $N_Y = 3$, are considered. The PMF of X_i is assumed to be

$$P_{X_i}(x) = \begin{cases} \frac{R_x}{\sum_x R_x} & x = 7, \cdots, 12, \\ 0 & otherwise. \end{cases} \tag{16}$$

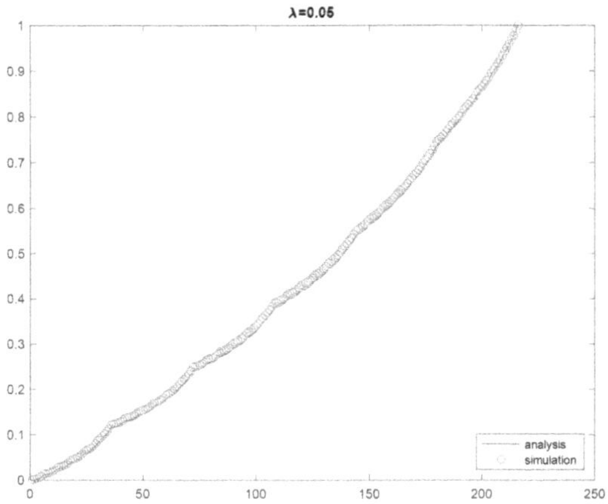

Figure 2. A comparison between analytical and simulation results for the cumulative probability distribution (CDF) of steady-state probabilities.

According to Eq. (16), in a homogeneous system where all SFs are selectable by each ED i, the traffic load can be balanced between SFs (R. B. Sorensen et al. 2017, F. Cuomo et al. 2018).

As can be seen in Figure 2, analytical results all match closely to simulation results under $\lambda_i = 0.05$ for the steady-state probability $\boldsymbol{\pi}$ obtained from solving simultaneously (D. Gross et al. 1985).

$$\boldsymbol{\pi} = \boldsymbol{\pi}\boldsymbol{P}, \sum_i \pi_i = 1, \tag{17}$$

where $\boldsymbol{\pi}_i$ is the ith element of $\boldsymbol{\pi}$. The results confirm that our proposed analysis is very accurate.

4 CONCLUSION

In this paper, an analysis is proposed in which an embedded Markov chain (EMC) is utilized to save the number of states, so that our proposed analysis can be more tractable. From numerical results, it is demonstrated that our proposed analysis is very accurate, compared with simulation results.

REFERENCES

F. Cuomo, M. Campo, E. Bassetti, L. Cartella, F. Sole and G. Bianchi, 2018. *IEEE Med-Hoc-Net.*
D. Gross and C. M. Harris, 1985. *John Wiley & Sons.*
D. Garlisi, I. Tinnirello, G. Bianchi and F. Cuomo, 2021. *IEEE Trans. Wirel. Commun.* 20(3) 2019–2033.
LoRa Alliance Technical Committee, 2017. *LoRa Alliance.*
R. B. Sorensen, D. M. Kim, J. J. Nielsen and P. Popovski, 2017. *IEEE Wireless Communication Letters* 6(5) 566–569.
Semtech, 2015. http://www.semtech.com/.

System Innovation for a Troubled World – Tak Lam et al. (Eds)
© 2023 The Author(s), ISBN: 978-1-032-45525-9

Amine grafted functionalized mesoporous silica nanomaterials for efficient environmentally friendly building materials for indoor humidity control

Kae-Long Lin*
Department of Environmental Engineering, National Ilan University, Yilan City, Taiwan, P.R. China

Ya-Wen Lin & Wei-Hao Lee
Institute of Mineral Resources Engineering, National Taipei University of Technology, Taipei City, Taiwan, P.R. China

Bo-Yi Kuo
Department of Environmental Engineering, National Ilan University, Yilan City, Taiwan, P.R. China

ABSTRACT: In this study, the silicon source and aluminum source rich in stone mud were extracted, and the best alkali fusion process for different alkali fusion temperatures and alkali agent ratios was determined by the alkali fusion salt method. The surface of the material was grafted by 3-aminopropyltriethoxysilane to obtain an amine grafted functionalized groups MCM-41 material (APTES/MCM-41). The APTES/MCM-41 material was analyzed and water adsorption/desorption experiment was performed. Utilizing stone sludge to obtain APTES/MCM-41, which has a high specific surface area (25.19 m^2/g).An adjustable pore size and an easy structural surface grafted can use different functional groups for the development of new materials. Over 24 hours at high humidity, the moisture adsorption capacity of the APTES/MCM-41 with 2.5 wt.% amine grafted functionalized groups can reach 37.02 g/m2, meeting the JIS A 1470-1 specifications. Comparing the moisture adsorption (18.6 g/m^2) of diatomaceous earth and inorganic materials in humidity control coatings, the performance of APTES/MCM-41 materials is relatively excellent. The APTES/MCM-41 can effectively compensate for the shortage of mechanical dehumidifiers due to its high energy consumption and inability for use in small spaces, replacing mechanical dehumidification as an emerging humidity control method.

1 INTRODUCTION

To make stone slabs ideal building materials, they are polished with natural lines and gloss, as well as having hardness and fire resistance performance. These have been widely used for floor and wall paving in modern buildings. However, steel grit and lime milk must be added to the sawing process of granite in steel grit. The granite waste obtained from the granite cutting industry is in the form of sludge, specifically, stone sludge that contains fine particles with small diameters has an excellent water permeability of less than 10-7 cm/s, and has a low dehydration rate. According to statistics, the amount of stone sludge produced in Taiwan is approximately 370,000 metric tons per year. If disposed of or recycled, stone sludge will cause environmental pollution (Environmental Protection Administration, 2020). Several studies have investigated issues related to sludge management and potential new product production. Stone sludge is used industrially as a substitute for fine aggregates in cement and concrete, bricks, and tiles (S.V. Ribeiro et al. 2014). Due to the high content of useful silica and alumina in stone sludge, recovery of these minerals for useful industrial applications has complete economic significance.

Many researchers have studied other chemical techniques using different silica precursors to produce mesoporous silica, especially MCM-41. Mesoporous MCM-41 is a hexagonal, one-dimensional material with mesopores in a uniform capillary system. The capillary size ranges from 2 to 10 nm. Large surface areas and excellent acidity have been widely used in support materials (S.T. Pham et al. 2021). However, the high cost of synthesizing mesoporous MCM-41 materials with pure chemicals limits their industrial applications. Therefore, it is necessary to find low-cost alternative sources as raw materials for synthesis. Other alternative materials for synthetic MCM-41 include kaolin (X. Li et al. 2010), waste glass (M.A. Salam et al. 2020), and rice husk ash (J. Andas et al. 2021). Matsumoto et al. determined that the introduction of amorphous silicon dioxide into tetrahedrally coordinated aluminum in its structure is a method to increase the Bronsted acid sites and ion exchange performance, which helps in increasing the stability of hydrothermal synthesis and the MCM-41 structural stability (A. Matsumoto et al. 1999). In addition, Rizzi

*Corresponding Author: klllin@niu.edu.tw

DOI 10.1201/9781003377399-9

et al. used amine grafted mesoporous silica nanoparticles to remove the blue dye and found that the amine group is beneficial for enhancing adsorption and catalysis (V. Rizzi et al. 2019).

Therefore, this article proposes a development and economical method that not only recycles waste as a starting material but also uses a simple yet effective method to produce stone sludge MCM-41 mesoporous material. The functionalization of MCM-41 via amine grafting is also discussed. In addition, the study reports how the MCM-41 material utilizes its material properties and the principle of the capillary phenomenon after amine grafting, thereby improving its adsorption characteristics as a water vapor adsorbent. The environmental humidity regulation performance is expected to have the potential for the sustainable conversion of stone sludge waste as an environmental humidity control material.

2 EXPERIMENTAL

The fusion process was completed by using a ratio of 1:1.25 of original stone sludge powder to sodium hydroxide. The mixture was then heated for several hours at different alkali fusion temperatures (650°C), cooled, and ground. Afterward, the obtained alkali fusion stone sludge powder and the distilled deionized water were mixed at different ratios (1:5) and aged under stirring conditions, and then the mixture was filtered to obtain an aluminosilicate sodium solution. In a typical synthesis process, the sodium aluminosilicate solution was put into a high-temperature autoclave, and a hydrothermal reaction was conducted at a hydrothermal temperature of 120°C for 48 hours. The obtained solid product was washed, filtered, dried, and calcined, hereinafter referred to as MCM-41.

However, 1 g of MCM-41 was added to dry toluene for washed, and different proportions of the amine functional groups (2.5 and 7.5 vol.%) were added to the toluene solution. The solution was then stirred and refluxed in a dry anhydrous toluene solution. Finally, the product was filtered and dried to form amine grafted functionalized mesoporous silica nanomaterials, hereinafter referred to as APTES/MCM-41.

3 RESULTS AND DISCUSSION

3.1 *Crystal phase of APTES/MCM-41*

Figure 1 shows the small-angle XRD pattern of APTES/MCM-41 with different amine grafted functionalized group. Observing the small-angle XRD pattern in Figure 4, two to three characteristic peaks are appearing at 2θ of 2°–5°. The typical MCM-41 characteristic shows that the strong peaks at 2θ of 2.46° and 4.52° represent a typical hexagonal ordered structure, which corresponds to the $d_{(100)}$ and $d_{(110)}$ planes (S.P. Samudrala et al. 2020). When the addition APTES/MCM-41 with 2.5 and 7.5 vol.% amine grafted functionalized groups is calculated, figure shows that as the addition of the amine functional groups increases, 2θ of the main characteristic peak and the second characteristic peaks tend to shift to a lower angle, and the crystallinity of the main characteristic peak decreases with an increasing amine functional groups concentration. In addition, when APTES/MCM-41 with different amine grafted functionalized, the main characteristic peaks $d_{(100)}$ 2θ are 2.48° and 2.24°, and from the Bragg diffraction formula, the interplanar spacings of the $d_{(100)}$ plane are 3.56 and 3.95 nm, respectively; and the 2θ of the second characteristic peak $d_{(110)}$ are the 4.42° and 3.92° positions, respectively. The APTES/MCM-41 with different amine-grafted functionalized groups belongs to the typical MCM-41 structure.

Figure 2 shows TEM observations of APTES/MCM-41. When the APTES/MCM-41 with 2.5 and 7.5 vol.% amine grafted functionalized groups are calculated, the results show that the particle size of the samples are all less than 1 μm, and the particle size tended to increase as the APTES/MCM-41 with amine grafted functionalized groups increased. It can be determined that the particle appearance of APTES/MCM-41 with 2.0 vol.% amine grafted functionalized groups is a relatively incomplete sphere (Figure 2(a)), and its particle size is approximately 0.15–0.32 μm. When the APTES/MCM-41 with 7.5 vol.% amine grafted functionalized groups (Figure 2(b)) was examined, the TEM observation showed that the particles had a larger particle size after grafted, and the particle size is approximately 0.18–0.32 μm. The results showed that the uniform mesoporous arrangement of APTES/MCM-41 was maintained, and no aggregation was observed (M.A. Mannaa et al. 2021).

Figure 3 shows the nitrogen adsorption and desorption curve of APTES/MCM-41. Nitrogen adsorption and desorption measurements showed that they resulted in a decrease in the BET surface area and the pore volume of the APTES/MCM-41 with amine grafted functionalized. When the APTES/MCM-41 with 2.5 and 7.5 vol.% amine grafted functionalized, the nitrogen adsorption and desorption curves of APTES/MCM-41 were classified by IUPAC as the second type of adsorption isotherm curve. When the relative pressure increases to 0.8–1.0, capillary condensation begins to occur and then becomes external. For multilayer adsorption on the surface, the highest nitrogen adsorption capacity of APTES/MCM-41 with 2.0 vol.% and 7.5 vol.% amine grafted functionalized groups is 40.40 cm^3/g and 55.22 cm^3/g STP, respectively. In addition, APTES/MCM-41 has a hysteresis loop similar to H3 when the relative pressure is 0.8-1. This type of adsorption curve is an upward curve at high relative pressure, usually caused by the formation of slit-shaped pores between particles.

Figure 4 shows the pore size distribution of APTES/MCM-41. The BET and BJH model simulations of the nitrogen adsorption and desorption test can calculate the specific surface area, the average pore volume, and the average pore diameter. The specific surface areas of APTES/MCM-41 with 2.0 vol.% and 7.5 vol.% amine grafted functionalized groups simulated by the BET model are approximately 25.19 m^2/g and 16.92 m^2/g, respectively. The average pores calculated for APTES/MCM-41 by the BJH method decreased from 0.086 cm^3/g to 0.063 cm^3/g. At this time, the specific surface area, the pore volume, and the pore diameter of APTES/MCM-41 are significantly reduced after graft. It is possible that the aminopropyl group is located in the mesopores, and the structural properties of materials with smaller pores are also significantly reduced (V. Zeleňák et al. 2008). When the APTES/MCM-41 with 2.5 and 7.5 vol.% amine grafted functionalized groups, the average pore diameters are 11.83 nm and 12.09 nm, respectively. The thickness of the pore wall of APTES/MCM-41 with amine grafted functionalized groups is thinner. This result also indicates that the amine functional groups were successfully grafted on the APTES/MCM-41.

Figure 5 shows the water vapor adsorption characteristics of APTES/MCM-41. The humidity control performance is characterized by a water adsorption capacity of 24 h and a water decomposition and adsorption capacity of 24 h. Obviously, when RH = 75%, the moisture adsorption in the 12th hour shows a downward trend. When the APTES/MCM-41 with 2.5 vol.% and 5.0 vol.% amine grafted functionalized groups, the moisture adsorptions at the 12th hour are 36.13 and 32.68 g/m^2, respectively. However, when the APTES/MCM-41 with 7.5 vol.% amine grafted functionalized groups, the moisture adsorption decreased to 19.45 g/m^2. In addition, under the same conditions, the moisture adsorption capacity of APTES/MCM-41 with 2.5 vol.% amine grafted functionalized groups is almost 1.83 times that of APTES/MCM-41 with 7.5 vol.% amine grafted functionalized groups under the same adsorption conditions. Based on the above observation, the APTES/MCM-41 with 2.5 vol.% amine grafted functionalized groups have the best water adsorption and desorption performance of 37.02 g/m^2. Compared with JIS A 1470 (Japanese Industrial Standards Committee 2008), the sample not only reached a very high value but also higher than the standard value.

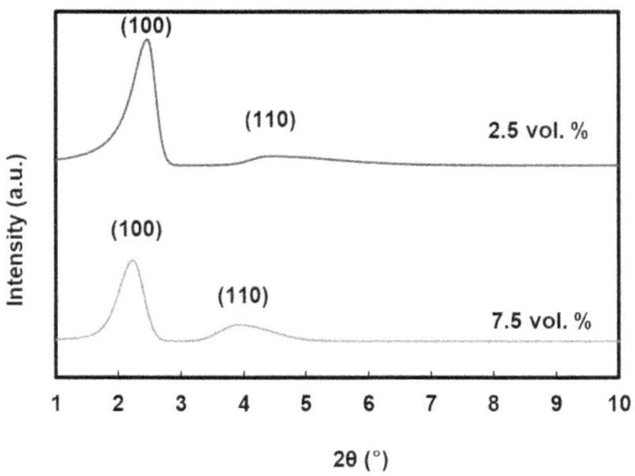

Figure 1. Small-angle XRD pattern of APTES/MCM-41.

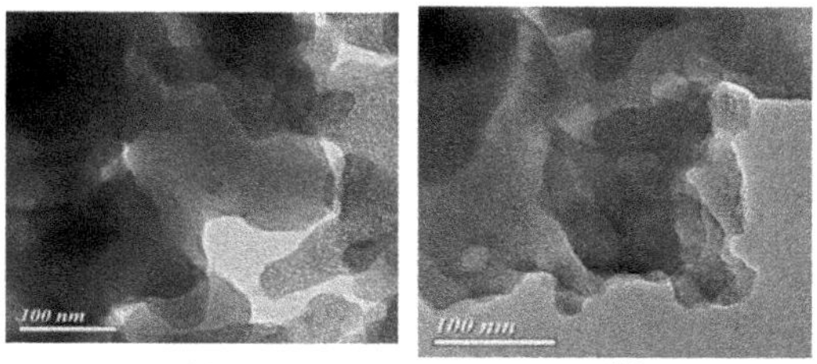

(a) (b)

Figure 2. TEM observations of APTES/MCM-41. (a) 2.5 vol.% and (b) 7.5 vol.%.

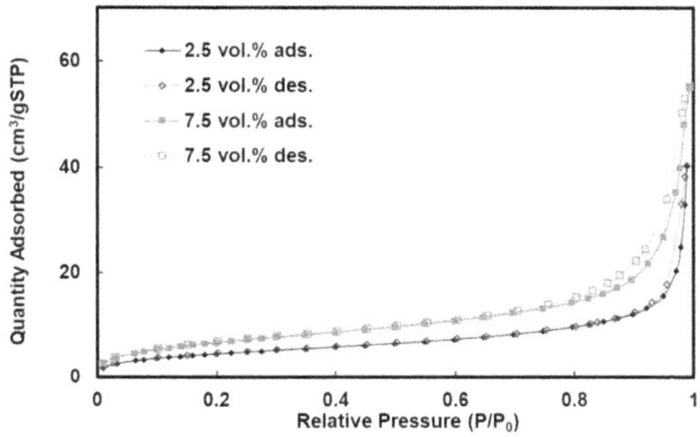

Figure 3. N_2 adsorption and desorption curve of APTES/MCM-41.

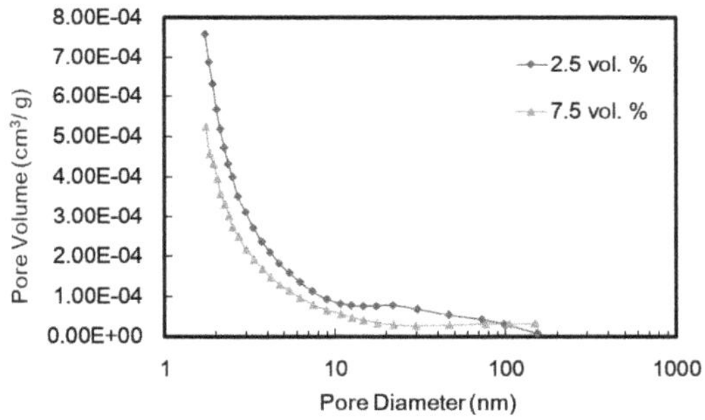

Figure 4. Pore size distribution of APTES/MCM-41.

4 CONCLUSION

In this study, stone sludge, as a raw material for the synthesis of APTES/MCM-41, is used to explore the use of alkali fusion treatment in order to extract Si and Al sources of this material. The experimental results confirmed that the use of SS can effectively synthesize APTES/MCM-41 and improve its structural stability and surface activity. After a small-angle XRD analysis, the APTES/MCM-41 shows that the pores are arranged in an orderly manner. The internal pores of the APTES/MCM-41 are arranged in parallel withthe hexagonal surface and the average pore diameter is approximately 11.83 nm. When the APTES/MCM-41 with 2.5 vol.% amine grafted functionalized groups is examined, the number of surface silanol groups is

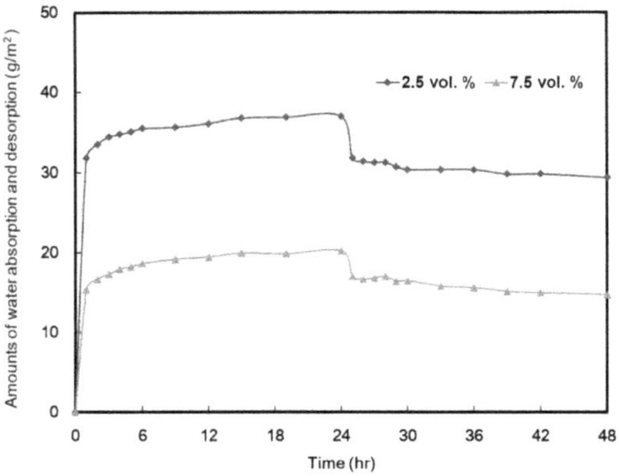

Figure 5. Water vapor adsorption characteristics of APTES/MCM-41.

more than the number of amine functional groups, resulting in a decrease in the average pore volume, average pore diameter, and the specific surface area. In addition, when the APTES/MCM-41 with 2.5 vol.% amine grafted functionalized groups, the best moisture adsorption of the humidity control test was 37.02 g/m^2, which met the referenced JIS A 1470-1 specifications and standards for humidity control building materials.

REFERENCES

J. Andas, S.H. Ekhbal, T.H. Ali, 2021. *Environ Technol Innov.* 21 101308.

Japanese Industrial Standards Committee, 2008. *JIS-A-1470-1-2008. Test method of adsorption/desorption efficiency for building materials to regulate an indoor humidity-part 1: response method of humidity.* Tokyo, Japan.

X. Li, B.S. Li, J.Q. Xu, Q. Wang, X.M. Pang, X.G. Gao, Z.Y. Zhou, J.R. Piao, 2010. *Appl Clay Sci.* 50 81–86.

A. Matsumoto, H. Chen, K. Tsutsumi, M. Grün, K. Unger, 1999. *Microporous Mesoporous Mater.* 32 55–62.

M.A. Mannaa, H.M. Altass, R.S. Salama, 2021. *Environ Nanotechnol Monit Manag.* 15 100410.

S.T. Pham, M.B. Nguyen, G.H. Le, T.D. Nguyen, C.D. Pham, T.S. Le, T.A. Vu, 2021. *Chemosphere.* 265 129062.

S.V. Ribeiro, J.N.F. Holanda, 2014. *Int J Eng Technol Innov.* 3 401–408.

V. Rizzi, E.A. Prasetyanto, P.K. Chen, J. Gubitosa, P. Fini, A. Agostiano, L. De Cola, P. Cosma, 2019. *J Mol Liq.* 273 435–446.

M.A. Salam, M.R. AbuKhadra, A.S. Mohamed, 2020. *Environ Pollut.* 259 113874.

S.P. Samudrala, S. Kandasamy, S. Bhattacharya, 2020. *Renew Energy.* 156 883–892.

Taiwan Environmental Protection Administration, Executive Yuan. 2020. https://www.epa.gov.tw. Accessed 26 June 2020.

V. Zeleňák, M. Badaničová, D. Halamová, J. Čejka, A. Zukal, N. Murafa, G. Goerigk, 2008. *Chem. Eng.* 144 336–342.

System Innovation for a Troubled World – Tak Lam et al. (Eds)
© 2023 The Author(s), ISBN: 978-1-032-45525-9

Research on the design of home auxiliary readers for the elderly base on QFD and TRIZ

Jeng-Chung Woo*
School of Design • Straits Institute of Technology, Fujian University of Technology, Fuzhou, China
Design Innovation Research Center of Humanities and Social Sciences Research Base of Colleges and Universities in Fujian Province, Fuzhou, China

Yu-Tong Chen, Qin Liu & Feng Luo
School of Design • Straits Institute of Technology, Fujian University of Technology, Fuzhou, China

ABSTRACT: According to the WHO, there will be 1.4 billion people over the age of 60 by 2030. Successful aging promotes the physical and mental health of the elderly, and lifelong learning exerts a positive influence on the elderly's successful aging. Reading habits play a critical role in this process, and the demand for auxiliary readers among the elderly grows with their physical deterioration. Combining quality function deployment (QFD) and the theory of inventive problem solving (TRIZ), based on the elderly's demand for the auxiliary reader, this study conducts a text analysis through web crawler on the online user evaluation (JD.com/About 400,000 words, Amazon/About 1.1 million words) of auxiliary readers in the consumer markets at home and abroad, and supplements customer needs with scenario analysis. KJ (Kawakita Jiro) is adopted to classify and turn it into engineering requirements to build an interaction matrix. The importance of customer requirements is then calculated through a questionnaire survey of 102 elderly people (average age is about 70 years), and experts are invited to evaluate the relationship matrix to determine the importance of engineering characteristics, so as to construct the house of quality (HoQ) of home auxiliary readers for the elderly (HARE). Based on the QFD study, it employs TRIZ to identify and resolve contradictions in engineering characteristics. By applying the systematic method that combines QFD and TRIZ, it finally summarizes and proposes the design principles and innovative plans for HARE, with specific application references for related research and design.

Keywords: Web crawler; Elderly assistance design; Auxiliary readers; QFD; TRIZ.

1 INTRODUCTION

The statistics from the WHO show that the number and proportion of people aged 60 years and older in the population are increasing. This number will increase to 1.4 billion by 2030 and 2.1 billion by 2050. In terms of demographic transition, the process of aging may produce an important impact as it has multiple and wide-ranging social and economic implications for society (Lee et al. 2011). Different age groups tend to have different levels of productivity and different economic demands; therefore, their effects on economic growth attract great attention from policymakers and researchers worldwide (Pham et al. 2021). To address challenges posed by population aging, the Member States of the WHO deliberated and approved the plan for a Decade of Healthy Aging 2020-2030 at the 73rd World Health Assembly in May 2020. The plan lists relevant sustainable development goals required for healthy aging, including quality education. As a medium of lifelong learning, quality education enables older people to retain the ability to make decisions and preserve their purpose, identity, and independence. In the design industry, auxiliary product for the elderly is one of senior design themes. Relevant researches on auxiliary product design for the elderly propose that the design should cater to physiological changes, be more adaptive to complicated environment, and consider internal and external factors of product for the design (Li X. et al. 2020). Besides, vision loss among the elderly is a major healthcare problem (Quillen 1999). This study focuses on the design of home auxiliary readers for the elderly (HARE) from the perspective of vision loss in older adults. HARE can compensate for vision loss to make reading easier for the elderly. In terms of healthy aging, HARE conforms to universal values of lifelong learning because it can ensure older adults' active reading and their access to external information and community links. Based on the reading requirements of the elderly, the study analyzes the survey by integrating quality function deployment (QFD) with theory of

*Corresponding Author: wwwjc2020@outlook.com

DOI 10.1201/9781003377399-10

inventive problem solving (TRIZ), and discusses how to address older adults' reading problems caused by physical deterioration, and proposes design principles of HARE.

2 LITERATURE REVIEW

2.1 *Home auxiliary readers of the elderly*

Auxiliary reading means to assist people to read in some ways. However, there is limited research on HARE. Alonzo et al. (2020) studied the auxiliary readers for deaf and hard-of-hearing individuals and proposed comments on the design of automatic text simplification (ATS) auxiliary reading tools. Based on children's reading difficulties and insufficient levels of reading proficiency, Tobias Lunte presented that gaze-contingent assistance was promising in terms of addressing reading difficulties (Lunte T. et al. 2020). As for the negative effects of the elderly's physical deterioration on reading, the study targeting the elderly and home environment proposes a design of HARE to improve the elderly's reading experience is meaningful.

2.2 *Quality function deployment (QFD)*

QFD is a method to help convert the voice of the customer into product design specifications (Akao Y. 1994). House of quality (HoQ) is the core tool of QFD, which combines customer requirements with product characteristics (Xi Z. et al. 2021). Initially, QFD was conceived to develop higher-quality products to meet or exceed customer requirements based on collecting and analyzing the voice of customers. In addition, products and services can be determined by QFD according to user needs (Ginting et al. 2020). Later, QFDs functions were applied to other fields such as design, planning, decision-making, engineering, and management. Essentially, there was no definite boundary for QFDs potential fields of applications (Chan et al. 2002). Babar et al. (2021) applied the QFD to a management decision to study hybrid electric vehicles through multi-regression analysis and fuzzy quality function deployment. Vimal et al. (2021) researched a new remanufacturing quality function deployment (RQFD) underpinned by QFD, which was applied to improve the design and sustainability of a product. Lapinskienė et al. (2021) presented a newly developed technology of Integrated Building Design by integrating QFD with axiomatic design (AD). In this study, QFD can achieve a deeper exploration of customer and engineering requirements about HARE, and pave the way for TRIZ to propose a design plan that satisfies the needs of elderly people.

2.3 *TRIZ (Theory of inventive problem solving)*

TRIZ is developed to support engineers and natural scientists to solve inventive problems by using the knowledge of former inventors (Moehrle, 2005). TRIZ provides a complete set of tools for analysis and solutions to problems in different fields (Ekmekçi et al. 2015). In the course of TRIZ research, algorithm of inventive problem solving (ARIZ) is a special analytical tool to help address inventive problems, among TRIZ the most commonly used tool is the contradictions matrix (Ekmekci et al. 2019). TRIZ can work alone via the algorithm of ARIZ or work with other methods. TRIZ is applied to fields of technology, design, management (Navas et al. 2015), education (Fan 2010), and others. Ramírez-Rios et al. (2021) used TRIZ to develop a plantar orthosis system of standardization, customization, and simultaneous actions. By studying smart head-neck helmets with KANO-QFD and TRIZ, Zhao et al. (2021) proposed an innovation that could improve products and verify the reasonableness of design through finite element analysis. After studying the feasibility of the synergistic effect of human-computer interaction (HCI) and TRIZ, Chen (2021) put forward rapidly developing design and evaluation, and interaction strategies. This study utilizes TRIZ to resolve potential product conflicts, improve product functions, and create innovative designs.

3 RESEARCH AND DESIGN

3.1 *Research on QFD*

QFD is imported to analyze the customer and engineering requirements of HARE. Customer requirements are conducted by text analysis and scenario analysis. The text analysis takes the web crawler technology. The web crawler is an organized, automated computer system for systematically navigating web pages residing on the web and for automatically searching information (Yang et al. 2019). It is possible to collect and analyze useful information by crawling the data of the website server. The scenario analysis is used to predefine and analyze elderly people in different usage scenarios. Then, KJ is used for the hierarchical classification, and for turning into engineering requirements to build the correlation matrix. Next, the questionnaire determined the importance of customer requirements and a relationship matrix by experts evaluation. Finally, the QFD study completed the HoQ construction. This phase of the study will identify product problems and lay the foundation for TRIZ research.

3.2 *Research on TRIZ*

Based on the QFD, TRIZ is introduced to propose suggestions and solutions for product problems. According to the correlation relationship in QFD, there exists a negative correlation among engineering requirements, which can be described as contradictions in the product. By analyzing contradictions with TRIZ, the contradictions are translated into the 39 general engineering parameters and correspond to the improving parameter and worsening parameter in the matrix. The author translated the 40 inventive principles into concrete solutions, thus putting forward innovative design.

3.3 *Integrated model of QFD and TRIZ*

QFD provides an objective direction for design on the basis of collecting user needs, whereas TRIZ provides the solution for the potential problems of product conflicts through matrix analysis. The integration of QFD and TRIZ brings out reasonable design concepts of HARE (Figure 1).

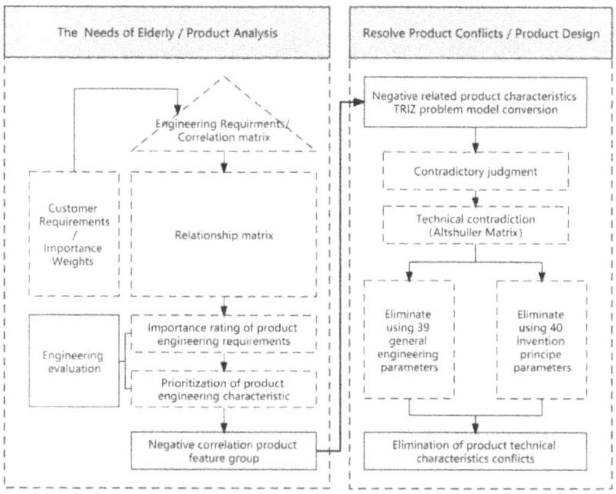

Figure 1. Integrated Model of QFD and TRIZ.

The integrated model of QFD and TRIZ can be divided into two parts. The first one is to identify the customer requirements and to carry out the product analysis. Second, TRIZ is introduced to eliminate the conflicts and propose a specific solution. Therefore, the concrete and improved design principles and scheme of HARE can be put forward by the integrated model of QFD and TRIZ.

4 RESULTS AND DISCUSSION

4.1 *Construction of the HoQ*

4.1.1 *Customer requirements*

The study of customer requirements uses text analysis by web crawlers and scenario analysis. The keywords "elderly reading" were retrieved from JD.com, and Python is used to capture the evaluation vocabulary of the online shopping platform. A total of more than 80 items from YiMoo and other brands are retrieved and collected, and online reviews are more than 400,000 words. The instant data scraper is used to collect foreign product reviews on Amazon for HARE. The author searches "Visual Impairment Aids", and retrieves more than 50 items from LANCOSC and other brands. In total, there are more than 13,000 reviews of online products, forming nearly 1,200,000 characters of text. Then, through the collation of the research data, the scenario analysis was a comprehensive analysis of the factors in the usage scenarios of HARE. It takes the usage scenario as the starting point of the phase research and analyzes user needs and expectations of the product according to the literature supplement and mining user evaluation data. Finally, the customer requirements for text analysis and scenario analysis are compared. Afterward, through the KJ for hierarchical classification and sorting, complete customer requirements were listed (Table 1).

Based on the items and hierarchical classification of customer requirements, the questionnaire is formed. Through offline and online questionnaires, a total of 110 questionnaires are collected and 102 questionnaires are valid, with 53 males and 49 females. Among these 96.1 percent of respondents are over 60 years old, with an average age of 70.2 years. The questionnaire is carried out in the form of multiple-choice questions. Moreover, based on the choice frequency of the KJ and the item's frequency of the corresponding sub-level, the importance proportion of customer requirements is calculated. The statistical results show that the ratio of the items ranges from 5% to 30%. Then, according to the maximum difference in the importance ratio of the items of customer requirements, five grades are set up to separate them and weights are given to each item, including 1 (5%–9%), 2 (10%–14%), 3 (15%–20%), 4 (21%–25%), and 5 (26%–30%). As a result, the importance weights of the items of customer requirements are obtained.

4.1.2 *Engineering requirements*

Engineering requirements are listed according to the corresponding technical characteristics of consumer requirements and the design elements of new products are planned. The customer requirements are converted into engineering requirements and classified by KJ to form the engineering requirements for HARE (Table 2).

4.1.3 *Correlation matrix*

After correlation matrix analysis, the conflicting engineering requirements can be solved by TRIZ analysis. In HoQ construction, the "roof" part is filled in the research of this stage and the correlation is marked according to the positive and negative relationship. If the two are positively correlated, "+" will be marked; otherwise, "−" will be marked. Additionally, if there is no obvious relationship between the two, it will not be marked. The engineering requirements of HARE are compared in pairs that show there were nine negative and eight positive correlations between engineering requirements (Figure 2).

4.1.4 *Relationship matrix*

The relationship matrix between customer requirements and engineering requirements is built. The relationship matrix is analyzed in three tiers: strong, medium, and weak, and the results are then populated into "rooms" in the HoQ. It is indicated by three symbols, namely "◎" for strong, "○" for medium, and "△" for weak. The relationship matrix in this phase is evaluated by three invited experts: a professor from

Table 1. Items and KJ classifications of HARE customer requirements.

First-Level Classification	Second-Level Classification	Customer Requirements
Common basic requirements	Reading clarity and improved visual adjustment	Moderate magnification Clear magnified image Magnification adjustment Reading distance adjustment Reading angle adjustment
	Improved material properties, design and workmanship	Wear-resistant lens materials Reading windows shapes (Circular, rectangular and others) Quality materials Sophisticated workmanship Good exterior design
	Improved user experience and practical functions	Shading security (avoid accidental fires) Smooth operation Easy operation Practical functions Long life span
Improved home-basedrequirements	Improved home-based basic functions	Overall dimensions (big) Overall dimensions (extendable) Heavy weight Stable frame structure Simple frame structure
	Improved home-based auxiliary functions	Improved lighting functions Video screen projection Audio reading system Low power consumption

Table 2. HARE Engineering Requirements Transformation and KJ Classification.

First-Level Classification	Engineering requirements			
Lens performance	Lens power	Lens resolution	Wear resistance	
Appearance properties	Size	Weight	Shape	Material rigidity
	Material utilization	Technical process	Exterior design	
Auxiliary functions	Regulation and control	Lighting	Video projection	Audiobook function
Usage and operation	Smooth operation	Easy operation		
Frame structure	Sable support	Simple structure		
User experience	Practical functions	Power performance	Security protection	Life span

the mechanical department of the university, a general manager of the instrumentation company, and a senior optician from the optical company. The analysis results are shown in Figure 2.

4.1.5 *Importance conversion*

Importance conversion is the process of converting the weight of customer requirements into the weight of engineering requirements. In this phase, the importance is converted through the independent collocation method, which is a method of multiplying the importance weights of customer requirements with the proportional values of ◎, ○, △ and then summing them vertically. The proportional values of the symbols ◎, ○, △ are ◎: ○: △ = 5:3:1. The importance's priority of engineering requirements of HARE is determined according to the data (Table 3).

4.1.6 *QFD Summary*

The QFD study investigates and classifies the needs of HARE. With reference to the importance of engineering requirements and customer requirements, the design of HARE focuses on improving the clarity of reading images, the user experience, and practical functions. Based on the feedback from the elderly on reading function, the HoQ of HARE is constructed (Figure 2).

Table 3. Ranking of the Importance of HARE Engineering Requirements (Top 5).

No.	Engineering Requirements	Importance rating	Priority (%)
1	Practical functions	68	8.6%
2	Lens resolution	65	8.2%
3	Regulation and control	64	8.1%
4	Lens power	58	7.3%
5	Video projection	55	7.0%
	Technical process	55	7.0%
	……		

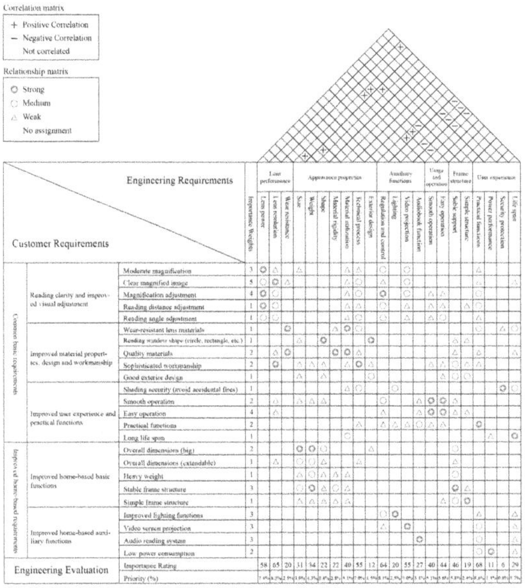

Figure 2. The HoQ of HARE.

4.2 *Model transformation of TRIZ problem*

4.2.1 *Product analysis*

According to the correlation matrix, the relevance of the first to third categories of contradictions is converted into technical contradiction. As the video projection and audiobook functions belong to functional configuration, the technical contradiction is divided into two groups (Table 4). The fact that product performance and practical functions should be improved because enhancing easy operation is an issue that needs to be addressed.

4.2.2 *Technical contradictions*

After analyzing and summarizing a large number of patents, Altshuller derives 39 general engineering parameters and 40 inventive principles, and links them to establish a 39×39 contradiction matrix. The technical contradictions are then grouped into two sets of conflicts. The inventive principles of resolving the contradictions can then be derived by consulting the contradictions matrix of the general engineering parameters (Table 5).

4.2.3 *Analysis and solution of design problems*

The focus of innovative design research is to take advantage of inventive principles to eliminate or offset the impact of technical contradictions on products. The analysis is given as follows.

First is the contradiction between regulation and control and stable support. The contradiction can be solved according to the analysis of the above-mentioned technical contradictions through the principle of "Blessing in disguise" or "Turn Lemons into Lemonade". In the new design, the stability depends on the reading platform contact area and the rigidity of the brackets. The connection of the sliding bracket on the back of the screen can ensure that the height and angle of the lifting screen can be adjusted. Meanwhile, the bracket structure is less likely to cause hand obstructions and the overall structure is more stable. Therefore, in combination with the adjustment of sliding bracket construction, it is possible to achieve structural flexibility while maintaining a certain level of stability.

The second is the mismatch between functional configuration and operation. Technical contradiction analysis adopts the principle of Intermediary to deal with conflicts. In the new design, buttons are designed as intermediaries to realize the single operation process, which is applied to the operation and control of different functions, thus reducing the costs of learning how to use buttons. The control buttons ensure multifunctional switching of auxiliary functions and the controlling of various functions. The clear key rebound feedback and the guiding icon notes, in addition to automatic focusing, can make reading easier for the elderly.

5 CONCLUSION AND RECOMMENDATION

The core of this research lies in the use of QFD to construct HoQ, which translates the needs of the elderly into engineering characteristics. Besides, TRIZ is used to analyze and solve the potential contradiction to propose solutions to design difficulties. The results of this study are as follows:

The web crawler can obtain big data of cross-domain consumers' comments to provide objective and complete research data.

The research and analysis of QFD provide in-depth product analysis and design reference direction for translating customer requirements into engineering requirements.

TRIZ contradiction analysis proposed innovative design from contradictions between regulation control and stable support, and function configuration and simple operation.

Based on systematic research, TRIZ ensures the objectivity of the research and the rationality of the design. The research puts forward the innovative

Table 4. Three negative correlations of HARE engineering requirements and their correlation reasons.

No.	Negative correlations			Reasons for the negative correlation
1	Regulation and control		Stable support	The regulation and control can meet the requirements of different people and environments while stable support can ensure a good reading experience. The controllable and adjustable support structure requires the assistance of mechanical parts, which results in reduced stability.
2	Video projection	Functional configuration	Easy operation	The video projecting function enables fast enlargement of the reading screen. But at the same time, this function results in the increase of operation steps and the use of computer system.
3	Audiobook function			The audiobook function enables an auditory reading experience in addition to visual reading. However, in addition to promoting the intelligence of the product to meet overall demand, this new function also reduces the usability of the product.

Table 5. Contradiction matrix of 39 general engineering parameters and its corresponding 40 inventive principles.

No.	Category	Specific parameters	39 general engineering parameters	40 invention principles
1	Improving Parameter	Regulation and control	No.35 Adaptability	35. Parameter changes 22. "Blessing in disguise" or "Turn Lemons into Lemonade 39. Inert atmosphere 23. Feedback
	Worsening Parameter	Stable support	No.13 Stability of object	
No.	Category	Specific parameters	39 general engineering parameters	40 invention principles
2	Improving Parameter	Functional	No.36 Complexity of device configuration	27. Cheap short-living object 9. Preliminary anti-action 26. Copying 24. 'Intermediary'
	Worsening Parameter	Easy operation	No.33 Convenience of use	

design of HARE and can provide a specific reference for related researches and designs. However, the current innovative design scheme does not include each project. In this context, this paper will combine AHP research methods and conduct an in-depth study to analyze the weight index of the current design scheme, to provide a more complete and detailed innovative design scheme.

ACKNOWLEDGMENTS

This work was supported by Fujian University of Technology [grant numbers GY-S21081, 2021], and Design Innovation Research Center of Humanities and Social Sciences Research Base of Colleges and Universities in Fujian Province.

REFERENCES

Akao, Y. (1994). Development history of quality function deployment. The Customer Driven Approach to Quality Planning and Deployment, 339, 90.

Alonzo, O., Elliot, L., Dingman, B., & Huenerfauth, M. (2020, October). Reading experiences and interest in reading-assistance tools among deaf and hard-of-hearing computing professionals. In The 22nd International ACM SIGACCESS Conference on Computers and Accessibility (pp. 1–13).

Babar, A. H. K., & Ali, Y. (2021). Enhancement of electric vehicles' market competitiveness using fuzzy quality function deployment. Technological Forecasting and Social Change, 167, 120738.

Chan, L. K., & Wu, M. L. (2002). Quality function deployment: A literature review. European journal of operational research, 143(3), 463–497.

Chen, S., Kamarudin, K. M., & Yan, S. (2021). Analyzing the Synergy between HCI and TRIZ in Product Innovation through a Systematic Review of the Literature. Advances in Human-Computer Interaction, 2021.

Ekmekçi, İ., & Koksal, M. (2015). Triz methodology and an application example for product development. Procedia-Social and Behavioral Sciences, 195, 2689–2698.

Ekmekci, I., & Nebati, E. E. (2019). Triz Methodology and Applications. Procedia Computer Science, 158, 303–315.

Fan, J. (2010, August). Application idea for TRIZ theory in innovation education. In 2010 5th International Conference on Computer Science & Education (pp. 1535–1540). IEEE.

Ginting, R., Ishak, A., Malik, A. F., & Satrio, M. R. (2020, December). Product development with quality function deployment (QFD): a literature review. In IOP Conference Series: Materials Science and Engineering (Vol. 1003, No. 1, p. 012022). IOP Publishing.

Lee, R. D., & Reher, D. S. (2011). Introduction: the landscape of demographic transition and its aftermath. Population and Development Review, 1–7.

Li, X., Wang, H., Liu, Y., Wang, Y., & Liu, Z. (2020, December). An Investigation of Assistive Products for the Elderly. In 2020 IEEE 19th International Conference on Trust, Security and Privacy in Computing and Communications (TrustCom) (pp. 1509–1514). IEEE.

Lunte, T., & Boll, S. (2020, October). Towards a gaze-contingent reading assistance for children with difficulties in reading. In The 22nd International ACM SIGACCESS Conference on Computers and Accessibility (pp. 1–4).

Lapinskienė, V., & Motuzienė, V. (2021). Integrated building design technology based on quality function deployment and axiomatic design methods: a case study. Sustainable Cities and Society, 65, 102631.

Moehrle, M. G. (2005). What is TRIZ? From conceptual basics to a framework for research. Creativity and innovation management, 14(1), 3–13.

Navas, H. V., Tenera, A. M., & Machado, V. A. C. (2015). Integrating TRIZ in project management processes: an ARIZ contribution. Procedia engineering, 131, 224–231.

Pham, T. N., & Vo, D. H. (2021). Aging population and economic growth in developing countries: a quantile regression approach. Emerging Markets Finance and Trade, 57(1), 108–122.

Quillen, D. A. (1999). Common causes of vision loss in elderly patients. American family physician, 60(1), 99–108.

Ramírez-Rios, L. Y., Camargo-Wilson, C., Olguín-Tiznado, J. E., López-Barreras, J. A., Inzunza-González, E., & García-Alcaraz, J. L. (2021). Design of a modular plantar orthosis system through the application of TRIZ methodology tools. Applied Sciences, 11(5), 2051.

Vimal, K. E. K., Kandasamy, J., & Duque, A. A. (2021). Integrating sustainability and remanufacturing strategies by remanufacturing quality function deployment (RQFD). Environment, Development and Sustainability, 23(9), 14090–14122.

Xi, Z., & Meng-di, Y. (2021). Research on Wheelchair Design for the Disabled Elderly Based on QFD/TRIZ. In Journal of Physics: Conference Series (Vol. 1750, No. 1, p. 012062). IOP Publishing.

Yang, H. C., & Kim, C. G. (2019). An Implementation of Python Web Crawler Using Thread. In Proceedings of the Korea Information Processing Society Conference (pp. 70–72). Korea Information Processing Society.

Zhao, D., & Chen, W. (2021). Design and Research of Smart Neck Helmets Based on the KANO-QFD Model and TRIZ Theory. Security and Communication Networks, 2021.

System Innovation for a Troubled World – Tak Lam et al. (Eds)
© 2023 The Author(s), ISBN: 978-1-032-45525-9

An advanced decision support architecture by joint utilization of contextual topic modelling and data envelopment analysis

Sin-Jin Lin*
Department of Accounting, Chinese Culture University, Taipei, Taiwan

Te-Min Chang
Department of Information Management, National Sun Yat-sen University, Kaohsiung, Taiwan

Ming-Fu Hsu
Department of Business Management, National United University, Miaoli, Taiwan

ABSTRACT: Financial statements have been widely viewed as a main channel to transmit inherent messages from corporate to current/potential market participants to form their own investment decisions. The information in the financial statements can be categorized into two parts: one is numerical information, and the other is non-numerical information (such as texts, figures, and tables). The numerical information accounts for around 10 to 15% of financial statements, and the remaining (85~90%) of financial statements are non-numerical information, such as texts, figures, and tables. Due to the nature of numerical information that is easy to realize and intuitive, this type of information turns out to be the main trigger for decisión-makers to judge the performance of each corporate and then formulate their own investment portfolios. However, merely adopting numerical information is unable to appropriately explain the corporate's current and future operating performance, especially in today's highly volatile business environment. One of the possible reasons is lack of considering the non-numerical information. To combat this, contextual topic modeling approach that can be applied to exploit the valuable information embedded into a large number of texts is considered. This study not only performs a contextual topic modeling approach to alleviate the time lag problem introduced by considering numerical information only but also equips the model with forecasting ability. The analyzed results are then inserted into the rotation forest to construct the forecasting model. By doing that the introduced model can be viewed as a decision support architecture to assist investors to form their own judgments under anticipated risk exposures as well as maximize their personnel wealth.

1 INTRODUCTION

Global financial disturbances, such as the subprime mortgage crisis, the European debt crisis, and recently the COVID-19 pandemic have attracted tremendous attention in corporate risk management. A financial crisis shows that the value of most financial properties drops together or one big company's financial trouble could propagate to other companies, which lead to financial instability, resulting in the collapse of the whole financial system as well as breakdown of the functioning of financial market to channel the resources to suitable places (Ouyang et al. 2021), that is, global financial disturbance can be viewed as a contagious risk (Baranoff et al. 2019). This statement is echoed by the European Central Bank which pointed out that global financial disturbance is a widespread risk for financial stability, which could ruin the establishment of financial systems/infrastructures and impede economic growth, and further result in economic recession and welfare of whole society losses (Bisias et al. 2012). To address this point, so many academics and practitioners reconsider their inherent risk pre-warning systems to quickly respond to the market and survive in this highly competitive business environment (Chen et al. 2021; Hsu & Lin 2021).

The company's risk management performance is relevant to its risk-absorbing ability. The company's risk-absorbing ability depends heavily on how many resources it owns or controls. A company with superior operating performance normally poses superior risk-absorbing ability. It is because the efficient company is good at utilizing assets to generate profits under an anticipated risk exposure. Hence, the more profit the company generates, the stronger the risk-absorbing power it has.

How to appropriately evaluate the company's operating performance turns out to be an urgent requirement of the market (Hsu 2019). Previous studies merely concentrated on one-input and one-output

*Corresponding Author: annman1204@gmail.com

DOI 10.1201/9781003377399-11

architecture and adopted numerical messages to form their decision. Unfortunately, the performance evaluation is a multidimensional task, and a simple model (i.e., one-input and one-output) is incapable to illustrate the real status of business operation. To combat this, much more advanced architecture for performance evaluation is taken. Data envelopment analysis has been demonstrated its worldwide applications in performance evaluation and gained so many positive results. Apart from prior parametric models, one of the nonparametric approaches, called data envelopment analysis (DEA), can handle multiple inputs and multiple outputs without pre-determining a production function and is considered to yield much deeper insights into performance evaluation.

Although numerical messages pose intuitive and comprehensive characteristics, it also comes with an obstacle of lacking the ability to transmit messages timely (Huang et al. 2014). To overcome this obstacle, text messages have been adopted and executed.

Text mining (TM) has attracted tremendous attention from both industrial and academic fields in recent years. It is because the textual messages contain valuable information that cannot be explicitly addressed in numerical data. Compared to numerical messages, textual messages normally do not organize in a predefined manner. This specific characteristic results in irregularities and ambiguities that make it hard to realize and comprehend via traditional approaches. To combat this, topic modeling is taken to group a large number of textual messages into a higher conceptual hierarchy so as to increase their comprehensibility. Latent Dirichlet allocation (LDA), one of the topic modeling algorithms, not only has an efficient probability inference method but also yields highly interpretable topics in an unsupervised fashion (Blei et al. 2003). LDA falls into the realm of "bag-of-words" where it cannot yield the synthetical and semantic meanings of textual messages (Saheb et al. 2022). To overcome this, an emerging word embedding approach called bidirectional encoder representation from transformer (BERT) is taken into consideration. By joint utilization of LDA and BERT (herein, contextual topic modeling), the contextual messages embedded into the texts can be extracted so as to gain deeper insights as well as assist users to deploy resources to the appropriate place (Miles et al. 2022).

The numerical and textual information derived from contextual topic modeling and performance ranks derived from DEA are then inserted into the twin support vector machine (TWSVM) to construct the model for performance forecasting. The model, examined by real cases, is a promising alternative for performance forecasting. The users can consider the model's outcome to adjust the policy as well as to reach the goal of sustainable development.

This study is structured as follows: we begin with a description of our introduced architecture and then go on with the findings. Finally, we finish the paper with a conclusion.

2 AN ADVANCED DECISION SUPPORT ARCHITECTURE

The advanced architecture encompasses three steps: numerical messages calculation, contextual topic determination, and forecasting model construction, as displayed in Figure 1.

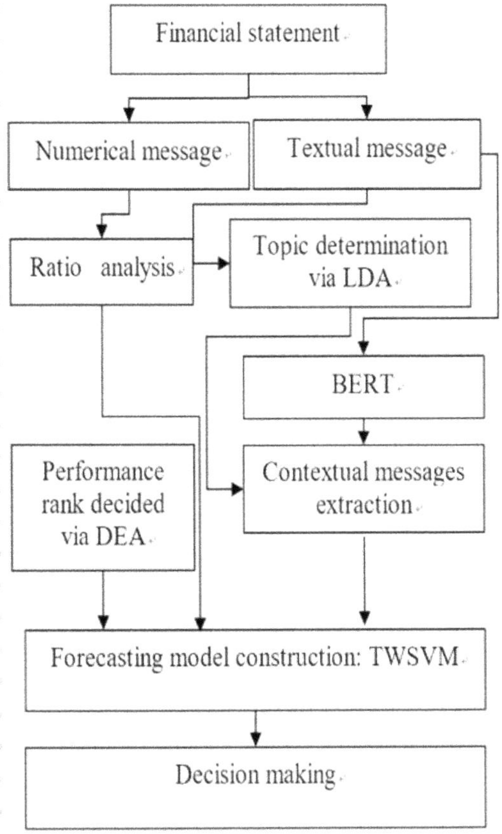

Figure 1. The advanced decision support architecture.

Ratio analysis is implemented to extract the numerical messages from financial statements. For example, the debt ratio (i.e., total debt to total assets) is used to represent the company's capital structure. The higher the value of the debt ratio, the bigger the financial burden the company confronted. The financial statement is the main channel for corporate to transmit their inherent messages to potential/current investors. However, too many messages surrounding users will mislead or influence their decision-making process. To overcome this issue, the textual message summation turns out to be an urgent requirement. The LDA is used to divide the large amount of textual information into some topics. However, the LDA belongs to the WOM realm, where it cannot represent the contextual and semantic information embedded in the texts. To avoid this, the BERT is considered. By joint utilization of LDA and BERT, we can develop a contextual topic algorithm to gain deeper insight into the text. DEA is conducted to determine the performance rank of each company. The analyzed outcome are then injected into a TWSVM to construct the forecasting model. The decision-makers can consider the potential implication of the outcome and adjust/modify their operating strategies so as to survive in this high turmoil atmosphere.

3 RESULTS AND DISCUSSION

The aim of this paper is to propose an effective model for corporate operating performance forecasting. Corporate operating performance is highly related to financial failures. Thus, the independent variables adopted in financial failure forecasting are conducted as predictors in this study. The textual messages derived from contextual topic modeling are four essentials: (1) business-related topics, (2) stock market-related topics, and (3) macroeconomic-related topics. Table 1 displays the adopted predictors.

Table 1. The predictors.

Symbol	Description
X1: S/TA	Sales to total assets
X2: TA/TL	Total assets to total liabilities
X3: NI/S	Net income to sales
X4: COGS/I	Cost of goods sold to inventory
X5: CA/CL	Current assets to current liabilities
X6: GP/S	Gross profit to sales
X7: WC/CA	Working capital to current assets
X8: B_topic	Business related topic
X9: S_topic	Stock market-related topics
X10: M_topic	Macroeconomic-related topic

To demonstrate the usefulness of textual information, this study considers two scenarios: (1) the model with textual information, and (2) the model without textual information. Five-fold cross-validation (CV) is taken to avoid the challenge of overfitting. Most recent relevant studies merely consider one assessment measure to conclude their finding. To reach a more robust outcome, the other two assessment measures are taken. The forecasting accuracy, sensitivity, and specificity are taken as assessment measures and the confusion matrix is displayed in Table 2.

Table 3 shows the forecasting performance under two scenarios. We can see that the model with textual information performs better than the model without textual information. This finding is in accordance with the work done by Hsu et al. (2021) who indicated that textual information can bring salient and incremental messages beyond a numerical aggregation from financial reports.

To demonstrate the effectiveness of the introduced model, this study further takes the TWSVM as a benchmark and compares it with the other three artificial intelligence models, such as back-propagation neural network (BPNN), classification and regression tree (CART), and rough set theory (RST). Table 4 expresses the outcome. We can see that the introduced model outperforms the other three models under all assessment measures.

Table 2. The confusion matrix.

Predicted↓/ Actual→	Efficient	Inefficient
Efficient	True Positive (TP)	False Positive (FP)
Inefficient	False Negative (FN)	True Negative (TN)

TP: It is the number of precisely predicted efficient companies. It tests how well a prediction model can recognize efficient companies.
FP: It is the number of inefficient companies that is imprecise and predicted as efficient companies. It tests the percentage of inefficient companies that are imprecisely predicted.
TN: It is the number of precisely recognized inefficient corporates. It tests how well a model can recognize an inefficient company.
FN: It is the number of efficient companies that are imprecisely recognized as inefficient. It tests the percentage of efficient companies that are imprecisely recognized.
Forecasting accuracy: (TN+TP)/(TP+FP+TN+FN)
Sensitivity: TP/(TP+FN)
Specificity: TN/(TN+FP)

Table 3. The prediction outcome under two scenarios.

Model	TWSVM	
	(1) With textual messages	(2) Without textual messages
Assessment measure: Accuracy		
CV-1	89.06	67.97
CV-2	85.94	68.75
CV-3	83.59	68.75
CV-4	85.16	71.09
CV-5	82.81	75.78
AVG.	85.31	70.47
Assessment measure: Sensitivity		
CV-1	84.38	68.75
CV-2	87.50	62.50
CV-3	81.25	65.63
CV-4	75.00	71.88
CV-5	75.00	65.63
AVG.	80.63	66.88
Assessment measure: Specificity		
CV-1	90.63	67.71
CV-2	85.42	70.83
CV-3	84.38	69.79
CV-4	88.54	70.83
CV-5	85.42	79.17
AVG.	86.88	71.67

Table 4. The prediction outcomes.

Mode	Prediction model			
	TWSVM	BPNN	CART	RST
Assessment measure: Accuracy				
CV-1	89.06	80.47	74.22	70.31
CV-2	85.94	72.66	73.44	67.19
CV-3	83.59	73.44	77.34	72.66
CV-4	85.16	73.44	82.81	71.88
CV-5	82.81	76.56	64.06	74.22
AVG.	85.31	75.31	74.38	71.25
Assessment measure: Sensitivity				
CV-1	84.38	78.13	71.88	62.50
CV-2	87.50	75.00	65.63	65.63
CV-3	81.25	71.88	68.75	71.88
CV-4	75.00	71.88	81.25	75.00
CV-5	75.00	62.50	59.38	75.00
AVG.	80.63	71.88	69.38	70.00
Assessment measure: Accuracy				
CV-1	90.63	81.25	75.00	72.92
CV-2	85.42	71.88	76.04	67.71
CV-3	84.38	73.96	80.21	72.92
CV-4	88.54	73.96	83.33	70.83
CV-5	85.42	81.25	65.63	73.96
AVG.	86.88	76.46	76.04	71.67

4 CONCLUSION

Compared to well-examined studies, such as financial crisis prediction or credit default forecasting, work on performance forecasting is quite scarce. To fill the gap, this study introduces an emerging model for performance forecasting. The introduced model consists of three main processes: the numerical message is handled by ratio analysis, the textual message is extracted via contextual topic modeling, and the performance rank is decided by DEA. Sequentially, the analyzed outcome are then injected into TWSVM to construct the model. The outcome shows that textual information can bring valuable information beyond the numerical message. The decisión-makers can consider the potential implication of this study and formulate their future policies as well as gain more competitive edge against future rivals' competition.

ACKNOWLEDGMENT

The authors would like to thank for Ministry of Science and Technology (MOST) for financial support under No. 109-2410-H-034-034-MY2 and No. 110-2410-H-239-017.

CONFLICT OF INTEREST

No conflict of interest exists in this study.

REFERENCES

D. Bisias, M. Flood, A. W. Lo, S. Valavanis (2012). A survey of systemic risk analytics. *Annual Review of Financial Economics*, 4(1), 119–131.

D. M. Blei, A. Y. Ng, M. I. Jordan, (2003), Latent Dirichlet Allocation, *Journal of Machine Learning Research* 3, 993–1022.

E. G. Baranoff, P. Brockett, T. W. Sager, B. Shi, (2019). Was the US life insurance industry in danger of systemic risk by using derivative hedging prior to the 2008 financial crisis? *Quantitative Finance and Economics*, 3(1), 145–164.

F. H. Chen, M. F. Hsu, K. H. Hu, (2021), *Enterprise's internal control for knowledge discovery in a big data environment by an integrated hybrid model.* Inf Technol Manag https://doi.org/10.1007/s10799-021-00342-8.

M.F. Hsu, S. J. Lin, (2021), A BSC-based network DEA model equipped with computational linguistics for performance assessment and improvement. *Int. J. Mach. Learn. & Cyber*. 12, 2479–2497.

M. F. Hsu, Y. S. Hsin, F. J. Shiue, (2021). Business analytics for corporate risk management and performance improvement. *Ann Oper Res* https://doi.org/10.1007/s10479-021-04259-x

M. F. Hsu, (2019), A fusion mechanism for management decision and risk analysis, *Cybernetics and Systems* 50 (6), 497–515.

S. Miles, L. Yao, W. Meng, C. M. Black, Z. B. Miled (2022), Comparing PSO-based clustering over contextual vector embeddings to modern topic modeling, *Information Processing & Management*, 59 (3), 102921,

Z. Ouyang, X. Yang, Y. Lai, 2021, Systemic financial risk early warning of financial market in China using Attention-LSTM model, The North *American Journal of Economics and Finance*, 56, 101383.

T. Saheb, M. Dehghani, T. Saheb (2022), Artificial intelligence for sustainable energy: A contextual topic modeling and content analysis, Sustainable Computing: *Informatics and Systems*, 35, 100699.

System Innovation for a Troubled World – Tak Lam et al. (Eds)
© 2023 The Author(s), ISBN: 978-1-032-45525-9

A preliminary study on the centennial evolution of penglaitu lacquerware technology in Taiwan

Shun-Yao Chiang
Graduate School of Design, Doctoral Program Graduate Student, National Yunlin University of Science & Technology, Yunlin, Taiwan

Shyh-Huei Hwang*
Professor, National Yunlin University of Science & Technology, Yunlin, Taiwan

ABSTRACT: "Penglaitu" is a lacquerware from the Japanese occupation period. It was designed and produced by a Japanese named Mr. Yamanaka Tadasu. He designed lacquer products with Taiwanese imagery based on Taiwan's scenery, fruits, habits, and aboriginal life. This study focuses on the Penglaitu lacquerware developed in Taiwan since the Japanese occupation period. Through literature research and interviews with existing creators, it studies and sorts out the development of Taiwan's unique lacquerware over the past century. The results of the study found: (1) Penglaitu lacquerware can be described as a "Japanese-made Taiwanese cultural creation": it started with the Japanese habit of using the lacquerware, and developed the best cultural souvenirs – Penglaitu lacquerware from Taiwan for Japanese to bring back to Japan and promote the "beauty of Taiwan"; (2) The Taiwanese elements of Penglai tu are relatively simple: with rough carvings, strong tropical colors, and rich Taiwanese characteristics, such as rich fruit crops, and vibrant aboriginal songs and dances, which shows that people who live in the northern Japan are yearn for the Penglai fairyland; (3) Penglaitu is authentic enough to distinguish the visual flavors of Taiwan, Japan and China: as far as the techniques and themes of Penglaitu are concerned, there is less Japanese flavor, less Chinese flavor, and more authentic Taiwanese flavor. (4) Modern interpretation of Penglaitu: Master Yamanaka Tadasu returned to Japan after the war, and Penglaitu lacquerware was almost vanished in Taiwan, but the Taichung City Government was successfully registered Wu Shu-Fa as the preserver of Penglaitu lacquerware, and Chen Ching-Hui as the preserver of lacquer craft, and left a hope for Penglaitu lacquerware to develop in Taiwan.

Keywords: Taiwanese crafts, lacquerware, Penglaitu lacquerware, craft design, souvenirs.

1 INTRODUCTION

Lacquerware is the most important daily necessity in Japanese life. In early Taiwan, the use of lacquerware was mainly for worship and celebration.

In daily use, it is only some more casual coating, and these lacquerwares are almost entirely from China. Lacquerware is not produced in Taiwan.

In 1916, Mr. Yamanaka Tadasu from Takamatsu, Shikoku, Japan was making Lacquerware in Taichung. He also assisted the Taichung government to set up schools to teach the production of lacquerware.

Yamanaka loves Taiwan and often travels to aboriginal tribes, so he often creates his own lacquerware with the theme of Taiwan aborigines. In 1933, Mr. Yamanaka established "Taichung Arts & Crafts Co., Ltd." in Taichung City and started producing lacquerware.

These lacquerwares are used as souvenirs for friends when the Japanese come to Taiwan visit and return to Japan. Because of its Taiwanese style, it was also selected as the most representative gift of Taichung at that time.

The lacquerware which Mr. Yamanaka Tadasu designed was different from traditional Japanese lacquerware. It was called "Penglaitu lacquerware" by the media at that time.

After the end of World War II and Japanese defeat in the war, the Republic of China ruled Taiwan, and lacquer craft education was ignored and sent back to the beginning. Penglaitu had also lost its market and no one continued to make it. After 1990s, Taiwan's economy became stable and the living standards of its citizens improved. The government began to pay attention to Taiwanese culture and craftsmanship, so the creators of lacquer art started to find back "Penglaitu lacquerware" again.

*Corresponding Author: jonamai@gmail.com

 DOI 10.1201/9781003377399-12

Today, Taiwan lacquer art has shown a glorious scene of success in these years, and Penglaitu lacquerware, which has a special Taiwanese image, has also begun to attract attention. The government also has the certification of Penglaitu lacquerware conservators, but there is no clear definition and interpretation of technical preservation methods and techniques.

This study is expected to achieve the following objectives:

(1) Discuss the era and significance of Penglaitu lacquerware with Japanese-made Taiwanese cultural and creative products.
(2) Use Taiwanese elements to construct the characteristics of Penglaitu lacquerware craftsmanship.
(3) Discuss the Penglaitu lacquerware made in Taiwan without using the Japanese style.
(4) Discuss the modern interpretations that give Penglaitu lacquerware a new hope.

2 CASE STUDY

(1) The regional craft characteristics of the Taiwan-Japan fusion of Penglaitu lacquerware.

The lacquerware designed by Yamanaka is called Penglaitu lacquerware (Figure 1). He uses Taiwanese elements to design lacquerware, such as the aboriginal Thao pestle dance, the scenery of Sun Moon Lake, bananas, pineapples, papayas, tropical fruits. etc., and then use carving techniques similar to Kamakura-bori (Figure 2) and Sanuki-bori (Figure 3) or use the drawing technique of Yakumo-Nuri (Figure 4). The main body of Penglaitu lacquerware is wood. At that time, electricity began to be popularized, and the body was mostly made of machinery, which could be mass-produced and the price was reduced. It was mainly used in three ways: turning, plate and milling machines (Wong, C., & Huang, L. 2010). Made with bright colors and powerful carving techniques, it is surrounded by Taiwan-based themes and is full of southern feelings.

Figure 1. Penglaitu lacquerware (photographed by author).

(2) Penglaitu lacquerware as a souvenir

During the Japanese occupation, Penglaitu lacquerware was used as a business trip souvenir from Taiwan

Figure 2. Kamakura-bori (photographed by author).

Figure 3. Sanuki-bori (photographed by author).

Figure 4. Yakumo-nuri (photographed by author).

for the Japanese to bring back to Japan, and the purchase of souvenirs could satisfy the consumer psychology of Japanese tourists in Taiwan. For travel experience, souvenirs have meanings such as "authenticity", "regional", and "story". An indispensable element to be used as an evocative of memories. This has a connecting purpose to the storytelling of the region being told by the giver. Let the recipient feel its authenticity. Especially in the process of building a regional brand in regional promotion, it is very important to discover, cultivate and develop new concept productsv(Suzuki, M 2021). Undoubtedly, Penglaitu lacquerware is a commodity that combines culture, life, and industry.

(3) Cultural propaganda that won the hearts of the people during the colonial period.

The themes of Penglaitu lacquerware include aboriginal songs and dances, farming by farmers, and various fruits. From here, it can be seen that Taiwan is a rich and beautiful island.

Japan's colonial assimilation policy of ruling Taiwan is a microcosm of the "tradition" shaped by the Penglaitu lacquerware phenomenon. It includes different historical, political, economic, and cultural aspects (Gao, Z. 2018). The terroir of Taiwan shows the effect of Japan's colonization of Taiwan. A rich southern country is a place that makes people want to go.

The existence of Taiwan is a special existence for the citizens of Japan.

(4) A century of industrial evolution.

The lacquerware industry shrank with the end of Japanese rule in Taiwan after World War II, and Taiwan's more representative lacquerware factories and schools were closed due to the turbulent times and the unintentional management of the rulers.

In the 1970s, when Japan turned to Taiwan for OEM work due to high wages and forest protection policies, Taiwan's lacquerware industry reached another peak. However, Penglaitu lacquerware did not have market demand and no one made it, so it almost disappeared in the development of lacquerware, in the whole of Taiwan.

Until the lacquer artist Huang Lih-Shwu discovered the relevant materials and cultural relics of Penglaitu lacquerware in the house of collector Liang Zhizhong, he asked Chen Huoqing to confirm that this Taiwan-style lacquerware is called "Penglaitu lacquerware" (National museum of history (Ed.). 1997).

Nowadays, Penglaitu lacquerware and Taiwan's lacquerware are also being valued, and the new generation of multi-stimulation is coming, so it would be necessary for traditional lacquer crafts to think about the future.

3 METHODS

This study focuses on the Penglaitu lacquerware developed in Taiwan since the Japanese occupation period. Through literature research and interviews with existing creators, it studies and resolve the development of Taiwan's unique lacquerware over the past century, and have an in-depth understanding of the preservation of lacquerware in Taiwan, and comprehend the context of lacquer art education (Figure 5). The interviewees include Penglaitu lacquerware preservers in Taichung City (Interviewer A), lacquer technology preservers (Interviewer B), and Penglaitu lacquerware collectors (Interviewer C). Based on the interview results, the "Centennial Evolution of Penglaitu lacquerware in Taiwan" was discussed and concluded.

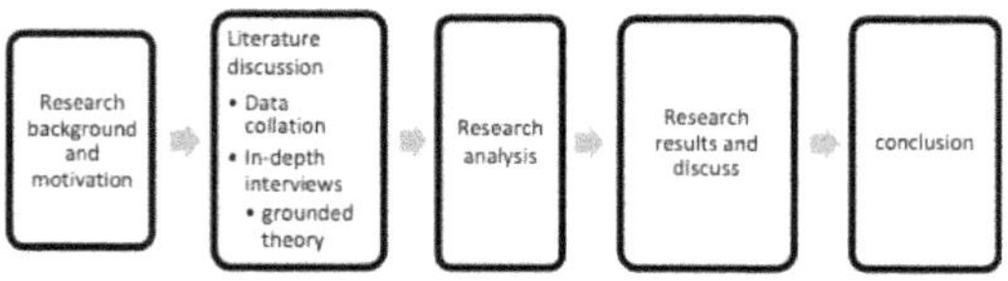

Figure 5. Research processes.

4 RESULTS AND DISCUSSION

This study takes Penglaitu lacquerware as the theme to research the development of Taiwanese lacquerware through literature review and interviews with creators.

Based on the research results and the interviewer's interview content, the following conclusions are drawn:

(1) Traditional innovation

Culture is the way of life. During the colonial period, Taiwan and Japan had different customs and habits. Mr. Yamanaka used the Taiwanese elements that Japan lacked, such as fruits and aboriginal people, and the Japanese loved lacquerware, which combined into Taiwan's unique cultural creativity.

The Japanese came to Taiwan to travel and returned to Japan to promote the beauty of Taiwan. "*Although the techniques used in Penglaitu lacquerware are similar to those of Japanese lacquerware, the most important part of craftsmanship is the presentation of Taiwanese culture through image composition.*" (Respondent C, interviewed for this study). Penglaitu lacquerware has been reborn under the traditional framework and has embarked on a road of innovation.

(2) Unique Craft Aesthetics

Crafts will exist over time, and sometimes they will be expanded. In addition to the visual sense of craftsmanship, unique craftsmanship aesthetics also exist in the technical construction method and cultural context. "*When I first came into contact with Penglaitu lacquerware, I always felt that it was different from Japanese Maki-e, because my carving skills were not proficient, but it was like the style of Penglai tu, and I had the foundation skills of oil painting, so I used oil painting techniques to create Penglaitu lacquerware.*" (Respondent B, interviewed for this study).

The uniqueness of the Penglaitu lacquerware is the rough carvings, strong tropical colors, strong Taiwanese characteristics, diverse fruits, lively aboriginal songs and dances, which make the Japanese living in the northern country yearn for the Penglai wonderland.

(3) The regional uniqueness of Taiwanese lacquerware from Penglaitu lacquerware

Being careful about the place where you live, and using the uniqueness of Taiwanese handicrafts to create Taiwan's excellent craftsmanship brand, Penglaitu lacquerware is authentic enough to distinguish the visual image of Taiwan, Japan, and China.

"*The Japanese have already set the composition and color very accurately when designing the Penglaitu lacquerware, whether it is just a souvenir or a work of art. Now is the era of global village, it is important to highlight local characteristics, especially craftsmanship, which can be used as a representative of local characteristics. Local characteristics and personal style are the future trends.*" (Respondent C, interviewed for this study).

As far as the techniques and themes of Penglaitu lacquerware are concerned, there is less Japanese flavor, less Chinese flavor, and more authentic Taiwan flavor.

(4) A New Interpretation of Taiwanese Elements

The government actively promotes the craft of lacquerware. In 2015, Taichung City Government registered Wu Shu-Fa as the traditional lacquer art preserver of Penglaitu lacquerware. The surviving Penglaitu lacquerware appears to be alive. Chen Ching-hui inherited the work of his father, Chen Huo-ching, to promote Penglaitu lacquerware. In 2021, he was selected as the preserver of traditional lacquer craftsmanship in Taichung City. Later, two artists, Mr. Wu and Mr. Chen, also assisted the government in setting up relevant lacquer art inheritance courses.

Facing changes in the environment, "*It is no longer a long-term study. What students learn is not just a technique, but a preparation for this technology as a career.*" (Respondent A, interviewed for this study).

"*Now Penglaitu lacquerware has the promotion of teachers and the government's attention. Not only Penglai tu, but also the entire lacquerware has received attention. No matter what, these are all positive responses. On the other hand, Japan is not so enthusiastic about learning lacquer work.*" (Respondent C, interviewed for this study).

After a century of evolution, Taiwan has transformed from a Japanese colony into an important indicator of Asian democracy, with different scenery regardless of its cultural characteristics and landscapes. A hundred years ago, Penglaitu lacquerware existed as a cultural and creative product representing Taiwanese characteristics. Could the creators also think what is Taiwanese characteristics? Using culture to demonstrate Taiwan's visibility in the world.

5 CONCLUSION

Yamanaka Tadasu looked at Taiwan from the perspective of a Japanese, how much he loved Taiwan, and how much he hoped that Taiwan would be noticed by the world, so he used the most familiar lacquerware to promote the culture of Taiwan. Craftsmen should pay attention to their own culture and create value for Taiwanese culture. Penglaitu lacquerware clearly conveys the importance of products for cultural promotion. In addition, craft creators also have the responsibility to retain and pass on the techniques of traditional crafts, which is the major problem facing Taiwan.

This study presents the following findings:

(1) Penglaitu lacquerware can be described as a "Japanese-made Taiwanese cultural creation": it started with the Japanese habit of using the lacquerware, and developed the best cultural souvenirs—Japanese used to bring back Penglaitu lacquerware from Taiwan to promote the "beauty of Taiwan".

(2) The Taiwanese elements of Penglaitu are relatively simple: with rough carvings, strong tropical colors, and rich Taiwanese characteristics, such as rich fruit crops, and vibrant aboriginal songs and dances, which shows that people who live in northern Japan yearn for the Penglai fairyland.

(3) Penglaitu is authentic enough to distinguish the visual flavors of Taiwan, Japan, and China. As far as the techniques and themes of Penglaitu are concerned, there is less Japanese flavor, less Chinese flavor, and more authentic Taiwanese flavor.

(4) Modern interpretation of Penglaitu: Master Yamanaka Tadasu returned to Japan after the war, and Penglaitu lacquerware almost vanished in Taiwan, but the Taichung City Government successfully registered Wu Shu-Fa as the preserver of Penglaitu lacquerware, and Chen Ching-Hui as the preserver of lacquercraft, and left hope for Penglaitu lacquerware to develop in Taiwan.

In addition, we also need to think about how to add Taiwanese awareness and aesthetic values to traditional craftsmanship, form a transformation of cultural identity with historical context, and position Taiwan in the wave of globalization.

"Penglaitu lacquerware" began to appear in the Century Time Tunnel with a complete Taiwanese style, and now I look forward to a gorgeous turning point and creating another wave of Penglai beauty.

REFERENCES

Chu, L. (2016, 9). The Development and Evolution of Lacquerware During the Japanese Colonization Period. *Journal of Art Forum, 10*, pp. 1–38.

Gao, Z. (2018). Traditional Invention—The Production and Influence of "Penglaitu" Lacquer in Taiwan during Japanese Occupation Period. *Creative and Design, 58*, pp. 71–75.

National museum of history (Ed.). (1997). *Ching Lacquer Artifacts-with Early Taiwanese Works.* (N. m. history, Ed.) National Museum of History.

Suzuki, M. (2021). Souvenirs without Memories: Changes of Confectionery Souvenirs in Japan. *Humanities Forum, 4*, pp. 281–296.

Wong, Chun-yi; Huang, Lih-shuw. (2010). Analysis on the Development and Design Characteristics of Taiwan Lacquerware Penglaitu in the 1930s. *Taiwan Literature, 61*(2), pp. 9–34.

Wong, Cunu-yi; Ueda, Akira; Miyazaki, Kiyoshi;. (2006). Design Character of the Taiwan Lacquer-Ware-HOURAINURI-at 1930's. *BULLETIN OF JSSD, 52*(5), pp. 35–44.

System Innovation for a Troubled World – Tak Lam et al. (Eds)
© 2023 The Author(s), ISBN: 978-1-032-45525-9

Curriculum design in creativity education with a focus on entrepreneurship

Tienhua Wu*
Department of Management, Air Force Institute of Technology, Gangshan, Kaohsiung, Taiwan

ABSTRACT: Creativity significantly influences the process and value of innovation, which affects the possibility of successful entrepreneurship. Hence, creativity education (CE) is emphasized in business education systems. This study first reviewed relevant literature to clarify the definitions of creativity and creativity in education. Second, based on the literature review and in-depth interview results, this study proposed a CE curriculum that includes the four stages for students to move towards being creative, engaged, and productive in novel and valuable solutions for entrepreneurship. Finally, this research conducted a nine-week experimental test on the proposed CE curriculum. The results showed that this CE curriculum was feasible and enabled students to enjoy the process of creative development and exhibit their creativity self-efficacy. The discussions and suggestions are also provided.

1 INTRODUCTION

Creativity is vital to innovation and entrepreneurship. Creativity can be viewed as a cognitive and mental process in which an individual connects existing knowledge to new ideas and applications (Ward 2004 & Jackson et al. 2012). The literature stresses the role of creativity in entrepreneurship: the process and value of innovation (Ward 2004), the new firm formation (Lee et al. 2004), problem-solving at lower costs in smaller ventures (Fillis and Rentschler 2010), affecting organization performance and survival (Shin et al. 2012), and the development and growth of creative industries (de Klerk 2015). Yar Hamidi et al. (2008) suggest that creativity critically contributes to raising the students' entrepreneurial intentions. Hence, addressing creativity in education may benefit from cultivating a creative workforce for entrepreneurship.

However, creativity education (CE) issues are sparse (Craft 2001). Indeed, DeHaan (2011) points out that educational research on teaching higher-order cognitive skills such as creativity is limited. The literature suggests that students indeed have creative performance (Daly et al. 2014), and educating them to be creative has practical importance (Simonton 2012). Moreover, given the essence of creativity in innovation and new ventures (Ward 2004), this research explores the educational issues in creativity to propose a curriculum and examines the feasibility and effectiveness of this proposed CE curriculum.

This study first reviews the relevant literature to clarify the definitions of creativity and creativity in education. Second, this study interviews educational experts. The results concerning the literature review and interviews are used to design the CE curriculum. This study then conducts a nine-week experimental test on the proposed CE curriculum to examine this curriculum's feasibility and effectiveness. Last, the discussions and the implications are presented.

*Corresponding Author: tienhua5678@gmail.com

2 LITERATURE REVIEW

2.1 *Creativity and creativity education*

Creativity is a broad, complex, and diverse concept (Berglund and Wennberg 2006 & Plucker et al. 2004), and its conceptual operationalization differs between disciplines (Jackson et al. 2012, Kleiman 2008). Ward (2004) and Jackson et al. (2012) suggest that creativity is a cognitive and mental process in which an individual connects existing knowledge to new ideas and applications. Kazerounian and Foley (2007) and Ferrari et al. (2009) note that creativity concerns the psychological phenomenon of an individual who offers novelty and usefulness of an idea. Overall, the notions of novelty and originality with notions of utility and value are highlighted in the studies (Kleiman 2008). For the research objective, this study adopts the definition proposed in the study of Fillis and Rentschler (2010) from an entrepreneurship viewpoint: *creativity is potentially most beneficial within the smaller enterprise as a way of overcoming barriers to acquiring resources and in terms of deriving alternative and lower cost solutions to solving problems.*

Ferrari et al. (2009) argue that creativity should not be limited to natural ability; instead, it can be identified and taught from an educational viewpoint. Thus, many research efforts focus on the process of creativity that provides a basis for educators to design a teaching process for creativity in education. Following the notion of teaching for creativity (Simonton 2012, Ferrari et al.

 DOI 10.1201/9781003377399-13

Table 1. The process of creativity or creative thinking.

Perspective	Stages of process	Author
Cognitive	(1) Generating ideas; (2) digging deeper into ideas; (3) openness and courage to explore ideas; and (4) listening to one's inner voice	Treffinger et al. 2002
Social psychology	(1) Problem or task presentation; (2) preparation; (3) response generation; (4) response validation; and (5) outcome	Amabile and Pillemer 2012
Entrepreneurship	(1) Problem identification; (2) preparation; (3) response generation; and (4) validation and communication	Fillis and Rentschler 2010
Learning	(1) Relevance; (2) learning ownership of knowledge; (3) control of learning process; and (4) innovation	Jeffrey 2006

2009, Jeffrey and Craft 2004), this study suggests that creativity education (CE) relates to implementing pedagogies, contents, and assessment mechanisms that address creativity or foster creativity. Simonton 2012 emphasizes the novel and valuable instructional techniques involving teaching creativity. Moreover, the literature provides critical elements addressing a curriculum: resources (time, space, and money), learning goals, plans, contents, pedagogies, assessment, and classroom atmosphere and instructions (Craft 2001, Daly et al. 2014, Ferrari et al. 2009). This study integrates these components above into curriculum design. Table 1 summarizes the varying perspectives on the process of creativity or creative thinking.

2.2 *Creativity education and entrepreneurship*

Creativity is essential for innovation and entrepreneurship education (Craft 2001, Simonton 2012). The literature notes that entrepreneurship is strongly connected to creativity that may result in idea development, new product innovations, and existing innovation improvement (Berglund and Wennberg 2006), value-added design for competitiveness (Penaluna and Penaluna 2009), and new functional requirements (Daly et al. 2014). Strong relationships between creativity, innovation, and new venture development exist. Creativity in education thus becomes necessary for entrepreneurship. More essentially, Yar Hamidi et al. (2008) acknowledge that creativity may contribute to improving students' entrepreneurial intentions. Consequently, cultivating students to be creative is practical for the dynamic marketplace and future entrepreneurship choices (Yar Hamidi et al. 2008, Berglund and Wennberg 2006).

3 METHODOLOGY

This study adopted two methods: in-depth interviews and an experimental classroom test. First, this study conducted interviews with three business and design school educators. The interview meeting was held twice for each informant from February to June 2020. The interview protocol concerns the primary considerations suggested in the CE literature to design a CE curriculum focusing on entrepreneurship. In doing so, this study proposed a tentative CE curriculum. This research then examined the feasibility and effectiveness of this tentative curriculum through an experimental classroom test in September 2020. Fifteen college students participated in this nine-week CE class.

4 RESULTS AND DISCUSSIONS

4.1 *Proposed CE curriculum*

The informants emphasized three main areas: pedagogies, a teaching environment, and creative tasks. The pedagogies such as team-, experience- or problem-based methods that enable learners to foster engagement and interpersonal communications are recommended. Providing an open-ended learning classroom benefits promoting motivation and fostering creative thinking. The creativity tasks or cases that allow students to observe and synthesize would be better for students' divergent thinking for tentative solutions. Based on the literature review results and interviewers' suggestions, the proposed curriculum containing the four stages of fostering creativity is identified. The stages are addressed below:

- *Teaching for creativity.* Teachers offer fundamental knowledge to creativity and ventures, understand students' backgrounds, provide resources for students to be free and open-minded, and utilize experience- and problem-based pedagogies.
- *Warm-up for being creative.* Teachers offer activities and resources that enable learners to freely observe, manipulate, explore, and brainstorm for new ideas and motivation promotion.

- *Fostering creativity*. Educators provide preliminary concepts concerning a subject task and guide students to engage in structured and heuristics processes to derive solutions to solving problems.
- *Outcome and assessment*. Teachers or peers focus on creative processes rather than products, evaluate solutions with positive appraisal rather than scathing criticism, and present ideas and solutions in multiple forms instead of written quiz assessment.

4.2 *Results of an experimental classroom*

In the first stage, the educator aimed to understand students' backgrounds, used varying open-ended questions for discussion, and provided adequate time for openness and self-expression. Eighty percent of the participants showed their desire to be creative or involved in creative thinking for problem-solving if they could have creativity-related training or education. Concerning the open-ended question of creative concept, the students' responses depended on individuals' cognition, domain knowledge, and experience. Similar to the literature, the words "novelty," "uniqueness," "new," or "effectiveness" were addressed by participants. In line with Jeffrey (2006), learners believed they could exhibit their creativity differently, at different levels, at different times, and in different conditions.

Second, the teacher provided warm-up activities that enabled students to move toward being creative. The participants had to offer several forms of self-portray and comments on classmates' drawing works. Eleven students (73.3%) enjoyed drawing activities and appreciated classmates' creative works from a positive angle, suggesting that students became more engaged in the creative development process.

Following the suggestion of Berglund and Wennberg (2006), a business venture case may provide more commercial practices that, in turn, equip students with the realities of entrepreneurial life. In the third and the fourth stages, the teacher assigned a social enterprise task and asked the participants to propose solutions to promote sales and services. Among these two stages, learners were asked to involve in the subjective and exploratory processes of touching, listening, and experimenting by conducting a field trip to this social enterprise. They also conducted interviews with owners and employees for solution considerations. The participants held meetings to share their observations, discuss their tentative alternatives, and propose solutions combined with novelty and utility for sales increase and service visibility.

This study surveyed students after this nine-week class. 86.7 percent of students provided positive feedback to the proposed CE curriculum, 66.7 percent of the participants engaged in generating creative ideas for marketing strategies, and 12 learners (80%) were confident with their creative self-efficacy. However, some limitations exist. Most participants suggested idea generation and evaluation difficulties, particularly in assessing the feasibility of ideas with marketing, financial, or banking criterion. Additionally, experience and knowledge relevance to learning tasks were challenging when generating ideas. Time-consuming is another limitation for creativity because of ineffective discussions or divergent thinking.

5 CONCLUSION

This study aims to propose a CE curriculum and empirically examine the feasibility and effectiveness of the proposed curriculum to respond to sparse research on creativity in education. This research first reviewed the relevant literature and conducted an in-depth interview to propose a tentative CE curriculum. An experimental class test was then employed to evaluate whether this curriculum is feasible. Based on the literature review and interview results, this study proposed a curriculum containing the four stages to foster creativity: teaching for creativity, a warm-up for being creative, fostering creativity, and outcome and assessment. The experimental results showed that most participants engaged in creativity development and became more confident with their creative self-efficacy. Despite the individual experience and time-consuming difficulties, the findings showed that this proposed CE curriculum is feasible and effective.

The proposed curriculum in this study is not universal for all students with varying socio-demographic characteristics, domains, and learning forces. Furthermore, the assessment mechanism is an essential part of a teaching-learning process, and evaluating an individual creative performance in terms of behavior, process, and outcomes is limited. Therefore, these limitations could be the future research directions, providing advanced knowledge of teaching-learning dynamics and outcomes of creativity education.

REFERENCES

T. B. Ward, "Cognition, creativity, and entrepreneurship," *Journal of business venturing*, vol. 19, pp. 173–188, 2004.

L. A. Jackson, E. A. Witt, A. I. Games, H. E. Fitzgerald, A. Von Eye, and Y. Zhao, "Information technology use and creativity: Findings from the Children and Technology Project," *Computers in human behavior*, vol. 28, pp. 370–376, 2012.

S. Y. Lee, R. Florida, and Z. Acs, "Creativity and entrepreneurship: A regional analysis of new firm formation," *Regional studies*, vol. 38, pp. 879–891, 2004.

I. Fillis and R. Rentschler, "The role of creativity in entrepreneurship," *Journal of enterprising culture*, vol. 18, pp. 49–81, 2010.

S. J. Shin, T.-Y. Kim, J.-Y. Lee, and L. Bian, "Cognitive team diversity and individual team member creativity: A cross-level interaction," *Academy of Management Journal*, vol. 55, pp. 197–212, 2012.

S. De Klerk, "The creative industries: An entrepreneurial bricolage perspective," *Management Decision*, vol. 53, pp. 828–842, 2015.

D. Yar Hamidi, K. Wennberg, and H. Berglund, "Creativity in entrepreneurship education," *Journal of Small Business and Enterprise Development*, vol. 15, pp. 304–320, 2008.

A. Craft, "An analysis of research and literature on creativity in education," *Qualifications and Curriculum Authority,* pp. 1–37, 2001.
R. L. DeHaan, "Teaching creative science thinking," *Science,* vol. 334, pp. 1499–1500, 2011.
S. R. Daly, E. A. Mosyjowski, and C. M. Seifert, "Teaching creativity in engineering courses," *Journal of Engineering Education,* vol. 103, pp. 417–449, 2014.
D. K. Simonton, "Teaching creativity: Current findings, trends, and controversies in the psychology of creativity," *Teaching of Psychology,* vol. 39, pp. 217–222, 2012.
J. A. Plucker, R. A. Beghetto, and G. T. Dow, "Why isn't creativity more important to educational psychologists? Potentials, pitfalls, and future directions in creativity research," *Educational psychologist,* vol. 39, pp. 83–96, 2004.
H. Berglund and K. Wennberg, "Creativity among entrepreneurship students: comparing engineering and busines education," *International Journal of Continuing Engineering Education and Life Long Learning,* vol. 16, pp. 366–379, 2006.
P. Kleiman, "Towards transformation: conceptions of creativity in higher education," *Innovations in education and teaching international,* vol. 45, pp. 209–217, 2008.
K. Kazerounian and S. Foley, "Barriers to creativity in engineering education: A study of instructors and students perceptions," *Journal of Mechanical Design,* vol. 129, pp. 761–768, 2007.
A. Ferrari, R. Cachia, and Y. Punie, "Innovation and creativity in education and training in the EU member states: Fostering creative learning and supporting innovative teaching," *JRC Technical Note,* vol. 52374, 2009.
D. J. Treffinger, G. C. Young, E. C. Selby, and C. Shepardson, "Assessing Creativity: A Guide for Educators," *National Research Center on the Gifted and Talented,* 2002.
T. M. Amabile and J. Pillemer, "Perspectives on the social psychology of creativity," *The Journal of Creative Behavior,* vol. 46, pp. 3–15, 2012.
B. Jeffrey, "Creative teaching and learning: Towards a common discourse and practice," *Cambridge Journal of Education,* vol. 36, pp. 399–414, 2006.
B. Jeffrey and A. Craft, "Teaching creatively and teaching for creativity: distinctions and relationships," *Educational studies,* vol. 30, pp. 77–87, 2004.
A. Penaluna and K. Penaluna, "Creativity in business/business in creativity: transdisciplinary curricula as an enabling strategy in enterprise education," *Industry and Higher Education,* vol. 23, pp. 209–219, 2009.

System Innovation for a Troubled World – Tak Lam et al. (Eds)

A privacy protection method for camera-based elderly monitoring system using single pixel representation

Chang Yueh Wang* & Fang Suey Lin
Graduate School of Design, National Yunlin University of Science and Technology, Yunlin, Taiwan

ABSTRACT: The aging of population is becoming a global health concern. The elderly monitoring systems can assist senior citizens to live independently and identify emergency situations. However, adoption of camera-based monitoring systems also raises the issues of the privacy-preservation of the elderly users. This paper presents a monitoring architecture based on machine learning, edge-computing, and information encoding for remote user interface. The system comprises a top-view fisheye camera, edge-computing module and an android application. An edge computing module uses a machine learning model to identify the person in the fisheye camera images and encodes the person's activity into a colored pixel. This encoded information is transmitted to the remote user interface, which mitigates the risk of privacy leak. The experiments based on real-world images confirm the applicability of our proposed system.

1 INTRODUCTION

The aging population around the world is on the rise according to the United Nations (UN) (United Nations, Department of Economic and Social Affairs 2019). At least 720 million of the world's population has reached the age over 65 and is expected to double by 2050. Based on UN reports, the Asian region accounts for a large proportion of the elderly population. The statistical bureau of Taiwan (Statistical Bureau 2019) reports approximately 3.607 million elderly people residing in the country. With aging, there are many health-related issues and safety concerns that arise for senior citizens (Colombo et al. 2012) that require continuous cognitive support and round-the-clock monitoring of such subjects. Elderly care is one of the biggest concerns for the current healthcare system as the number of elderly subjects is rising. In the last decade, many monitoring systems are developed by engineers and researchers for the elderly population.

The elderly monitoring systems developed and used in recent years (Azimi et al. 2017; Stavropoulos et al. 2020) are mainly based on cameras and wearable sensors. While camera-based elderly monitoring systems provide high accuracy and reliability, it also raises privacy concerns among users. As the image data is transmitted directly to the database servers and to the monitoring user interfaces, such as mobile applications, this gives rise to a sense of being continuously watched (Kensbock & Stöckmann 2020; Power 2016; Sætra 2019). Also, many recent cybercriminal incidents and cybersecurity leaks (BBC 2021; Owaida 2020) make user privacy a critical requirement for such systems. Therefore, to increase the acceptance of monitoring systems among the elderly population and to mitigate privacy concerns for users, an innovative and practical approach is needed.

In this paper, a privacy protection method is discussed for camera-based elderly monitoring systems. This study presents an elderly monitoring system that uses a top-view fisheye camera for image acquisition, machine learning model for person and activity detection, an information encoding scheme for encoding the monitored scene, and a remote user interface for displaying the monitoring information. The information encoding scheme includes one colored pixel in the image, where pixel represents the location of person and color represents the person's activity such as standing, sitting, sleeping and falling. The study is built upon internet-of-things (IoT) technology such as wireless fidelity (Wi-Fi) for image data transmission from camera, edge computing on the computing module for people and activity detection with information encoding, and hypertext transfer protocol secure (HTTPS) for secure transfer of data to the remote user interface. This paper provides a brief relevant literature review in the related works section. The research method is described in the methodology section, followed by the results and conclusion.

2 LITERATURE REVIEW

Many researches are focused on person detection and elderly monitoring using camera images. In Kottari, Delibasis, & Maglogiannis (2020), a real-time fall-detection system is introduced using fisheye camera images using image processing techniques. However,

*Corresponding Author: d10930015@gemail.yuntech.edu.tw

 DOI 10.1201/9781003377399-14

most of the camera-based monitoring systems use a machine learning (ML) technique for person and activity detection. Such as in (Buzzelli et al. 2020) presents a vision-based elderly monitoring system that can effectively identify daily life actions, basic poses, and alerting situations using the ML model and IoT communications. Similarly, an eight-camera-based low-cost monitoring system is discussed in Shu & Shu (2021), which applies ML and IoT methods to the image feed of strategically placed cameras for monitoring the entire living space for fall detection. Vision-based recognition of human activity (Beddiar et al. 2020) is the backbone of IoT-based elderly monitoring systems which include features such as people detection, fall detection, pose detection, and location identification. Some systems also incorporate many advanced algorithms to improve the accuracy of monitoring systems for fall detection. Background subtraction, Kalman filtering and optical flow are also used, as in De Miguel, Brunete, Hernando, & Gambao (2017) for improving fall detection accuracy. Overall, several recent researches show that innovations to increase accuracy for activity detection and IoT technology for ease of use and reliability (Beddiar et al. 2020; Dolatabadi et al. 2019; He et al. 2018; Kim et al. 2020; Luo et al. 2018; Malik et al. 2019; Nguyen et al. 2014). Edge-computing (Hamdan et al. 2020; Zhu et al. 2020) is also an area where the research is focused on offloading the cloud and local computing tasks. Some of researches (Ahmed et al. 2021; Nikouei et al. 2018) uses lightweight ML models for person detection that enable such systems to operate on edge-computing devices/layer and offer real-time performance. Although many studies, including the ones cited above, focus on the area of technological innovation, they fail to address privacy concerns and user data security.

Since this research uses top-view fisheye camera images similar to Kottari et al. (2020), it is important to discuss the relevant works in this field. "Mono-Eye" is (Hwang et al. 2019) of the state-of-art models for estimating three-dimensional human pose using monocular fisheye camera. Similarly, (Chiang et al. 2020) introduces a pedestrian detection model for top-view fisheye images. "RAPiD" (Duan et al. 2020) is another state-of-art method for rotation-aware people detection ML model for top-view fisheye images which also addresses the computational complexity. Where much research uses bounding box methods for people detections, Mask-RNN-based segmentation approach is also used as in (Wang et al. 2019). Some researchers also provide the public dataset for other researchers to build upon the existing methods (Duan et al. 2020; Krams & Kiryati 2017; Scheck et al. 2020). However, many available public datasets do not include all relevant poses that are required for ML model training. This problem is addressed in Scheck et al. (2020), which provides a synthetic dataset for person detection using artificially generated images. Also, this dataset format is compatible with the popular bounding box object detection model known as "YOLO-v4" (Bochkovskiy et al. 2020). "YOLO-v4" is the successor of "YOLO-v3" (Redmon & Farhadi 2018) (a term used for "you only look once" type of ML model) with accuracy and speed improvements. YOLO-v4 utilizes a convolutional neural network (CNN) architecture for image feature extraction and bounding box regression. Due to the ease-of-use and availability of open-source code, YOLO-v4 is regarded as a popular model for object detection.

3 METHODOLOGY

In this paper, an elderly monitoring system is proposed based on a fisheye camera. As shown in Figure 1, the system architecture comprises a Wi-Fi fisheye camera that transmits the live image stream to the edge computing module using transmission control protocol/internet protocol (TCP/IP) over a Wi-Fi connection at a frame rate of 30 frames per second. The edge computing device (in this case a personal computer is used to simulate a server module) feeds the 15th frame of live image stream to the pre-trained YOLO-v4 ML model. The YOLO-v4 model predicts the location of the person in the image together with the activity they are performing. The person's location and activity are thereafter encoded using an information encoding scheme, and the encoded image is thereafter transmitted to a mobile application that acts as a remote user interface. The encoded image is transferred using an HTTPS server hosted on the edge computing module. The mobile application connects to the HTTPS server and retrieves the image and activity data. Thereafter, remote user interface on the mobile application displays the real-time activity location with respect to the camera's field of view and notifies the end user if an alerting situation (fall detection) occurs.

For this study, the public dataset provided by Scheck et al. (2020) is used. This dataset comprises 100,000 high-resolution artificially generated top-view fisheye images with 14 different classes. The images provided in the dataset comprise different indoor scenery. Since the dataset does not provide the pose for the persons in the image. For simplicity, a small subset of 2,000 images is selected and manually labeled for different poses for people in the image. The poses selected for this study are standing, sitting, sleeping, and falling. Thereafter, 1,500 images with labels are used for training the YOLO-v4 model and 500 images are used for testing. In Figure 2, the image processing and computation scheme are presented. Every 15th frame of the live image feed is used as the input for YOLO-v4 model. The original image is then masked and resized from the resolution of 800 pixels to 80 pixels. The center point of the bounding box values from YOLO-v4 prediction is used to draw pixels (one pixel per person) for persons detected in the image. The activity predictions are encoded using a color scheme shown in Figure 2. Where red color shows the fall, yellow represents standing, blue represents sleeping/lying and green is used for sitting action. The color encoding is then transferred to pixels in the resized images, thus color-encoded representations of monitored are generated. As it is clear from Figures 1 and 2, only encoded representation/image is transferred from the edge module to the remote user interface on mobile application, thus the end user can not have access to the real image data of the monitored area. This method protects the privacy of the elderly person being monitored by encoding the real image data into a low-resolution single-pixel representation while providing sufficient monitoring information to the guardian/caregiver using the remote user interface.

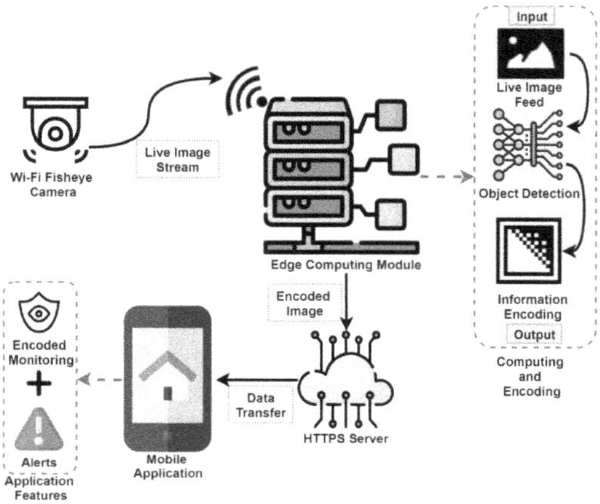

Figure 1. The system architecture of the proposed privacy-protected elderly monitoring system.

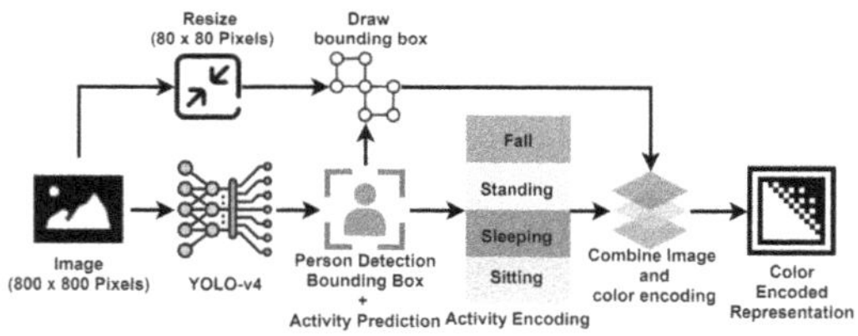

Figure 2. The process flowchart for person/activity detection and colored pixel encoding of image.

4 RESULTS AND DISCUSSION

The YOLO-v4 model trained on 1,500 synthetic images and tested on 500 images reaches a training accuracy of 81.69% while a prediction accuracy of 74.32% for activity recognition. The default settings for YOLO-v4 are used for training and prediction. The accuracy calculated above only takes the label prediction into account and bounding-box errors are ignored, since the final image is a simplified representation of the original image, which minimizes the effects of bounding-box errors. In Figure 3(a), (b), and (c), examples of input image (left column), bounding box prediction (center column, bounding boxes as green rectangles) and color encoded represented (right columns) are shown. In Figure 3(a), an image of a living room with three persons is used, the bounding box image shows the ML model predictions, and the encoded representation shows the activity (two standing and one sitting) in the single pixel format. Figure 3(b) shows the results of proposed method, where the image of a bedroom is used where four persons are present (two standing, one sitting and one sleeping). In Figure 3 (c), an image of a different living room is shown where three persons are present (one standing and two sitting). The results shown in Figure 3 clarify that the proposed privacy protection method can effectively encode the person's location and activity from the fisheye camera images without exposing the actual images to the end-user (the guardian or the caregiver).

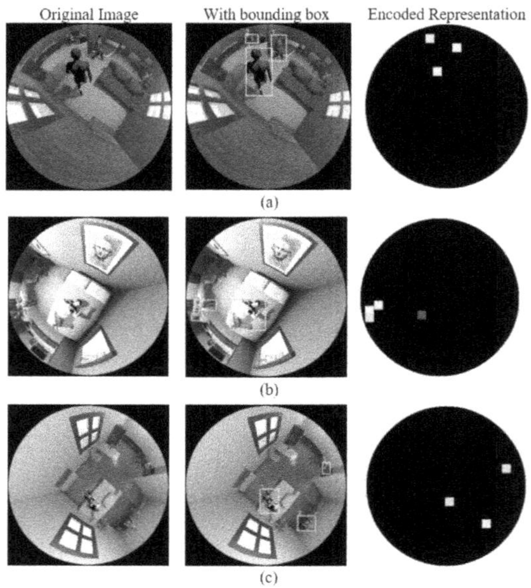

Figure 3. Results of person detection and activity information encoding using colored pixel using proposed privacy protection method.

In Figue 4(a), the log-in page of the remote user interface on the mobile application is shown. The guardian/caregiver can sign into the monitoring system using custom username and password. However, one must register to the edge computing device's internet protocol version 4 (IPv4) address and port before using the service. After logging into the user interface, the end user can access the monitoring information, such as the number of persons in the room, their location in the scene and their activity, as shown in Figure 4(b). It is also possible to select the monitoring scene from the top left corner of the user interface, for example, living room, bedroom, etc.

5 CONCLUSION

This paper presents a proof-of-concept for an elderly monitoring system with a privacy protection feature.

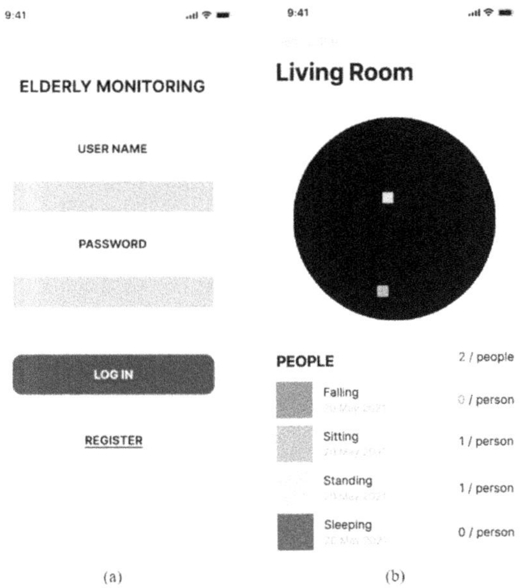

Figure 4. Image of remote user interface for privacy-protected elderly monitoring system.

A machine learning model is used to detect a person's location and activity from top-view fisheye camera images. The detected persons and their activities are encoded using a single colored pixel that represents their location in the image and their corresponding activities such as sitting, standing, sleeping, and falling. The encoded image and activity information is then transferred to a remote user interface on a mobile application that displays a privacy-protected representation of the monitored area. The main advantage of such a system includes the edge computation of the fisheye images that enables image encoding at the premises of monitored area (i.e., at home, care homes, etc.), therefore the actual image data is never exposed to the internet or cloud database. This effectively mitigates the risk of sensitive image data ever being available to the end-user (i.e., guardian, caregiver) and the database service provider (i.e., company managing the monitoring system). By using the proposed privacy protection method, one can gain the trust and acceptance of elderly persons for using monitoring systems.

The limitation of this study is that it uses synthetic images for training and verifying the proposed concept for elderly monitoring. However, due to the unavailability of public dataset on real-life images, synthetic images can provide insight on the effectiveness of the method. Future direction of this study includes the collection of real-life image dataset for improving the performance of the proposed system. The animated representation of persons and objects in the image instead of a single-pixel representation will also be considered in future work by following the privacy protection concept.

REFERENCES

Ahmed, I., Ahmad, M., Rodrigues, J. J. P. C., & Jeon, G. (2021). *Edge computing-based person detection system for top view surveillance: Using CenterNet with transfer learning. Applied Soft Computing*, 107, 107489. https://doi.org/10.1016/j.asoc.2021.107489

Azimi, I., Rahmani, A. M., Liljeberg, P., & Tenhunen, H. (2017). Internet of things for remote elderly monitoring: a study from user-centered perspective. *Journal of Ambient Intelligence and Humanized Computing*, 8(2), 273–289. https://doi.org/10.1007/s12652-016-0387-y

BBC. (2021, March 10). Hack of "150,000 cameras" investigated by camera firm. *BBC News Services*. Retrieved from https://www.bbc.com/news/technology-56342525

Beddiar, D. R., Nini, B., Sabokrou, M., & Hadid, A. (2020). Vision-based human activity recognition: a survey. *Multimedia Tools and Applications*, 79(41–42), 30509–30555. https://doi.org/10.1007/s11042-020-09004-3

Bochkovskiy, A., Wang, C.-Y., & Liao, H.-Y. M. (2020). YOLOv4: *Optimal Speed and Accuracy of Object Detection*. Retrieved from http://arxiv.org/abs/2004.10934

Buzzelli, M., Albé, A., & Ciocca, G. (2020). A Vision-Based System for Monitoring Elderly People at Home. *Applied Sciences*, 10(1), 374. https://doi.org/10.3390/app10010374

Chiang, S.-H., Wang, T., & Chen, Y.-F. (2020). Efficient Pedestrian Detection in Top-View Fisheye Images Using Compositions of Perspective View Patches. *Image and Vision Computing*, 105, 104069. https://doi.org/10.1016/j.imavis.2020.104069

Colombo, P. J., Crawley, M. E., East, B. S., & Hill, A. R. (2012). Aging and the Brain. Encyclopedia of Human Behavior: Second Edition, 53–59. https://doi.org/10.1016/B978-0-12-375000-6.00006-9

De Miguel, K., Brunete, A., Hernando, M., & Gambao, E. (2017). Home camera-based fall detection system for the elderly. *Sensors* (Switzerland), 17(12). https://doi.org/10.3390/s17122864

Dolatabadi, E., Zhi, Y. X., Flint, A. J., Mansfield, A., Iaboni, A., & Taati, B. (2019). The feasibility of a vision-based sensor for longitudinal monitoring of mobility in older adults with dementia. *Archives of Gerontology and Geriatrics*, 82(December 2018), 200–206. https://doi.org/10.1016/j.archger.2019.02.004

Duan, Z., Ozan Tezcan, M., Nakamura, H., Ishwar, P., & Konrad, J. (2020). RAPiD: *Rotation-Aware People Detection in Overhead Fisheye Images*. 2020 IEEE/CVF Conference on Computer Vision and Pattern Recognition Workshops (CVPRW), 2020-June, 2700–2709. https://doi.org/10.1109/CVPRW50498.2020.00326

Hamdan, S., Ayyash, M., & Almajali, S. (2020). Edge-Computing Architectures for Internet of Things Applications: A Survey. *Sensors*, 20(22), 6441. https://doi.org/10.3390/s20226441

He, Z., Lu, D., Yang, Y., & Gao, M. (2018). An Elderly Care System Based on Multiple Information Fusion. *Journal of Healthcare Engineering*, 2018. https://doi.org/10.1155/2018/4098237

Hwang, D.-H., Aso, K., & Koike, H. (2019). *Mono-Eye: Monocular Fisheye Camera-based 3D Human Pose Estimation. 2019 IEEE Conference on Virtual Reality and 3D User Interfaces* (VR), 988–989. https://doi.org/10.1109/VR.2019.8798267

Kensbock, J. M., & Stöckmann, C. (2020). "Big brother is watching you": surveillance via technology undermines employees' learning and voice behavior during digital

transformation. *Journal of Business Economics*, 91(4), 565–594. https://doi.org/10.1007/s11573-020-01012-x
Kim, J., Min, K., Jung, M., & Chi, S. (2020). Occupant behavior monitoring and emergency event detection in single-person households using deep learning-based sound recognition. *Building and Environment*, 181, 107092. https://doi.org/10.1016/j.buildenv.2020.107092
Kottari, K. N., Delibasis, K. K., & Maglogiannis, I. G. (2020). *Real-Time Fall Detection Using Uncalibrated Fisheye Cameras. IEEE Transactions on Cognitive and Developmental Systems*, 12(3), 588–600. https://doi.org/10.1109/TCDS.2019.2948786
Krams, O., & Kiryati, N. (2017). *People detection in top-view fisheye imaging*. 2017 14th IEEE International Conference on Advanced Video and Signal Based Surveillance (AVSS), 1–6. https://doi.org/10.1109/AVSS.2017.8078535
Luo, Z., Hsieh, J.-T., Balachandar, N., Yeung, S., Pusiol, G., Luxenberg, J., …Fei-Fei, L. (2018). Computer Vision-Based Descriptive Analytics of Seniors' Daily Activities for Long-Term Health Monitoring. *Proceedings of Machine Learning Research*, 85(August), 1–18.
Malik, R., Rastogi, K., Tripathi, V., & Badal, T. (2019). An efficient vision based elderly care monitoring framework using fall detection. *Journal of Statistics and Management Systems*, 22(4), 603–611. https://doi.org/10.1080/09720510.2019.1609185
Nguyen, V. D., Le, M. T., Do, A. D., Duong, H. H., Thai, T. D., & Tran, D. H. (2014). *An efficient camera-based surveillance for fall detection of elderly people*. Proceedings of the 2014 9th IEEE Conference on Industrial Electronics and Applications, ICIEA 2014, 994–997. https://doi.org/10.1109/ICIEA.2014.6931308
Nikouei, S. Y., Chen, Y., Song, S., Xu, R., Choi, B. Y., & Faughnan, T. R. (2018). *Real-time human detection as an edge service enabled by a lightweight CNN*. Proceedings - 2018 IEEE International Conference on Edge Computing, EDGE 2018 - Part of the 2018 IEEE World Congress on Services, 125–129. https://doi.org/10.1109/EDGE.2018.00025
Owaida, A. (2020). *Google disables Xiaomi smart home integration after camera bug*. Retrieved June 6, 2021, from ESET website: https://www.welivesecurity.com/2020/01/03/google-disables-xiaomi-smart-home-integration/
Power, D. J. (2016). "Big Brother" can watch us. *Journal of Decision Systems*, 25, 578–588. https://doi.org/10.1080/12460125.2016.1187420
Redmon, J., & Farhadi, A. (2018). YOLOv3: *An Incremental Improvement*. Retrieved from http://arxiv.org/abs/1804.02767
Sætra, H. S. (2019). Freedom under the gaze of Big Brother: Preparing the grounds for a liberal defence of privacy in the era of Big Data. *Technology in Society*, 58(June), 101160. https://doi.org/10.1016/j.techsoc.2019.101160
Scheck, T., Seidel, R., & Hirtz, G. (2020). *Learning from THEODORE: A Synthetic Omnidirectional Top-View Indoor Dataset for Deep Transfer Learning*. 2020 IEEE Winter Conference on Applications of Computer Vision (WACV), 932–941. https://doi.org/10.1109/WACV45572.2020.9093563
Shu, F., & Shu, J. (2021). An eight-camera fall detection system using human fall pattern recognition via machine learning by a low-cost android box. *Scientific Reports*, 11(1), 1–17. https://doi.org/10.1038/s41598-021-81115-9
Statistical Bureau, R. of C. (Taiwan). (2019). *General situation of population*. Retrieved June 7, 2021, from National Statistics, Taiwan website: https://eng.stat.gov.tw/ct.asp?xItem=41871&ctNode=6339&mp=5
Stavropoulos, T. G., Papastergiou, A., Mpaltadoros, L., Nikolopoulos, S., & Kompatsiaris, I. (2020). *Iot wearable sensors and devices in elderly care: A literature review*. Sensors (Switzerland), 20(10). https://doi.org/10.3390/s20102826
United Nations, Department of Economic and Social Affairs, P. D. (2019). *World Population Ageing* 2019. In World Population Ageing 2019. Retrieved from United Nations website: https://www.un.org/en/development/desa/population/publications/pdf/ageing/WorldPopulationAgeing2019-Report.pdf
Wang, T., Hsieh, Y. Y., Wong, F. W., & Chen, Y. F. (2019). *Mask-RCNN Based People Detection Using A Top-View Fisheye Camera*. Proceedings – 2019 International Conference on Technologies and Applications of Artificial Intelligence, TAAI 2019. https://doi.org/10.1109/TAAI48200.2019.8959887
Zhu, R., Liu, L., Song, H., & Ma, M. (2020). Multi-access edge computing enabled internet of things: advances and novel applications. *Neural Computing and Applications*, 32(19), 15313–15316. https://doi.org/10.1007/s00521-020-05267-x

A design for client association and resource allocation in mmWave 5G networks

Jeng-Ji Huang* & Yi-Rui Li
Department of Electrical Engineering, National Taiwan Normal University, Taipei, Taiwan

David Shiung
Department of Electronics Engineering, National Changhua University of Education, Changhua, Taiwan

ABSTRACT: A 5G millimeter wave network is able to deliver speeds of Gbps through the use of beam-forming and the dense deployment of small cell access points. In this paper, we consider both the associations of clients and allocations of channel resources, where demands from clients are taken into account. The problem can be formulated into an integer nonlinear optimization. To solve this, a dual decomposition method can be utilized, and then a distributed algorithm is developed. Numerical results demonstrate that our proposed algorithm can substantially outperform the previous work in which no client demand is taken into account.

1 INTRODUCTION

In this paper, a millimeter wave network is considered, where clients are uniformly distributed in a service area and access points (APs) are deployed for clients to access through. Let $\mathcal{C}$ be the set including all clients, and $\mathcal{A}$ be the set consisting of all APs. A client is associated with an AP, and resources are allocated by an AP to all the clients associated with it.

A binary association indicator is denoted by x_{ik}. If client i is associated with AP k, $x_{ik}=1$; otherwise, $x_{ik}=0$. the resource allocated by AP k to client i is denoted by x_{ik}, which amounts to the portion of time that AP k will direct its antenna to client i. Let r_{ik} be the bit rate between client i and AP k, which is basically calculated by $r_{ik}=B\cdot log_2(1+SNR_{ik})$, where B is the bandwidth and SNR_{ik} is the signal-to-noise ratio of client i seen at AP k.

When proportional fairness is taken into account, the problem can be formulated as (Y. Xu et al. 2016)

$$\max_{x,y}\sum_{i\in\mathcal{C}}\sum_{k\in\mathcal{A}}x_{ik}\cdot\log\left(r_{ik}y_{ik}\right)$$

s. t.

$$\begin{aligned}&\sum_{k\in\mathcal{A}}x_{ik}=1,\ x_{ik}\in\{0,1\},\ \forall i\in\mathcal{C},k\in\mathcal{A},\\&\sum_{i\in\mathcal{C}}y_{ik}\le 1,\ \ y_{ik}\ge 0,\qquad \forall i\in\mathcal{C},k\in\mathcal{A},\end{aligned}\tag{1}$$

where $\boldsymbol{x},\boldsymbol{y}$ are vectors of x_{ik},y_{ik}, respectively.

$\boldsymbol{y}$ in Eq. (1) can be solved by first letting x_{ik} is known. That is, by assuming $\boldsymbol{x}$ a feasible solution, $\boldsymbol{y}$ can then be found from r_{ik}. The steps of finding $\boldsymbol{y}$ are illustrated as follows. First of all, the objective function in Eq. (1) can be rewritten as $\max_y\sum_{i\in\mathcal{C}}\sum_{k\in\mathcal{A}}x_{ik}\log\left(r_{ik}y_{ik}\right)$

$$=\max_y\sum_{i\in\mathcal{C}}\sum_{k\in\mathcal{A}}x_{ik}(\log r_{ik}+\log y_{ik})$$

$$=\max_y\sum_{i\in\mathcal{C}}\sum_{k\in\mathcal{A}}x_{ik}\log r_{ik}+x_{ik}\log y_{ik}\tag{2}$$

Let $n_k=\sum_{i\in\mathcal{C}}x_{ik}$. As x_{ik} is known, Eq. (2) is equivalent to

$$\max_y\sum_{i\in\mathcal{C}}\sum_{k\in\mathcal{A}}x_{ik}\log y_{ik}=\max_y\sum_{i\in\mathcal{C}_k}\sum_{k\in\mathcal{A}}\log y_{ik},\tag{3}$$

where $\mathcal{C}_k=\{i|x_{ik}=1\}$ is the set consisting of those clients that are associated with AP k. Eq. (3) can further be expressed as

$$\max_y\sum_{k\in\mathcal{A}}\log\left(\prod_{i=1}^{n_k}y_{ik}\right).\tag{4}$$

As $\log(\cdot)$ is an increasing function, Eq. (4) can be rewritten as

$$\begin{aligned}&\max_y\prod_{i=1}^{n_k}y_{ik}\\&s.t.\ \sum_{i\in\mathcal{C}_k}y_{ik}\le 1.\end{aligned}\tag{5}$$

*Corresponding Author: hjj2005@ntnu.edu.tw

DOI 10.1201/9781003377399-15

According to arithmetic-geometric mean inequality, $\prod_{i=1}^{n_k} y_{ik}$ is maximized when $y_{ik} = \frac{1}{n_k}$. From this and Eq. (2), Eq. (1) can be rewritten as

$$\max_{x,n} \sum_{i \in \mathcal{C}} \sum_{k \in \mathcal{A}} x_{ik} b_{ik} - \sum_{k \in \mathcal{A}} n_k \log n_k,$$

s. t.

$$\begin{aligned} &\sum_{k \in \mathcal{A}} x_{ik} = 1, \ x_{ik} \in \{0, 1\}, \ \forall i \in \mathcal{C}, k \in \mathcal{A}, \\ &\sum_{i \in \mathcal{C}} x_{ik} = n_k, \qquad \forall k \in \mathcal{A}, \end{aligned} \tag{6}$$

where $b_{ik} = \log r_{ik}$, $\boldsymbol{n}$ is a vector formed by n_k.

In order to solve Eq. (6), Lagrange multipliers are utilized. Furthermore, x_{ik} is relaxed to be a real number between 0 and 1. Let $\boldsymbol{\lambda}$ be a vector formed by λ_k, where $k \in \mathcal{A}$. The Lagrangian function can then be expressed as

$$\begin{aligned} &\mathcal{L}(\boldsymbol{x}, \boldsymbol{n}, \boldsymbol{\lambda}) \\ &= \sum_{i \in \mathcal{C}} \sum_{k \in \mathcal{A}} x_{ik} b_{ik} - \sum_{k \in \mathcal{A}} n_k \log n_k + \sum_{k \in \mathcal{A}} \lambda_k \left(n_k - \sum_{i \in \mathcal{C}} x_{ik} \right) \\ &= \sum_{i \in \mathcal{C}} \sum_{k \in \mathcal{A}} x_{ik} \left(b_{ik} - \lambda_k \right) + \sum_{k \in \mathcal{A}} n_k (\lambda_k - \log n_k). \end{aligned} \tag{7}$$

Eq. (7) can be solved by using the dual decomposition method. First, Eq. (7) is rewritten into its Lagrangian dual function as

$$\min_{\lambda} \mathcal{L}(\boldsymbol{x}, \boldsymbol{n}, \boldsymbol{\lambda}) = \min_{\lambda} f(\boldsymbol{n}, \boldsymbol{\lambda}) + g(\boldsymbol{x}, \boldsymbol{\lambda}), \tag{8}$$

where

$$f(\boldsymbol{n}, \boldsymbol{\lambda}) = \sum_{k \in \mathcal{A}} n_k \left(\lambda_k - \log n_k \right), \tag{9}$$

$$g(\boldsymbol{x}, \boldsymbol{\lambda}) = \sum_{i \in \mathcal{C}} \sum_{k \in \mathcal{A}} x_{ik} (b_{ik} - \lambda_k). \tag{10}$$

By using the dual decomposition method, the minimization of Eq. (7) can be implemented in a distributed manner. That is, according to current $\boldsymbol{\lambda}$, Eqs. (9) and (10) seek to search for an optimal solution for $\boldsymbol{n}$ and an optimal solution for $\boldsymbol{x}$, respectively, in each round. Since

$$\frac{\partial \mathcal{L}}{\partial \lambda_k} = n_k - \sum_{i \in \mathcal{C}} x_{ik}, \tag{11}$$

$\boldsymbol{\lambda}$ is updated for the next round by

$$\lambda_{k+1} = \lambda_k - \delta \cdot \left(n_k - \sum_{i \in \mathcal{C}} x_{ik} \right). \tag{12}$$

where δ is a predefined step size. Note that the term for adjustment, i.e., $-\delta \cdot \left(n_k - \sum_{i \in \mathcal{C}} x_{ik} \right)$, is in the reverse direction of $\frac{\partial \mathcal{L}}{\partial \lambda_k}$ in Eq. (11), meaning that in each round a new λ lets $\mathcal{L}$ get smaller. The algorithm stops only when $n_k = \sum_{i \in \mathcal{C}} x_{ik}$, $\forall k \in \mathcal{A}$.

The optimization of Eq. (9) is found by

$$\frac{\partial f}{\partial n_k} = \lambda_k - (\log n_k + 1), \tag{13}$$

or

$$n_k = e^{(\lambda_k - 1)}, \quad \forall k \in \mathcal{A}. \tag{14}$$

On the other hand, the optimization of Eq. (10) is given as

$$\frac{\partial g}{\partial x_{ik}} = b_{ik} - \lambda_k, \quad \forall i \in \mathcal{C}. \tag{15}$$

Note that Eq. (15) is calculated using the current λ_k at each client i. Since each client i has its own limited capacity of contributing to the maximization of Eq. (10), it is clear that the best policy for client i is to let $x_{ik} = 1$ and $x_{ik'} = 0$ for $k' \neq k$. It is because $\frac{\partial g}{\partial x_{ik}}$ is the steepest, and, by doing so, client i has done its utmost for maximizing Eq. (10).

After Eq. (15) is performed at each client i, x_{ik} is then determined, which can be made known to each AP k. Then, at AP k, $n_k = e^{(\lambda_k - 1)}$ calculated by Eq. (14) for current λ_k. When the calculated n_k is different than the actual number of associated clients, i.e., $\sum_{i \in \mathcal{C}} x_{ik}$, the algorithm is yet to finish and has to continue its next round by updating λ_k in Eq. (12).

2 THE PROPOSED SCHEME

In this section, the demand from each client i is taken into account in the problem, which manifests itself as a weight w_i. The problem, like Eq. (1), can be written as

$$\max_{x,y} \sum_{i \in \mathcal{C}} \sum_{k \in \mathcal{A}} x_{ik} \cdot w_i \cdot \log (r_{ik} y_{ik}).$$

s. t.

$$\begin{aligned} &\sum_{k \in \mathcal{A}} x_{ik} = 1, \ x_{ik} \in \{0, 1\}, \ \forall i \in \mathcal{C}, k \in \mathcal{A}, \\ &\sum_{i \in \mathcal{C}} y_{ik} \leq 1, \ \ y_{ik} \geq 0, \qquad \forall i \in \mathcal{C}, k \in \mathcal{A}, \end{aligned} \tag{16}$$

Similar to the process in Section 1, given that x_{ik} is known, y_{ik} is first solved, which is detailed as follows. First of all, Eq. (16) can be rewritten as

$$\begin{aligned} &\max_{y} \sum_{i \in \mathcal{C}} \sum_{k \in \mathcal{A}} x_{ik} \cdot w_i \cdot \log (r_{ik} y_{ik}) \\ &= \max_{y} \sum_{i \in \mathcal{C}} \sum_{k \in \mathcal{A}} x_{ik} \cdot w_i \cdot (\log r_{ik} + \log y_{ik}) \end{aligned}$$

$$= \max_{y} \sum_{i \in \mathcal{C}} \sum_{k \in \mathcal{A}} x_{ik} w_i \log r_{ik} + x_{ik} w_i \log y_{ik} \quad (17)$$

Second, let $n'_k = \sum_{i \in \mathcal{C}} w_i x_{ik}$ be the total weight of all clients i associated with AP k. Since $b_{ik} = \log r_{ik}$ is a constant, the first term in Eq. (17) can be removed and Eq. (17) can be reduced to

$$\max_{y} \sum_{i \in \mathcal{C}} \sum_{k \in \mathcal{A}} x_{ik} w_i \log y_{ik}$$

$$= \max_{y} \sum_{k \in \mathcal{A}} \sum_{i \in \mathcal{C}_k} w_i \log y_{ik}$$

$$= \max_{y} \sum_{k \in \mathcal{A}} \log \prod_{i=1}^{n'_k} y_{ik}^{w_i}. \quad (18)$$

As $\log(\cdot)$ is an increasing function, Eq. (18) can be rewritten as

$$\max_{y} \prod_{i=1}^{n'_k} y_{ik}^{w_i}.$$

$$\text{s.t.} \sum_{i \in \mathcal{C}_k} y_{ik} \leq 1. \quad (19)$$

Suppose w_is are positive integers. $y_{ik}^{w_i}$ can be viewed as y_{ik} multiplied by itself w_i times. By letting n'_k be the total number of imaginary clients associated with AP k, the optimal y_{ik} can be found as

$$y_{ik} = \frac{w_i}{n_k}. \quad (20)$$

Note that the optimal solution for y_{ik} can be applied to cases that w_is are real numbers, rather than integers.

Third, like Eq. (7), the Lagrangian function can then be expressed as

$$\mathcal{L}(\boldsymbol{x}, \boldsymbol{n}', \boldsymbol{\lambda})$$

$$= \sum_{i \in \mathcal{C}} \sum_{k \in \mathcal{A}} x_{ik} w_i (b_{ik} - \lambda_k)$$

$$+ \sum_{k \in \mathcal{A}} n'_k w_i (\lambda_k - \log n'_k). \quad (21)$$

Eq. (21) can then be minimized via the dual decomposition method. Similar to Eqs. (8), (9), and (10), the problem in Eq. (16) can be expressed as

$$\min_{\lambda} \mathcal{L}(\boldsymbol{x}, \boldsymbol{n}, \boldsymbol{\lambda}) = \min_{\lambda} f(\boldsymbol{n}, \boldsymbol{\lambda}) + g(\boldsymbol{x}, \boldsymbol{\lambda}), \quad (22)$$

where

$$f(\boldsymbol{n}, \boldsymbol{\lambda}) = \sum_{k \in \mathcal{A}} n_k (\lambda_k - \log n_k), \quad (23)$$

$$g(\boldsymbol{x}, \boldsymbol{\lambda}) = \sum_{i \in \mathcal{C}} \sum_{k \in \mathcal{A}} x_{ik} (b_{ik} - \lambda_k). \quad (24)$$

As before, x_{ik} would be either 1 or 0 when minimizing g for a given $\boldsymbol{\lambda}$, which is updated by

$$\lambda_{k+1} = \lambda_k - \delta \cdot \left(n'_k - \sum_{i \in \mathcal{C}} w_i x_{ik} \right). \quad (25)$$

The minimization of f for a given λ, similar to Eq. (14), can be obtained as

$$n'_k = e^{(\lambda_k - 1)}, \quad \forall k \in \mathcal{A}. \quad (26)$$

On the other hand, the minimization of g for a given λ, similar to Eq. (15), can be achieved as

$$\frac{\partial g}{\partial x_{ik}} = w_i \cdot (b_{ik} - \lambda_k), \quad \forall i \in \mathcal{C}. \quad (27)$$

Note that for client i, w_i is simply the same. Therefore, Eq. (27) is equivalent to Eq. (15), in which AP k with the largest $(b_{ik} - \lambda_k)$ is selected.

3 NUMERICAL RESULTS

In this section, numerical results are demonstrated from our experiments. In our experiments, the locations of 150 clients are uniformly distributed in a circular service area of radius 150 m, while the locations of 5 APs are uniformly distributed within a concentric circular area of radius 80 m. An example is shown in Figure 1, where blue circles are used to denote the locations of clients and black triangles are used to denote the locations of APs. The path loss model in (T. S. Rappaport et al. 2015) is employed between a client and an AP.

Figure 2 shows a result obtained from solving Eqs. (8)–(10). As can be seen, almost every client chooses

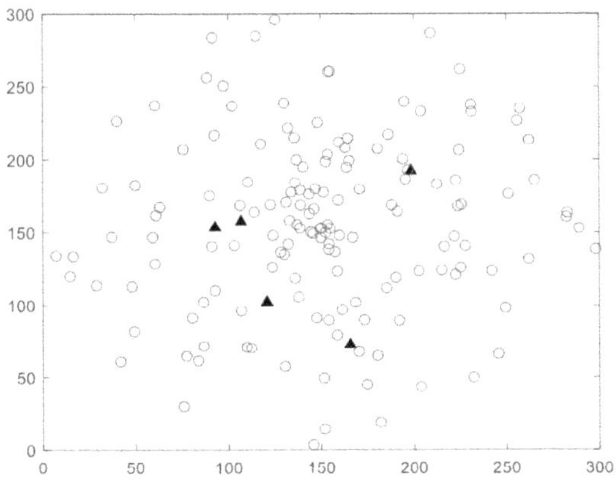

Figure 1. A circular service area of radius 150 m in our experiments, where 150 clients and 5 APs are uniformly distributed in concentric circular areas of radii 150 and 80 m, respectively.

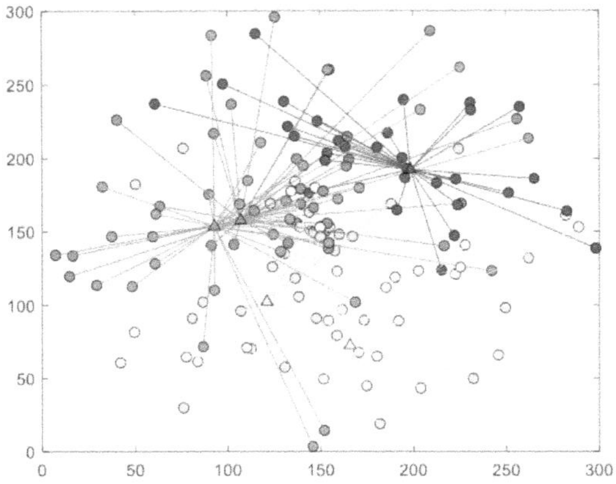

Figure 2. A result of client associations, where different colors indicate different APs with which clients associate.

to make an association with the nearest AP. However, there are still exceptions if a client chooses to make associations with an AP that is not the nearest one. The main reason is that, from Eq. (14), λ_k tends to grow when n_k is large. However, when n_k becomes too large, a large λ_k makes AP k less likely to be chosen by a client, from Eq. (15).

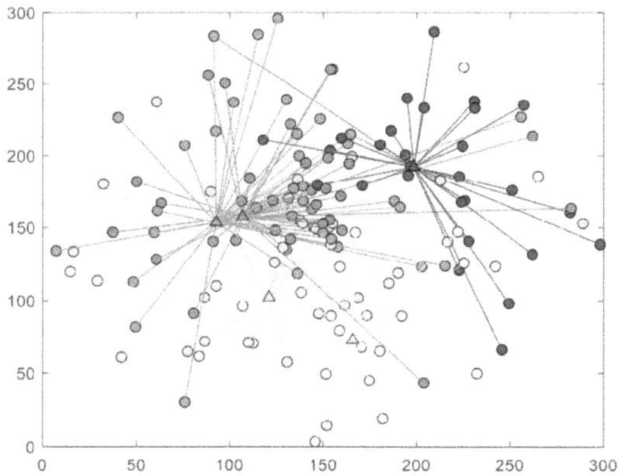

Figure 3. An overlap of two results for client associations, with/without demands from clients.

Figure 3 shows an overlap of two results for client associations, when there are no demands from clients. As can be seen, most clients make associations with the same APs, no matter if there are demands from clients. However, there are still exceptions that some clients make associations with different APs when there are demands from clients, which are shown by those having two different colored lines connected to APs. This shows that our proposed scheme is able to reflect different weights from different clients and let some clients make different association decisions in order to meet their demands.

4 CONCLUSION

In this paper, a client association problem is discussed for a millimeter wave 5G network. In particular, demands from clients are explicitly incorporated into an optimization problem, in which a dual decomposition method is utilized. Numerical results show that our proposed scheme is able to let demands from clients be reflected on clients' decision about their associations with APs.

REFERENCES

T. S. Rappaport, G. R. MacCartney, M. K. Samimi and S. Sun, 2015. IEEE Trans. Commun. 63(9) 3029–3056.

Y. Xu, H. Shokri-Ghadikolaei and C. Fischione, 2016. *IEEE Trans. Wirel. Commun.* 15(12) 7955–7970.

System Innovation for a Troubled World – Tak Lam et al. (Eds)
© 2023 The Author(s), ISBN: 978-1-032-45525-9

Research on operating modes of industrial design studios from qualitative and quantitative perspectives

Jeng-Chung Woo*
School of Design · Straits Institute of Technology, Fujian University of Technology, Fuzhou, China
Design Innovation Research Center of Humanities and Social Sciences Research Base of Colleges and Universities in Fujian Province, Fuzhou, China

Guo-Qing Pan, Zhe-Hui Lin & Yu-Tong Chen
School of Design · Straits Institute of Technology, Fujian University of Technology, Fuzhou, China

ABSTRACT: Studio teaching is remarkably effective in the education of industrial design. However, in the previous studies in relation to it, neither systematic analysis nor specific guiding principles can be observed. In response to it, this study conducted interviews with 5 experts using a semi-structured interview questionnaire formed from the analysis of the reletive literatures. Then, the audio files of the interviews were converted into verbatim transcripts (with 68,870 words in total) which were analyzed and coded using the grounded theory. After that, 245 categories, 82 open coding structures, 22 axial coding structures, and 6 selective coding structures were formed, all of which were split into three-grade indicators as the initial questionnaire for the Delphi method (110 questions in total) based on the division of the studio operating mode into teaching, management, operation, atmosphere, appraisal, and university-enterprise cooperation. By consulting 11 experts with more than two years of experience in hosting studios for three-round revision, finally, 17 clauses were modified with 10 added and 12 deleted. In this way, a set of effective guidelines for studio operation has been finalized through systematic analysis and generalization using qualitative and quantitative methods, which will provide reference and paradigms for the establishment of industrial design studios in universities, propelling the disciplinary development of the industrial design.

Keywords: Design studio; Industrial design; Grounded theory; Delphi method

1 INTRODUCTION

In the history of industrial design education, Bauhaus has exerted far-reaching impacts on design education and thinking. (Cross A. 1983) By integrating art with technology and offering workshop-style teaching, Bauhaus has successfully incorporated the studio system into the cultivation of students' creativity and practical competencies and established a set of sophisticated models and frameworks for design education. Specifically, the studio system has been introduced to core courses aiming to foster students' design competencies in many colleges and universities. For example, the duration of design studio classes in Malaysian universities is about 6 to 8 hours per week (Heskett et al. 1997).

With the development of industrial design, the studio model as a core course for cultivating specialized talents has been broadly recognized by many universities. Schön, a scholar, defines design studios as "reflective practices in design" to illustrate the effectiveness of the studio system in fostering students' creativity and practical competencies. (Schön et al. 1998).

However, given a late start and the constraints over design education due to China's national conditions, the studio system in Chinese universities has not been well established and students enrolled in the industrial design field have been considered as having low practical skills and creativity (Fan X et al. 2011). Although many universities have adopted the studio system in industrial design education, a lack of guidance by specific studio operation models has adversely affected the outcomes of studio teaching. Using grounded theory and the Delphi method, this research examines the operation model of industrial design studios. A systematic analysis based on both qualitative and quantitative approaches is conducted, with a view of summarizing the guiding principles for the operation model of industrial design studios and thus improving studio teaching and facilitating the construction and reform of the industrial design specialty field in universities.

*Corresponding Author: wwwjc2020@outlook.com

DOI 10.1201/9781003377399-16

2 LITERATURE REVIEW

2.1 *Current status of design studio research*

As the teaching model aimed at cultivating industrial design talents, the studio system has been intensively researched. An analysis reveals that existing studio research mainly focuses on topics such as school-enterprise cooperation, studio management, studio teaching, performance evaluation, design methods, and procedures, with studio management and teaching receiving most of the attention. (Sawyer R K. 2017)

In the body of studio teaching research, researchers start with teaching and learning to analyze such topics as cross-disciplinary integration, interaction modes between teachers and students, global design trends, and co-design. For example, in her work addressing the interaction modes between teachers and students, Philippa Kethro elaborates on the connection between teacher's feedback and students' autonomous thinking in design evaluation, highlighting different cognitions between teachers and students towards problem research, design expression, and design purpose (Kethro P. 2013). The line of research on studio management is mostly focused on studio environment and cross-disciplinary integration. For example, in carrying out a study of the studio model from perspectives of environment and spatial arrangement, open subjects, design evaluation, and assessment, scholar Wayne C. Chung argued that studios should create their own learning and thinking atmosphere and should maintain mutually separated and well-functioning working, gaming, meeting, classroom, and presentation spaces through reasonable organization. (Chung W C. 2019)

In their studies of school-enterprise cooperation models, scholars like Chen et al. elaborated the necessity of school-enterprise cooperation and how to implement the cooperation models from the perspective of basic characteristics of design universities. However, these studies only set out the authors' opinions without providing convincing validation about the proposed school-enterprise models (Chi S et al. 2013). In studies of performance evaluation systems, Alagbe O et al. explored different performance evaluation systems implemented by design studios. (Alagbe O et al. 2015).

As can be found from the above-mentioned discussions on studios, researchers mostly focused on some parts of the studio model with little attention paid to the overarching guiding principle; a few scholarly studies that address the operation model of studios tend to adopt a method of view description or literature review without rigorous analysis. Thus, researching studio operation criteria using qualitative and quantitative methods stands to fill the void in such a line of research and generate more objective and rigorous studio operation models.

2.2 *Grounded theory*

As a bottom-up method, the grounded theory emphasizes that theoretical research should not be based on existing hypotheses but on a process starting with direct observations of practical activities, collection, and organization of raw data, followed by systematic data analysis and generalization of theories capable of reflecting social practices, that is, "establishing theories by the qualitative method" (Xu R et al. 2021) Coding analysis is a key step and fundamental process toward theoretical construction. Studies based on grounded theory generally entail three levels of coding analyses: open coding, axial coding, and selective coding (Ahangama N et al. 2021) Employing the ground theory, Tiffany Lee Hooda and Darcy Copeland discussed key events experienced by nursing students in the clinical environment, advising all faculty members to step up their attention to students' psychological health (Hood et al. 2022) Ayla Kaya utilized the grounded theory to explore the experience of Turkish teenagers exposed to the risk of addiction to digital games, thereby increasing awareness among schools and parents of the issue and preventing digital game addiction (Kaya A et al. 2021). In this research, the grounded theory is applied to exploring the relevant experience of guiding design studios; 3-level coding analyses of the initial data of expert interview texts are conducted to preliminarily distill studio operation criteria.

2.3 *Delphi method*

The Delphi method is a process used to collect opinions on specific subjects of interest (De Villiers M R et al. 2005). In essence, it is a questionnaire research to collect anonymous feedback. It is a non-face-to-face method to collect expert opinions and allow for the concentration, collision and integration of experts' social resources. (Shen L et al. 2019) The Delphi method often involves multiple rounds of opinion solicitation until the process ends with a consensus reached between experts. Using the Delphi method, Amanda A. Olsen studied more effective curriculum design criteria in pharmaceutical education (Olsen A A et al. 2021). Muzaffer Berna Dogan used the Delphi method to explore the standard method model for claims in medical accidents (Doğan M B et al. 2020). In this research, the Delphi method is applied to conduct three rounds of solicitation of expert opinions concerning studio operation criteria, which are modified based on suggestions from multiple experts until a consensus is reached within the expert panel, thereby giving rise to a set of scientific, rigorous studio operation criteria.

3 METHODS

3.1 *Data collection*

To collect true and effective data on studio operations to discuss studio guiding principles, this research employed the grounded theory to assemble an expert panel of 5 university teachers who have presided over studio projects (concerning industrial design, visual communication, product design, and other fields) and conduct expert interviews using semi-structured questionnaires. An interview outline was prepared based on organized and summarized literature, which encompasses five aspects: studio management, studio teaching, school-enterprise cooperation, performance evaluation, and design methods and process. A total of 19 items were prepared to guide the expert interview and collect their experience and opinions on how to guide studios. The interview was conducted in both online and offline manners. Experts' permission to record the interview was sought and all records were transcribed into texts. In the end, a transcription of 68,870 characters was obtained. Subsequently, the transcription was coded word-by-word and sentence-by-sentence using the grounded theory and then organized into preliminary studio operation criteria.

3.2 *Grounded theory*

3.2.1 *Open coding*

Open coding requires researchers to carry out in-depth analyses of raw data with an open mentality. In the analysis, researchers needed to categorize semantic information related to the guiding principles of studio operation and integrated highly correlated semantic information into initial labels; then, initial labels were further conceptualized and abstracted into more generalized concepts. This can be divided into three stages: labelization, conceptualization, and categorization (Walker D et al. 2006).

By repeatedly comparing and organizing interview data on guiding principles of studio operation, a total of 245 concepts were distilled. These concepts were further refined and analyzed by establishing necessary connections and categories, deriving a total of 82 open codes (as shown in Table 1).

3.2.2 *Axial coding*

Subsequent to open coding, it was necessary to obtain more generalized primary categories from initial ones to further explore the logical relations between different categories and concertize their attributes and dimensions, which is the key to axial coding (as shown in Table 2). A total of 22 axial codes were obtained through repeated comparison and generalization.

3.2.3 *Selective coding*

On the basis of axial coding, selective coding is implemented to sort out the relationships between primary categories, abstracting core categories capable of generalizing all aspects, and connecting the primary and secondary categories with a storyline. In this paper, a systematic analysis of relevant categories was conducted to obtain 6 selective codes, which are operation vision, teaching model, operation system, environmental atmosphere, assessment system, and school-enterprise cooperation.

3.3 *Delphi method*

3.3.1 *Construction of the expert panel*

In this research, 11 experts with more than 2 years' worth of experience (5 years on average) in presiding over studios related to the purpose and content of this study were invited to form an expert panel; specifically, the researched studios mainly involved areas of industrial design, art design, product design, marine culture and science popularization, visual communication design, and interior design.

3.3.2 *Questionnaire design*

Studio operation criteria were preliminarily built upon the original logical relations among 6 selective codes, 22 axial codes, and 82 open codes generalized using the grounded theory; on that basis, the first round of expert consultation table was prepared and a Likert five-point scale was used to invite experts to modify indicators of the operating criteria. Apart from assessing each indicator, the experts were also asked to express their opinions and thoughts.

Table 1. Concept cluster of the open code "serving the locality".

Labelization	Conceptualization	Open code
b1, b2, b3, b4 b5	B1 Be attentive to local specialty economy	Serving the locality
c11, c12	C5 Driven by local resources	
d27, d28	D16 Local resources and conditions promote studio development	
d32, d33	D19 Local economy constrains studio development	
d55	D36 Studios should serve the local economy and culture	

Table 2. Categories of the axial code "adherence to ideals".

Conceptualization	Open code	Axial code
A45, C27, D29, E10 and E12	Cross-disciplinary integration	Ideas upheld
D49:	Integration of teaching, learning, research and innovation	
C3, E25 and E26	Design enthusiasm	
A13, B14, A6, B9 and B13	Aptitude-based teaching	

3.3.3 *Calculation methods of evaluation indicators*
(1) The degree of expert enthusiasm, that is, the recovery rate of expert consultation tables, the magnitude of which indicates the level of concern extended by experts toward the project. The coefficient of expert enthusiasm can be expressed by the recovery rate of valid questionnaires, that is,

$$P = \frac{E}{T} x 100\%$$

where P denotes the coefficient of expert enthusiasm, E denotes the number of valid questionnaires collected, and T denotes the total number of questionnaires distributed. In this research, three rounds of expert consultation were conducted and no attrition rate was registered. Based on the aforementioned equation, the coefficient of expert enthusiasm was 100% in all three rounds of consultation, indicating that the experts were highly enthusiastic in scoring the indicator system (Dawood K A et al. 2021).
(2) Degree of expert authoritativeness
Expert authoritativeness is determined by two factors: First, the basis by which experts judge the plan; second, the familiarity of experts with the problems (Tang et al. 2014). The quantitative values of the level and category of each factor are shown in Tables 3 and 4.

Table 3. Quantitative table on the basis of judgment and its influencing degree.

	Degree of influence on expert judgment		
Basis of judgment	Major	Moderate	Minor
Theoretical analysis	0.3	0.2	0.1
Practical experience	0.5	0.4	0.3
Peer understanding	0.1	0.1	0.1
Personal intuition	0.1	0.1	0.1

Table 4. Coefficient of experts' familiarity with questions.

Degree of familiarity	Quantitative value
Very familiar	1
Familiar	0.8
Somewhat familiar	0.5
Not very familiar	0.2
Unfamiliar	0

The equation of the degree of expert authoritativeness is:

$$C_R = \frac{Cs + Ca}{2}$$

where C_R denotes the coefficient of expert authoritativeness, Cs denotes the coefficient of the degree of expert familiarity, and Ca denotes the coefficient of judgment (Liu G et al. 2009).

$$C_a = C_{a1} + C_{a2} + C_{a3} + C_{a4}$$

It can be seen from the calculation, the coefficient of expert authoritativeness averages, invariably, above 0.8, indicating that the experts in this research have a relatively high level of authoritativeness. As can be known from the functional relationship between the degree of expert authoritativeness and forecast precision, this research also has a relatively high level of forecast precision.
(3) The degree of expert-opinion concordance is expressed by Kendall's W, which is calculated by the importance scores of the three-level indicators:

$$\text{Kendall's W coefficient} = \frac{12}{m^2n^3 - n} \sum_{j=1}^{n} d_j^2$$

where m is the number of experts, n is the number of indicators, and d is the difference of importance scores of the jth indicator and all indicators. The degree of the concordance of expert opinions is used to measure the degree of consistency in attitudes held by different experts toward the same question (Sun B et al. 2015). All P-values of expert opinions in expert consultation of this research are smaller than 0.05, and the coefficients of Kendall's W were 0.201, 0.225, and 0.207 in the 1st, 2nd, and 3rd rounds, respectively.

3.3.4 *Expert consultation*
A total of three rounds of expert opinion solicitation were carried out and the preliminarily established studio operation criteria were modified based on the mean value, standard deviation, and coefficient of variation. The mean value denotes the level of concentration in expert opinions, the higher the mean value, the higher the importance of corresponding indicators. Standard deviation denotes the degree of concordance in expert opinions and indicates the degree of fluctuation in expert's opinion about the relative importance of the indicator. The smaller the standard deviation, the higher expert concordance in the indicator. The coefficient of variation reflects the degree of divergence in the expert evaluation of results. The smaller the index of variability, the smaller the degree of divergence (Tang et al. 2014). Generally, a mean value larger than or equal to 4, a standard deviation smaller than 1, and a coefficient of variation smaller than 0.25 is the fundamental criteria satisfying the indicator requirements. Through analyzing relevant data in combination with expert opinions and suggestions, a total of 17 items were modified, 10 extra items were added, and 12 items were deleted.

The expert opinion data was collected and processed, and the indicators whose mean values, standard deviations, and coefficients of variation failed to meet the basic criteria were ruled out and changed. Then, the suggestions and opinions of experts are modified in combination with data analysis, and the analysis is as follows:

Some experts asked, "whether it is better to put social responsibility under the indicator of talent cultivate and adherence to ideals". The opinion was accepted after a further review of the literature, and the social responsibility under the teaching model indicator was adjusted as a tertiary indicator under the secondary indicator of talent cultivation. Experts were also suggesting that "enterprise projects should be changed to crosswise subjects while project research could be changed to vertical projects". This research argues that enterprise project and project research are consistent with the wordings of crosswise and vertical research projects, and the formers are more easily understandable. Thus, the "enterprise project" and "project research" are retained.

In three rounds of expert opinion consultation, experts largely agreed upon the primary indicators but raised some suggestions and opinions on part of the secondary and tertiary indicators. Based on data analysis, changes to studio operating indicators are shown in Tables 5–7.

After three rounds of expert consultation, the expert panel had reached a consensus on all indicators for studio operation criteria, and thus expert consultation was concluded.

Table 6. Modifications to indicators in the second round of expert consultation.

Tertiary indicator	Modification results
2-4-1 Work evaluation	2-4-1 Work review
3-2-2 Data archiving	3-2-2 Data management
4-2-5 Teacher talents	4-2-5 Teacher resources
5-4-2 External visits	5-4-2 External visits and learning
Division of labor among multiple supervisors	Added
5-2-1 Supervisor counseling out system	5-2-1 Supervisor assistance
3-3-3 Studio librarization system	3-3-3 Study lounge form

Table 7. Modifications to indicators in the third round of expert consultation.

Tertiary indicator	Modification results
3-3-3 Study lounge form	Delete

Table 5. Modifications to indicators in the first round of expert consultation.

Secondary indicator	Modification results
2-4 Sense of social responsibility	Delete
2-4 Teaching mode	Added
3-2 Supervisor background	6-3 Background of supervisor at school
6-3 Enterprise supervisors	6-4 Background of enterprise supervisors
Tertiary indicator	**Modification results**
1-2-3 Design passion	1-2-3 Enthusiastic about design
1-3-5 Inter-disciplinary talents	Delete
1-4-4 Assist classmates with their self-development	1-4-4 Self-development
3-1-2 Multiple supervisor responsibility system	Delete
3-3-1 Team building activities	4-3-3 Team building activities
3-3-4 Autonomous work selection	Delete
4-4-3 Sense of accomplishment	4-4-3 Obtain a sense of accomplishment
5-2-3 Inability to concentrate on studio work	5-2-3 Inability to concentrate
5-2-2 Dismissal of the least competent system	Delete
5-2-1 Voluntary exit	Delete
Supervisor counseling out system	Added
5-3-3 Member evaluation	Delete
Awareness of responsibility	Added
5-4-3 Credit reward	Delete
5-4-1 Form of reward	Delete
Honor conferment	Added
6-1-3 Payback to society	6-1-3 Social evaluation
6-1-4 Enterprise payback willingness	6-1-4 Enterprise willingness
6-2-4 School-dominated management	Delete
Enterprise feedback and suggestion	Added
6-3-2 Diploma requirement	Delete
Teaching competence	Added
6-3-1 Business competence	6-3-1 Practical competence

Note: 2-4 Sense of social responsibility represents the fourth secondary indicator under the second primary indicator.
1-2-3 Design passion means the third third-level indicator under the second second-level indicator under the first first-level indicator.

Table 8. Studio operation indicators.

Primary indicator	Secondary indicator	Tertiary indicator
1 Operation vision	1-1 Development orientation	1-1-1 Following social dynamics 1-1-2 Specialties of the major
	1-2 Ideas upheld	1-2-1 Cross-disciplinary integration 1-2-2 Enthusiastic about design 1-2-3 Integration of teaching, learning, research and innovation 1-2-4 Aptitude-based teaching
	1-3 Talent cultivation	1-3-1 Autonomous learning 1-3-2 Practical competence 1-3-3 Design thinking 1-3-4 Design expression 1-3-5 Sense of social responsibility
	1-4 Functionality	1-4-1 Design practice 1-4-2 Serving the school 1-4-3 Serving the locality 1-4-4 Self-development
2 Teaching model	2-1 Practical projects	2-1-1 Disciplinary competition 2-1-2 Enterprise project 2-1-3 Enterprise research Projects
	2-2 Organizing activities	2-2-1 Field studies 2-2-2 Lectures on industrial knowledge 2-2-3 Diverse knowledge sharing meetings 2-2-4 Visits to practical bases 2-2-5 Introduction of expert guidance
	2-3 Guidance model	2-3-1 Team-based teaching 2-3-2 Communicative teaching 2-3-3 "Teaching, helping and guiding" model 2-3-4 One-on-one guidance
	2-4 Teaching mode	2-4-1 Work review 2-4-2 Teaching by example 2-4-3 Social practices 2-4-4 Project induction
3 Operation systems	3-1 Supervisor system	3-1-1 Single supervisor responsibility system 3-1-2 School-enterprise supervisor system 3-1-3 Division of labor among multiple supervisors
	3-2 Students affairs	3-2-1 Post determination and allocation 3-2-2 Data management
4 Environmental atmosphere	4-1 Spatial environment	4-1-1 Exhibitive space 4-1-2 Discussion and reading space 4-1-3 Work space 4-1-4 Sanitary environment
	4-2 Resource environment	4-2-1 Book zone 4-2-2 Equipment resources 4-2-3 Fund management 4-2-4 Cooperated enterprises 4-2-5 Teacher resources
	4-3 Team atmosphere	4-3-1 Learning atmosphere 4-3-2 Sense of belongingness in teams 4-3-3 Team building activities
	4-4 Students' enthusiasm	4-4-1 Honor incentives 4-4-2 Model building 4-4-3 Obtain the sense of accomplishment

(*continued*)

Table 8. Continued

Primary indicator	Secondary indicator	Tertiary indicator
5 Assessment system	5-1 Approach to new member recruitment	5-1-1 Brief reporting and face-to-face discussion
		5-1-2 Feedback from specialized teachers
		5-1-3 Specialized qualities
	5-2 Approach to termination	5-2-1 Supervisor assistance
		5-2-2 Inability to concentrate
	5-3 Assessment method	5-3-1 Attendance
		5-3-2 Attitude to learning
		5-3-3 Awareness of responsibility
		5-3-4 Competition achievements
		5-3-5 Project progress
		5-3-6 Learning goals for each grade
	5-4 Reward method	5-4-1 Honor conferment
		5-4-2 External visits and learning
6 School-enterprise cooperation	6-1 Enterprise selection	6-1-1 Enterprise scale
		6-1-2 Enterprise orientation
		6-1-3 Social evaluation
		6-1-4 Enterprise willingness
	6-2 Enterprise cooperation	6-2-1 Practical base model
		6-2-2 Internship recommendation opportunity
		6-2-3 Introduction of enterprise projects
		6-2-4 Externally hired enterprise experts
		6-2-5 Enterprise feedback and suggestion
	6-3 Background of enterprise supervisors	6-3-1 Practical competence
		6-3-2 Teaching competence
	6-4 Background of supervisor at school	6-4-1 Theoretical competence
		6-4-2 Enterprise resources
		6-4-3 Supervisor willingness

3.4 *Constructing the indicators of studio operation system*

After three rounds of modifications to studio operation indicators using the Delphi method, the final studio operation criteria are as follows:

4 CONCLUSION

Starting with relevant literature and focusing on research segments, semi-structured questionnaires were devised to interview 5 experts who had presided over university studios. Based on grounded theory, three levels of coding were conducted on interview texts, deriving 245 categories, 82 open codes, 22 axial codes, and 6 selective codes, followed by preliminary construction of studio operation criteria. Using the Delphi method, 11 experts with experience (at least two years) in presiding studios were invited to participate in three rounds of expert opinion consultation and modification. Eventually, 17 provisions were modified, 10 provisions were added and 12 provisions were deleted. Six categories, including operation vision, teaching model, operation system, environmental atmosphere, assessment system, and school-enterprise cooperation, related to design studios were distilled, giving rise to a complete set of studio operation criteria. The following conclusions are drawn from this research:

1. The Delphi method can validate the scientificness and rigorousness of indicators established based on the grounded theory, and can compensate for the insufficiency in indicator definition and completeness.
2. In this research, a set of design studio operation criteria was established, providing references for universities intending to create and develop studios. For example, when setting studio operation vision, considerations can be given to aspects like "development orientation", "ideas upheld", "talent cultivation" and "functionality". Additionally, the development orientation can be determined through "following social dynamics" and "specialties of the major".
3. Combining grounded theory with the Delphi method helps sufficiently distill experts' guidance experience and construct operation criteria, thus adding objectivity and accuracy to the research of the studio operation model.

When constructing a studio operation model, the analysis of the weights of operation indicators stands to derive effective references and suggestions for universities. Therefore, in future studies, the analytic hierarchy process may be introduced to analyze the weights of studio operation criteria with a view of better helping universities construct their own studio models and facilitating the development of the industrial design discipline.

ACKNOWLEDGMENT

This work was supported by the Fujian University of Technology [grant numbers GY-S21081, 2021], and the Design Innovation Research Center of Humanities and Social Sciences Research Base of Colleges and Universities in Fujian Province.

REFERENCES

Ahangama, N., & Prasanna, R. (2021). International Journal of Disaster Risk Reduction, 65, 102569.

Alagbe, O., Aderonmu, P., Opoko, P., Adeboye, A., Akinjare, O., & Izobo-Martins, O. (2015). Global Journal on Humanities and Social Sciences, 1(1).

Chung, W. C. (2019). Springer International Publishing.

Chi, S., Chen, Z., & Wu, T. (2013, March). In Proceedings of the 2013 International Conference on Information, Business and Education Technology (ICIBET 2013) (pp. 240-245). Atlantis Press.

Cross, A. (1983). Design studies, 4(1), 43–52.

Dawood, K. A., Sharif, K. Y., Ghani, A. A., Zulzalil, H., Zaidan, A. A., & Zaidan, B. B. (2021). Information and Software Technology, 130, 106453.

Design Studio (Doctoral dissertation, Rhodes University).

De Villiers, M. R., De Villiers, P. J., & Kent, A. P. (2005). Medical teacher, 27(7), 639–643.

Doğan, M. B., Yükseloğlu, E. H., Doğan, M., & Uğraþ, S. D. (2020). Journal of Forensic and Legal Medicine, 71, 101932.

H., & Seitamaa-Hakkarainen, P. (2014). Journal of Learning Design.

Heskett, J. Industrial Design.

Hood, T. L., & Copeland, D. (2021). Journal of Professional Nursing, 37(5), 885–893.

Fan, X. (2011). University of Cincinnati.

Kaya, A., Karataş, N., Gürcan, M., & Dalgıç, A. İ. (2021). Journal of Pediatric Nursing.

Kethro, P. (2013). Pedagogical Ways-of-knowing in the Lahti,

Liu, G., Wang, Y., Wang, X., & Pan, Y. (2009, August). In 2009 Sixth International Conference on Fuzzy Systems and Knowledge Discovery (Vol. 2, pp. 603-607). IEEE.

Olsen, A. A., Wolcott, M. D., Haines, S. T., Janke, K. K., & McLaughlin, J. E. (2021). Currents in Pharmacy Teaching and Learning, 13(10), 1376–1385.

Schön, D. A. (1999). Classic Readings in Architecture, 374.

Sawyer, R. K. (2017). Educational research review, 22, 99–113.

Shen, L., Yang, J., Jin, X., Hou, L., Shang, S., & Zhang, Y. (2019). Nurse education today, 79, 67–73.

Sun, B., Jiang, J., Zheng, F., Zhao, W., Liaw, B. Y., Ruan, H., ... & Zhang, W. (2015). Journal of Power Sources, 282, 146–157.

Tang, Y., Sun, H., Yao, Q., & Wang, Y. (2014). Energy, 75, 474–482.

Walker, D., & Myrick, F. (2006).Qualitative health research, 16(4), 547–559.

Xu, R., Luo, F., Chen, G., Zhou, F., & Abdulahi, E. W. (2021). International Journal of Industrial Ergonomics, 86, 103228.

System Innovation for a Troubled World – Tak Lam et al. (Eds)
© 2023 The Author(s), ISBN: 978-1-032-45525-9

A hybrid model for recommender system in e-commerce

Teng-Chun Yu* & Shie-Jue Lee*
Department of Electronic Engineering, National Sun Yat-sen University, Kaohsiung, Taiwan

ABSTRACT: With the popularity of the Internet, e-commerce has grown rapidly. Many online applications use recommender system to predict user preferences and make predictions on items based on user preference information to provide a better service experience and increase sales of items. Since online applications often add new items into the system, it is difficult to recommend new items to users without any feedback from them. Similarly, when a new user registers to the system, it becomes very difficult to recommend items to the new user because there is no previous purchase history of the new user. This condition is known as cold-start problem. There are two types of cold-start problem: user cold-start problem and item cold-start problem. In order to solve the cold-start problem, we propose a hybrid approach that combines traditional machine learning methods and neural network techniques. For the user cold-start problem, we extract additional information between users and items and convert the information into latent features as a basis. Experiments will be done with real-world datasets to verify the effectiveness of this method on the user cold-start problem.

1 INTRODUCTION

Nowadays, there are many online platforms for users to conduct electronic trading activities, but it is difficult for online shopping users to find the target in a large number of items in a limited time. In order to enable users to find the items they are interested in more quickly, the recommender system has begun to receive attention and has an important position in e-commerce. Many academic researchers have made breakthroughs and contributions in recommender systems, and even online platforms have held competitions to encourage various recommendation methods to compete with each other. However, the cold-start problem is still a challenge in the field of recommender systems.

The rating matrix, as shown in Figure 1, is an important basis for prediction. Assuming there are n users and $\mathbf{m}$ items, $r_{u,i}, 1 \leq u \leq n, 1 \leq i \leq m$, represents the rating of item i by user u. Usually, $r_{u,i}$ is an integer greater than 0; if there is no rating then $r_{u,i} = 0$.

$$\mathbf{R} = \begin{bmatrix} r_{1,1} & \cdots & r_{1,m} \\ \vdots & \ddots & \vdots \\ r_{n,1} & \cdots & r_{n,m} \end{bmatrix}$$

Figure 1. A rating matrix.

The user reviews, as shown in Figure 2, form another basis for prediction. Most users express their preferences and emotions through text when giving reviews on items, and the semantic meaning of text can be used to analyze deeper hidden characteristics.

Figure 2. User reviews.

Collaborative filtering (C. Desrosiers 2011) is mainly based on the user's past preference information and the interaction between users to achieve personalized recommendation. Matrix factorization (S. Funk 2006) decomposes the user-item interaction matrix into two matrices and maps them to a common latent space. The receiver introduces the mapped matrices into the latent space and predicts the user's rating of the item. Content-based filtering is a method of recommendation that does not rely on user ratings. It is effective in recommending new items without any user feedback by using the information possessed by the items themselves, such as item attributes, item descriptions, keywords, etc. However, it does not provide recommendation diversity because it only recommends items with similar features. With additional information to increase the accuracy of recommendations, many hybrid recommender systems have been developed. Trust SVD (Guo 2015) strengthens the

*Corresponding Authors: tcyu@itlm.ee.nsysu.edu.tw and leesj@mail.ee.nsysu.edu.tw

DOI 10.1201/9781003377399-17

importance of user feedback on items through the information of trust value among users, which reduces the sparsity problem and enhances the analysis of hidden characteristics of users.

Due to the development of machine learning, some recent work has incorporated deep neural networks (DNNs), which have powerful nonlinear modeling capabilities, to perform recommendation tasks. Neural collaborative filtering (NCF) (He 2017) is a collaborative filtering model that combines the traditional approach of matrix factorization and multi-layer perceptron (MLP) for capturing linear and nonlinear interactions between users and objects. Item2Vec (Barkan 2016) learns the dense embedding for items from item sequences with which users have interacted. The wide and Deep model (Cheng 2016) captures features linearly and then uses deep neural networks to remember them for the recommendation. The deep factorization machine (Guo 2017) combines the factorization machine and MLP to perform feature crossover and extract latent factors to build predictive models.

We propose a combination of collaborative filtering and neural networks to solve the user cold-start problem and use both linear and nonlinear approaches to discover users' possible hidden preferences and latent features of items. This not only provides users with more choices but also increases their trust in the recommender system.

The remainder of this paper is organized as follows. Section II describes the dataset we worked on. Section III describes the proposed prediction model framework and details. Section IV presents experimental results. Finally, Section V concludes the paper.

2 DATASET

We use two different datasets, Yelp and Movielens-1M. The Yelp dataset contains user ratings, reviews, hours of operation, and environmental information for stores and restaurants, but here we only use ratings. The rating range is from 1 to 5, with 5 being the most satisfied and 1 being the opposite. The Movielens-1M dataset contains the rating of various users for movies, the rating range is from 1 to 5, with 5 being the most satisfied and 1 being the opposite. The number of users, the number of items, and the sparsity of the two datasets are listed in Table 1.

Table 1. Dataset.

Name	Number of Users	Number of Items	Sparsity
Yelp	23549	17135	99.66%
Movielens-1M	6040	3706	95.53%

3 THE PROPOSED METHODOLOGY AND MODEL

We propose a new hybrid recommender system to solve the user cold-start problem. This approach combines collaborative filtering and neural network techniques. When a new user enters the system, since the new user does not have any interaction records in the system, we leverage the attribute information which refers to the characteristics possessed by the user/item and can be regarded as the user's/item's self-introduction. These attributes can be divided into two different types, categorical and continuous. Categorical data can be expressed as a group containing a limited number of distinct objects, e.g., gender, horoscope, occupation, etc. On the contrary, data such as age, size, and price are continuous data.

To deal with attribute information, we first represent it in binary vectors. If the attribute is categorical, its value can be easily converted to a binary vector. For example, item i is converted to a vector of length N, if N is the total number of items, with a 1 at the ith entry and 0's at the other entries of the vector. As a result, the first item will be represented as $[1\ 0\ 0 \ldots\ 0]^T$, the second item will be as $[0\ 1\ 0 \ldots\ 0]^T$, and so on. If the attribute is continuous, the attribute values should be discretized into a finite number of intervals and each interval is then represented as a binary vector. Any value which belongs to an interval is represented by the same binary vector as the interval. For example, if the values of the price attribute are within 0 and 1,000, we divided the values into three intervals: 0-300, 301-600, and 601-1000, then each value is represented as a binary vector of length 3. The value 699 is represented as $[0\ \ 0\ \ 1]^T$. The binary vector of each item is called item vector (IV). We concatenate the binary vectors of all the user attributes into a binary vector called user attribute vector (UAV). Similarly, we concatenate the binary vectors of all the item attributes into a binary vector called Item Attribute Vector (IAV).

We obtain the attribute information of old items and old users, as well as the existing ratings in the rating matrix. We use the ratings of old users on old items as training data, and use traditional machine learning and deep neural network techniques to develop the recommender system model which consists of linear and non-linear components, as shown in Figure 3.

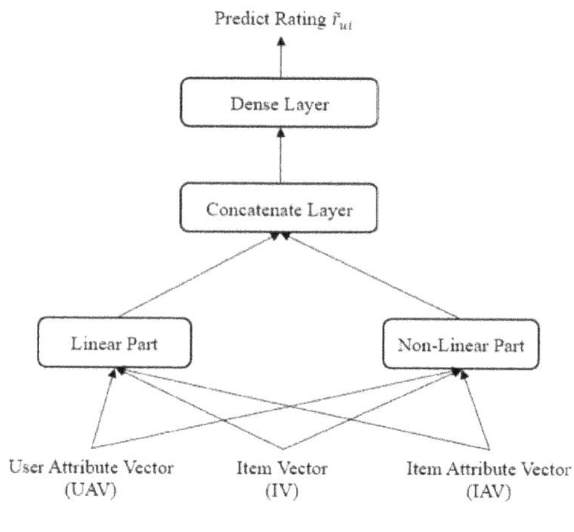

Figure 3. Block diagram of our proposed model.

First, the linear component uses matrix factorization to deal with the attribute information of users and items, and the interaction between users and items is extracted, linear component as shown in Figure 4.

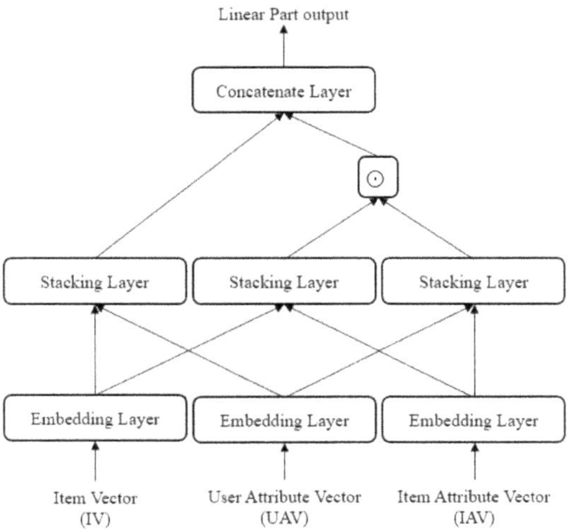

Figure 4. Block diagram of Linear component.

Next, the nonlinear component uses a multilayer perceptron to learn the nonlinear latent feature vectors (NLFVs). Finally, the outputs of the two components are concatenated and fed into a fully connected layer, and the weights and biases of the layer are adjusted. After the model is trained, the attributes of the new user and the item's feedback history are fed to the model to obtain the predicted user's score for the item. The user cold-start problem is then solved by successfully recommending old items to new users.

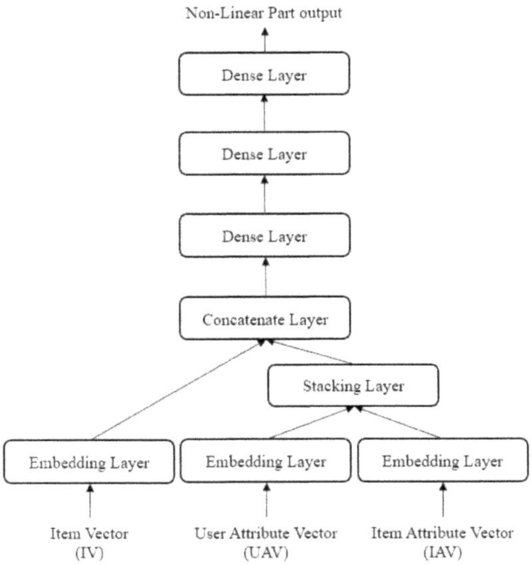

Figure 5. Block diagram of Non-linear component.

4 EXPERIMENT

In this experiment, we simulate the user cold-start problem, and we use 80% of the user attribute information in the dataset as the training set and the remaining 20% without rating data as the testing set. For our model, we use Adam as the optimizer. The training loss and test loss of the model are shown in Figures 6 and 7.

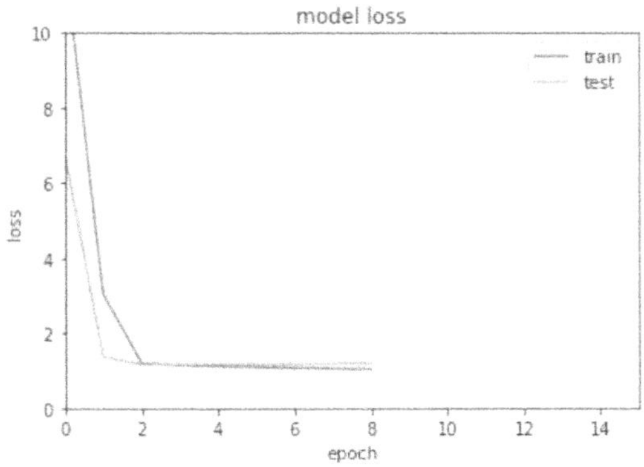

Figure 6. Model loss of yelp.

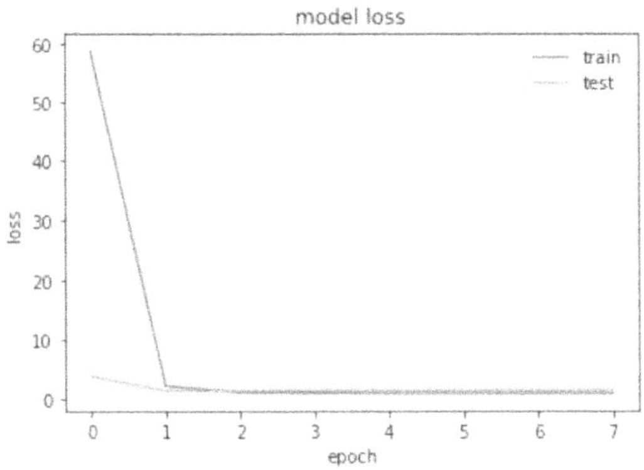

Figure 7. Model loss of Movielens-1M

The experimental results are expressed in terms of MAE and RMSE in Table 2.

Table 2.

Dataset	MAE	RMSE
Yelp	0.907	1.096
Movielens-1M	1.007	1.213

5 CONCLUSION

Our proposed model can handle the user cold-start problem very well and can make recommendations without any feedback from new users. By analyzing the interaction between user and item attribute information, we believe that attribute information can be used as important auxiliary information in the recommendation, and by analyzing the interaction between user and item attribute information, we can explore important potential features to increase the accuracy of recommendation, and if new users have these attribute information, they can also be recommended, and the experiment proves that it is achievable.

ACKNOWLEDGMENT

This work was supported by the grants MOST-108-2221-E-110-046-MY2 and MOST-110-2221-E-110-049, Ministry of Science and Technology, the NSYSU-KMU Joint Research Project (#NSYSUKMU 110-KN002), and the "Intelligent Electronic Commerce Research Center" from the Featured Areas Research Center Program within the framework of the Higher Education Sprout Project by the Ministry of Education in Taiwan. Ministry of Education.

REFERENCES

C. Desrosiers, G. Karypis, *A comprehensive survey of neighborhood-based recommendation methods, Recommender Systems Handbook*, 2011, p. 107–144, doi:10.1007/978-0-378-85820-3_4.

S. Funk, *Netflix update: Try this at home*, 2006, doi:10.1126/science.286.5440.681d.

G. Guo, J. Zhang, N. Yorke-Smith, *TrustSVD: Col laborative filtering with both the explicit and implicit influence of user trust and of item ratings*, 2015, in: Proceedings of the AAAI Conference on Articial Intelligence, Vol. 29, doi:10.1016/j.engappai.2017.10.006.

X. He, L. Liao, H. Zhang, L. Nie, X. Hu, T.-S. Chua, *Neural collaborative filtering*, 2017, in: Proceedings of the 26th International Conference on World Wide Web, p. 173-182, doi:10.1007/s12065-020-00409-5.

Oren Barkan, Noam Koenigstein, *Item2Vec: Neural Item Embedding for Collaborative Filtering*, 2016, in: Proceedings of the IEEE Workshop on Machine Learning for Signal Processing, doi:10.1109/MLSP.2016.7738886.

H.-T. Cheng, L. Koc, J. Harmsen, T. Shaked, T. Chandra, H. Aradhye, G. Anderson, G. Corrado, W. Chai, M. Ispir, et al., 2016, *Wide & deep learning for recommender systems, in: Proceedings of the 1st Workshop on Deep Learning for Recommender Systems*, p. 7-10, doi:10.1145/2988450.2988454.

H. Guo, R. Tang, Y. Ye, Z. Li, X. He, *DeepFM: a factorization-machine based neural network 545 for CTR prediction*, 2017, arXiv preprint arXiv:1703.04247, doi:10.24963/ijcai.2017/239.

System Innovation for a Troubled World – Tak Lam et al. (Eds)
© 2023 The Author(s), ISBN: 978-1-032-45525-9

Design and testing of a reversed intake for a turboprop engine on a subsonic pusher aircraft

Hung-Chieh Liu* & Yi-Ming Wang*
Department of Mechatronics Engineering, College of Engineering, National Changhua University of Education, Changhua, Taiwan

ABSTRACT: The objective of this paper is to design a new reversed intake for a pusher aircraft having dimension restriction of the nacelle and establish a testing platform to verify the results. For a pusher aircraft, the turboprop engine is mounted on the tail of the aircraft, and hence a reversed intake with U-elbow to guide airflow flowing through a 180° bend, maintaining excellent flow quality, prior to entering the engine is required. The U-elbow channel has a twisted angle to connect the throat and the inlet mouth of the engine. In the design and analysis, the ANSYS CFX was employed to examine the influence produced by the variation of intake geometry on two crucial performance parameters: total pressure recovery and flow distortion coefficient. In the experiment, the TPE331-10 engine was employed and a testing platform was built to carry out the test.

1 INTRODUCTION

When a turboprop engine is mounted at the rear portion of the fuselage of a subsonic pusher aircraft, the suction of the engine is in the reverse direction of the airflow; under such condition, a reversed intake, which brings the airflow to turn reversely and flow into the engine, is required. Meanwhile, when airflow flows through a curved duct, e.g., U-elbow, the pressure gradient may be developed between the inner and the outer walls if the radius of curvature of the duct and the Reynolds number of airflow are inappropriately designed and selected (Nirmal K Das t al. 2012; K. Sudo et al. 2000; Tong-Miin Liou et al. 1998). As known, the worse pressure gradient can induce secondary flow and generates nonuniform pressure load which causes the occurrence of airflow distortion and instability such that the performance of the engine is significantly reduced.

Papa et al. (2000, 2002) employed the artificial compressibility method to analyze the strength of secondary flow in a curved duct having a circular cross-section duct and 180° bend. They took the centrifugal and Coriolis forces into account to examine the rotating effects on the airflow quality. Their results showed that the occurrence of secondary flow strongly depends on the radius of curvature and the velocity of airflow. Similarly, Nobari et al. (2009) numerically studied 3D incompressible viscous flow and heat transfer in a rotating U-shaped duct. They pointed out that vortices and secondary flow were generated by improper radius of curvature. In other words, a significant pressure gradient between the inner and outer walls of the duct and high inlet flow distortion of the engine has occurred.

The occurrence of inlet flow distortion and its influence on the performance of turboprop engine were investigated by the teams of Selvanayagam, Pecinka, and Thomas et al, respectively. (Jeyatharsan Selvanayagam et al. 2017; Jiri Pecinka et al. 2017; Thomas Kächele et al. 2018). Their results disclosed that inlet flow distortion caused the flow separation, or called nonuniform flow distribution, in the intake; it resulted in the total pressure nonuniformity at the interface between intake outlet and the aerodynamic interface plane (AIP) of the engine, which had a significant impact to engine performance and operational safety of the aircraft.

Kim et al. (2001) studied the efficiency of the centrifugal compressor of the engine with a 90° curved duct with a circular shape. They pointed out that distortion intensity of the flow decreased quickly along the flow direction. Meanwhile, the downstream flow field was greatly affected by the bend curvature of the curved duct.

The above-mentioned studies show that when the bending angle of a duct is larger than 90°, secondary flow can be generated easily at the curved region; this may cause the occurrence of inlet flow distortion, total pressure loss, and poor engine performance. Hence, as mentioned previously, the purpose of this paper is to design a new reversed intake for a pusher aircraft. The new intake has to not only overcome the restriction of geometric dimension and shape of nacelle but also maintain excellent airflow quality, e.g., maximum mass flow rate, high-pressure recovery, and

*Corresponding Author: d0351005@mail.ncue.edu.tw

DOI 10.1201/9781003377399-18

low-pressure distortion coefficient. In addition, a testing platform was built and the TPE331-10 engine was employed to examine and verify the performance of the reversed intake.

2 ESTABLISHMENT OF THE GEOMETRIC MODEL OF A REVERSED INTAKE

In the modeling, the CATIA V5 CAD software is applied to construct the 3D model of the reversed intake. Figures 1 and 2 show the schematic mounting position and the profile of the reversed intake, respectively. Note that the growth of the reversed intake is along its axial center line under different cross-section and is symmetrically arranged on both sides of the fuselage. Meanwhile, the reversed intake is composed of three parts: inlet mouth, convergent guide duct, and twisted U-elbow channel with outlet. As shown in Figure 2, the freestream is sucked into the duct from the inlet mouth; the air then flows along the guide duct and moves upward into the U-elbow. Afterward, the air moves downward until reaching the inlet mouth of the engine. It is mentioned here that bad performance of engine and stall of the compressor can easily occur if the flow distortion coefficient at AIP is higher than the maximum value of engine requirement.

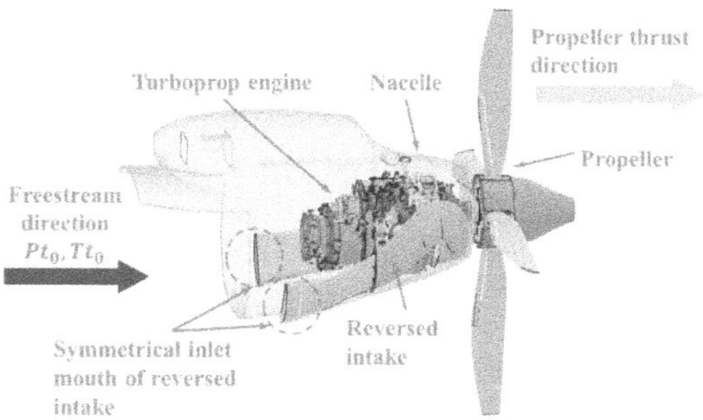

Figure 1. The schematic diagram of the reversed intake.

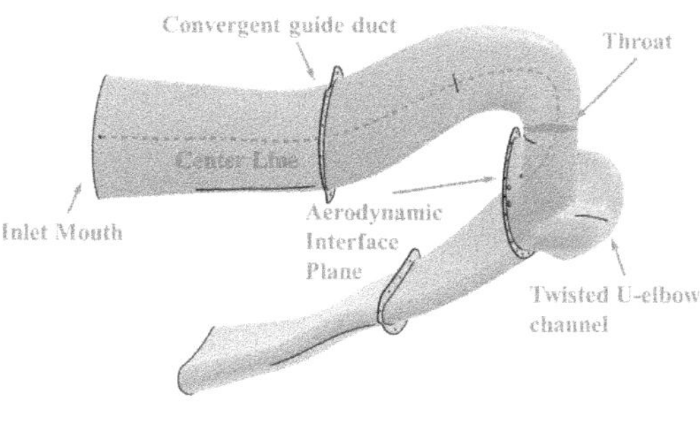

Figure 2. The profile of the reversed intake.

To break through the shape constraint of the nacelle and avoid interference with engine mounting system, one inlet mouth for each side of the fuselage is set out on the lower space. It is followed by a convergent guide duct which grows forward and upward to bypass the engine supporting system. The guide duct is also served as a convergent duct to prevent turbulent flow generated by nonuniform flow when the airflow moves upward in the duct, where the throat is at the terminal of the guide duct. Since the suction of the engine is in the reverse direction of the freestream and the velocity of airflow has to be reduced before entering the inlet mouth of the engine to fulfill engine specification, a twisted U-elbow channel having divergent cross-section is linked to the guide duct.

In the following, the design of the reversed intake is carried out. First, by referring to the main operational parameters of the TPE331-10 turboprop engine and the flight state of the aircraft, the design point of the intake can be determined. In addition, the design point in this study is that the aircraft takes off with maximum velocity at sea level. Table 1 indicates useful parameters of the design point.

Table 1. Main parameters of the design point.

Parameter	Symbol	Value
Air density	ρ_i	1.225kg/m^3
Outside total temperature	Tt_0	288.2k
Outside total pressure	Pt_0	101.3kPa
Flow velocity at inlet	V_i	66.88m/s
Mass flow rate	$\dot{m}$	4.54kg/s
Flow distortion coefficient	DC	<10%
Total pressure recovery	PR	≧90%
Max. flow velocity at throat	V_{th}	170.16m/s

To calculate the area of the inlet mouth of the intake, it is assumed that the air is under a frictionless and isentropic process. From Table 1 and the conservation of mass equation, $\dot{m} = A_i V_i \rho_i$ the total area of inlet mouth can be calculated and is equal to 0.065m^2 (A_i) the single area of total area is equal to 0.0325m^2 due to the symmetrically design, where $\dot{m}$, V_i, and ρ_i are shown in Table 1.

Assuming one-dimension flow, the subsonic airflow moves through the convergent duct and its velocity gradually increases and reaches the maximum when the flow is at the throat, where the velocity of airflow at throat is 170.16 m/s (Mach number = 0.5, Table 1) (Jack D. Mattingly 1996). The area of the throat can be calculated by the air-compressible function (Jack D. Mattingly 1996),

$$A_{th} = \frac{\dot{m}\sqrt{T_t}}{MFP \cdot P_t} \tag{1}$$

The function of the mass flow parameter *MFP* as follows:

$$MFP = M\sqrt{\frac{\gamma g_c}{R}}\left(1 + \frac{\gamma - 1}{2}M^2\right)^{-[\gamma+1/2(\gamma-1)]}$$

Where the *MFP* with M, γ, R, and g_c are the Mach number (=0.5), the specific heat ratio (=1.4), gas constant (=8.314Pa/m^3), and gravity, respectively. After some manipulations, the area of the throat (A_{th}) is equal to 0.0126m^2.

As mentioned previously, the U-elbow channel connects the throat and the inlet mouth of engine. Meanwhile, to obtain high total pressure recovery and small flow distortion coefficient, according to the engine specification, the velocity of airflow at AIP has to be reduced, i.e., the U-elbow is served as a divergent channel. It is noted here that the design restriction of the U-elbow is that the outlet of the U-elbow has to be fitted with the inlet mouth of the engine; therefore, the cross-sectional area of the U-elbow gradually increases until being equal to that of inlet mouth of the engine. Figure 3 indicates the location and numbering of the interface of each part of the reversed intake. The area of each number defined in Figure 3 is given in Table 2.

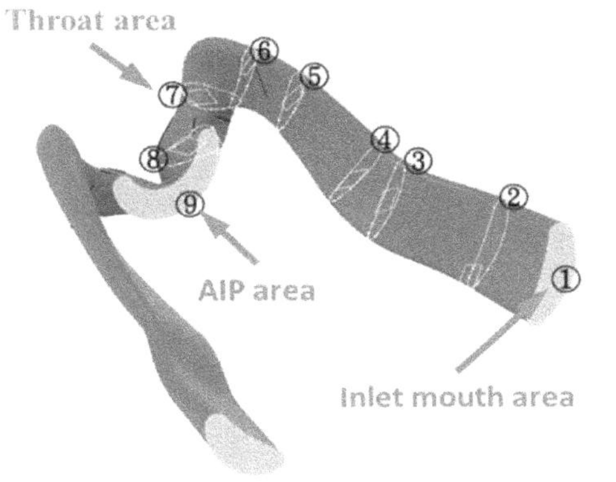

Figure 3. The location and numbering of the interface of each part of the model.

Table 2. The area of each number defined by Figure 3.

Number	Area value (m^2)	Note
1	0.0325	inlet mouth
2	0.0294	convert point of guide duct
3	0.0215	convert point of guide duct
4	0.0210	convert point of guide duct
5	0.0181	convert point of guide duct
6	0.0152	convert point of guide duct
7	0.0126	throat, outlet of guide duct
8	0.0204	divert point of U-elbow
9	0.0458	AIP (engine specification)

3 NUMERICAL SIMULATION FOR FLOW FIELD INSIDE THE REVERSED INTAKE

The ANSYS-CFX is employed to simulate the three-dimensional flow condition in the reversed intake under the following assumptions:

a. Steady-state flow field and sea level standard surrounding conditions are assumed.
b. Subsonic inlet flow and ideal gas property are taken into account.
c. Adiabatic, smooth, and impermeable boundary conditions of the inner wall of the intake are considered.
d. the K-ε model of CFX is chosen to take over the effects generated by turbulence.
e. Adjusting the static pressure at AIP to gain the inlet mass flow rate.

Recall that high total pressure recovery (PR) and small flow distortion coefficient (DC) are the two crucial parameters to guarantee the required output performance of engine and safe operation of the aircraft being fulfilled. By the fundamental theory of aerodynamics, it is known that high total pressure recovery can reduce the total pressure loss and maintains uniformly deliver flow to the engine with less turbulence and pressure differential. In addition, if the total pressure distribution on AIP is not uniformly distributed, a large pressure differential occurs; this implies the occurrence of flow separation on the upstream seriously. Referring to the requirement of DC percent of the TPE331 engine, the DC percent must be less than 10%. The requirements of TPE331 engine are PR $\geq$90% and DC<10% at the interface between the exit of the intake and the inlet mouth of engine (AIP) (TPE331 engine installation handbook 1997). Referring to Mattingly and Triantafyllou (Jack D. Mattingly 1996; T. Triantafyllou et al. 2015), and Figure 4, the definition of PR and DC are given by:

$$PR\% = 1 - \frac{Pt_0 - Pt_{AIP,avg}}{Pt_0} \times 100\% \tag{2}$$

$$DC\% = \frac{Pt_{AIP,max} - Pt_{AIP,min}}{Pt_{AIP,avg}} \times 100\% \tag{3}$$

where in Eq. (2), Pt_0 = the mean total pressure of freestream and $Pt_{AIP,avg}$ = the average total pressure at AIP. In Eq. (3), $Pt_{AIP,max}$, $Pt_{AIP,min}$, and $Pt_{AIP,avg}$ are the maximum, the minimum, and the average of total pressure at AIP, respectively.

Referring to Eqs. (2) and (3), the PR and DC of airflow are affected by the total pressure distribution at AIP. In Figure 4, there are two subplots and the parameters used are $\dot{m}$ = 4.54kg/s Pt_0 = 101.3kPa and Tt_0 = 288.2k (Table 1). Subplots 4(a) and 4(b) show the total pressure distribution at the U-elbow and AIP, respectively. Plot 4(a) clearly indicates that internal flow separation develops at the U-elbow and low total pressure zone exists on the wall of the corner. As we know, internal flow separation occurs if the airflow passes a rapidly expanding duct with corner (David C. Wilcox. 2007). Note that under this condition, we found that the PR is 8604%. Meanwhile, as shown in subplot 4(b), when airflow reaches the AIP, low total pressure zone occurs in the two flow channels at the top of the AIP. The value of DC is 29.07% at AIP, which does not meet the minimum requirement of engine specification. Hence, the reversed intake has to be redesigned under the restriction that the geometric shape of the intake keeps unchanged.

Since the minimum area of the intake is at throat and Bernoulli's law indicates that decreasing the velocity of airflow and increasing pressure can reduce the impact produced by a rapidly expanding duct to the occurrence of flow separation, the redesign procedures are as follows:

a. Taking the dimension restriction of available space of nacelle into account and increasing a bit of the throat area.
b. According to the new throat area, increasing the area of neighboring sections first and then gradually expanding to other regions.
c. Running CFX and iterating until the result being acceptable.

After some manipulations, the throat area $0.0185 m^2$ is chosen, where the velocity of airflow is reached 105.5m/s. The cross-sectional area of other sections is given in Table 3.

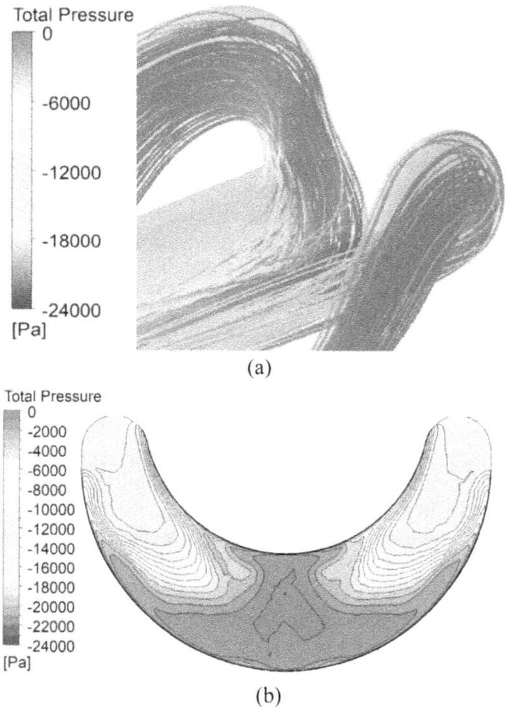

Figure 4. Subplot 4(a) and 4(b) present the total pressure distribution at U-elbow and AIP, respectively.

Figure 5 shows similar information as Figure 4, except that the redesigned intake is considered. Plot 5(a) reveals that the total pressure at U-elbow is high and the airflow separation near the corner disappeared. The PR of airflow becomes 95.67% ($\geqq$90%). Plot 5(b) discloses that the pressure in the two flow channels at the top of AIP is high. The value of DC is 7.3% ($<$10%) at AIP. Therefore, it is concluded that the remodeled reversed intake presents better dynamic performance than the minimum requirement of engine specification. In addition, referring to Table 3, a real reversed

Table 3. The new area of each number defined by Figure 3.

Number	Area value (m^2)	Note
1	0.0325	inlet mouth
2	0.0294	convert point of guide duct
3	0.0215	convert point of guide duct
4	0.0210	convert point of guide duct
5	0.0196	convert point of guide duct
6	0.0191	convert point of guide duct
7	0.0185	throat, outlet of guide duct
8	0.0204	divert point of U-elbow
9	0.0458	AIP (engine specification)

intake made by carbon fiber composite is built to verify the results.

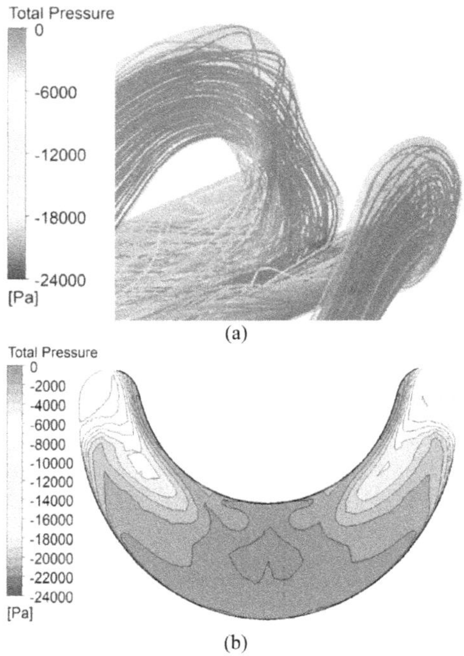

Figure 5. This figure shows similar information to that shown in Figure 4.

4 EXPERIMENT AND SOLUTION VALIDATION

A new test platform is built to perform the experiment. The test is performed at a sea-level engine test facility. It is known that rotational torque and axial thrust will be generated when the engine with a propeller is running. Hence, before the platform was constructed, structural analysis of the platform was executed to insure the structure of platform was strong enough. Meanwhile, the platform is firmly fixed on the ground by ground anchors and steel ropes to prevent the platform from being toppled over.

In order to let the testing condition is close to real situation, in addition to the reversed intake being mounted on the TPE331-10 engine (Figure 6(a)), nacelle covers, an exhaust pipe as well as a push-propeller (Figure 6(b)) are also included in the engine assembly. The freestream is sucked into the inlet mouth of the intake by engine self-sucking. The mass flow rate increases with the increase in engine operation speed (rpm); it reaches the maximum value (4.54kg/sec, Table 1) when the engine is under full power.

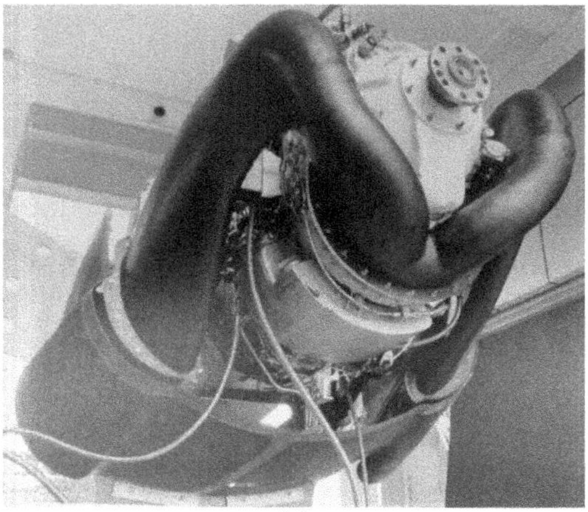

(a)

(b)

Figure 6. The pictures of the assembly and testing platform and engine assembly; subplots 6(a) and 6(b) show the view of the reversed intake and push-propeller, respectively.

As shown in Figure 7, the pressure ring is employed to measure the total pressure distribution of the interface between the outlet of intake and the inlet mouth of engine (AIP). There are 6 probes, 3 total pressure probes, and 3 static pressure probes. The position of those probes is given in Figure 7 and Table 4. The probes arrange that the 3 total pressure probes are placed near the center of the channel with a curvature similar to that of the inner wall having larger curvature of the pressure ring and the 3 static pressure probes are also equally evenly apart and taped on the inner wall with smaller curvature of the pressure ring. Note that the static pressure probes are used to calculate the mass flow rate of the airflow.

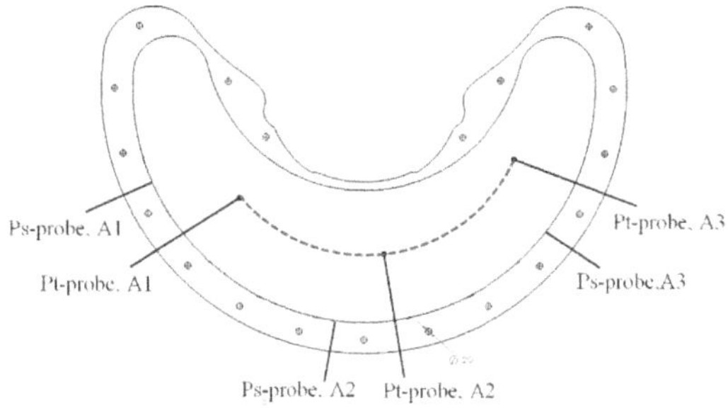

Figure 7. The schematic diagram of the pressure ring.

The comparisons of the results obtained by simulation (black line type with black hollow blocks) and experiment (red line type with blue points) are shown in Figures 8 and 9. Figure 8 indicates the variation in the percentage of total pressure recovery vs. the mass flow rate (kg/s) of the airflow. Figure 9 presents the change in the percentage of distortion coefficient vs the mass flow rate of the airflow. The two figures clearly indicate that the proposed intake can produce better airflow qualities, e.g., high PR% and low DC%, than engine specification. The two figures also show that the results of simulation and experiment are in agreement with each other. It is worth noting that the testing consequence is better than the numerical result. This is because the pressure data of test and simulation are acquired by the site of probes and the two flow channels at the top of AIP, respectively.

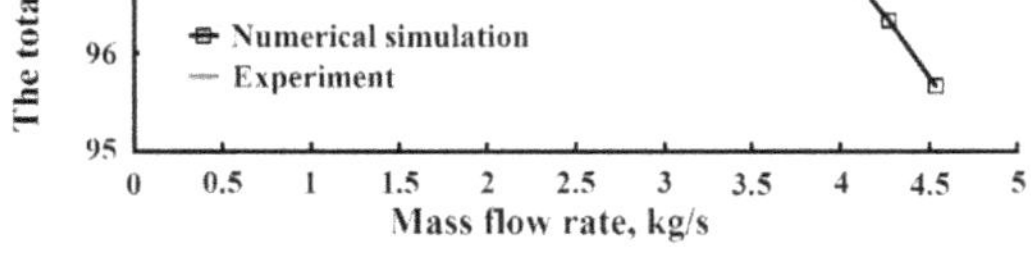

Figure 8. The variation of total pressure vs mass flow rate of airflow.

5 CONCLUSIONS

A new reversed intake with U-elbow for a pusher aircraft which has dimension limitation on nacelle is

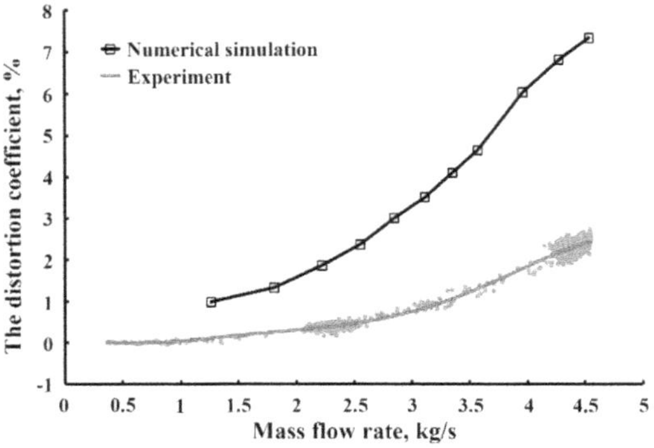

Figure 9. The variation of total pressure vs mass flow rate of airflow.

developed and studied. The U-elbow has twisted angle to fit the shape of nacelle and connect the throat and the inlet mouth of engine. Meanwhile, in order to obtain high total pressure recovery and small flow distortion coefficient of airflow at AIP, the U-elbow is also served as a divergent channel to let the airflow reduces its velocity prior to entering AIP. A testing platform and the TPE331-10 engine with nacelle covers, exhaust pipe, and a push-propeller are employed. The intake is mounted on the engine and the assembly is installed on the platform to perform the experiment and examine the dynamic performance of the reversed intake. The result indicates that internal airflow separation occurs if the intake has a rapidly expanding duct with corner; it results in low PR and high DC of the airflow, i.e., causing the inefficient dynamic performance of the engine. The drawbacks can be reduced by readjusting the throat area and then changing the area of neighboring sections and gradually expanding to other regions. Finally, running iteration until the solution is acceptable. The simulation solution is validated by experiment. It shows that both of them are in agreement with each other. Therefore, the results of this study should guide to better future design of the intake of pusher aircrafts having dimension restriction of nacelle and are vital to reliable and safe operations of these devices.

REFERENCES

David C. Wilcox, 2007. *Basic Fluid Mechanics. 3rd ed. Mill Valley: DCW Industries, Inc.* 664–668.

Garrett TPE331-14GR/HR turboprop engine installation handbook, 1994. pp. 13–11.

Jack D. Mattingly, 1996. *Elements of gas turbine propulsion. Int. ed. McGraw-Hill Education*. pp. 124–127, 156–161, 758–767.

Jeyatharsan Selvanayagam, Cristhian Aliaga, John Stokes, 2017. *Numerical Simulation of an Aircraft Engine Intake S-Duct Diffuser. AIAA Propulsion and Energy Forum*. DOI: 10.2514/6.2017-4837.

Jiri Pecinka, Gabriel T. Bugajski, Petr Kmoch, Adolf Jilek, 2017. *Jet engine inlet distortion screen and descriptor evaluation. Acta Polytechnica*. 57 22–31.

Y Kim, A Engeda, R Aungier and G Direnzi, 2001. *The influence of inlet flow distortion on the performance of a centrifugal compressor and the development of an improved inlet using numerical simulations. Proc. Instn. Mech Engrs*. 215 323–338.

Nirmal K Das, Ashoke K Raman, Amar N Mullick, B. Halder, P. Ray, B. Majumdar4, 2012. *Numerical Validation of Flow Through an S-shaped Diffuser. Int. Conf. on Fluid Dynamics and Thermodynamics Technologies*. 33 87–91.

M.R.H. Nobari, A. Nousha, E. Damangir, 2009. *A numerical investigation of flow and heat transfer in rotating U-shaped square ducts. Int. J. Thermal Sciences.* 48 590–601.

F. Papa, K. Vaidyanathan, T.J. Keith, K.J. Dewitt, 2000. *Numerical computations of flow in rotating ducts with strong curvature. Int. J. Numerical Methods for Heat & Fluid Flow.* 10 541–556.

F. Papa, K.J. Dewitt, T.J. Keith, K. Vaidyanathan, 2002. *Numerical calculation of developing laminar flow in rotating ducts with a 180-deg bend. Int. J. Numerical Methods for Heat & Fluid Flow.* 12 780–799.

K. Sudo, M. Sumida, H. Hibara, 2000. *Experimental investigation on turbulent flow through. Experiments in Fluids*. 28 51–57.

T. Triantafyllou, T. Nikolaidis, M. Diakostefanis and P. Pilidis, 2015. *Total pressure distortion levels at the aerodynamic interface plane of a military aircraft. The Aeronautical Journal*. 19 1147–1166.

Tong-Miln Llou, Yaw-Yng Tzeng, Chung-Chu Chen, 1998. *Fluid flow in a 180 deg sharp turning suct with different divider. Int. Gas Turbine & Aeroengine Congress & Exhibition*. 98-GT-189.

Thomas Kächele, Rudolf P. M. Rademakers, Tim Schneider, Reinhard Niehuis, 2018. *Numerical simulations of an intake-compressor system. J. Glob. Power Propuls. Soc.* 2 442–452.

System Innovation for a Troubled World – Tak Lam et al. (Eds)
© 2023 The Author(s), ISBN: 978-1-032-45525-9

Human alcohol residual detection application designed based on smart phone platform

Shou-Cih Chen*, Yen-Yen Kuo, Kai-Fu Ke & Yu-Chen Chen
Department of Management, Air Force Institute of Technology, Kaohsiung, Taiwan

ABSTRACT: In recent years in Taiwan, people are paying more and more attention to the issue of drunk driving. Many drunken driving accidents cause the tragedies and broken families. Even though drunk driving is life-threatening, it continues to happen. People who are involved in itare not only get punished under traffic laws but also cause car accidents, injuries, and even death. In order to amend the wrong behavior, the government has revised the law to make the penalties for drunk driving more severe,. Even if there are traffic laws to control the behavior of drunk driving, there are still many cases happened by ignoring the residual value of alcohol. The residual value of alcohol in the blood is undeniably the most common reason for drunk driving accidents and it will not simply be metabolized overnight, so people ignore that easily. Based on the residual value of alcohol, this research intends to design a low-cost personal portable alcohol sensor and application that includes the function of taxi service and a driving safety video display.The gas sensor, MQ-3, is used as the sensing hardware, with MIT App Inventor 2 as the software designing, and it is applied to the smart phone. After exhaling to the sensor designed in this research, the value will be displayed on application on the mobile phone, so that the alcohol concentration can be detected immediately after meals. This system is intended to increase the vigilance of traffic safety with this system and hopefully reduce the occurrence of drunk driving behavior.

Keywords: sobriety test, Aride-hailing, AMQ-3, AMIT App Inventor 2.

1 INTRODUCTION

Alcohol has been an irresistible drink for some people, especially when it comes to gatherings with friends or socializing, "alcohol" will never be absent. However, no matter who has had a drink, they will face the same problem of alcohol residual value. The common misconception if you sleep after alcohol is consumed, you can drive on the road right the next day. In fact, the time to metabolize alcohol is dependent on different body functions, and people easily ignore the residual value of alcohol when they think they are sober. Unfortunately, things happened in different ways.

In nowadays society, people get more and more aware of not driving after drinking, and traffic regulations have become stricter. Indeed, many people have developed the habit of avoiding driving after having alcohol. More people are alert to those who attempt to drink and drive and are trying to persuade them not to. However, the residual value of alcohol is often ignored in the body. Most people think that there is no problem in driving if you have a good sleep at night. According to the recommendation of the Taiwan Drink and Driving Prevention Social Care Association, alcohol needs at least 10 to 20 hours to be completely metabolized. The incidents of detecting drunk driving are endless and the main reason for many drunk driving cases is that people ignore the alcohol residual value after drinking, which leads to the occurrence of drunk driving. In Taiwan, the drinking and driving regulations are getting stricter to deter all the possibility of drinking and driving. In this research, we developed a low-cost portable alcohol detector (E. Schechtman et al. 2011) and paired application (H. Wilson et al. 2016).

Most of the alcohol detectors on the market use an MQ-3 gas sensor (P. Sahul et al. 2017) as the concentration sensing device, which has the advantages of low cost, high accuracy, and strong stability, and thus this study uses MQ-3 gas sensor. The concentration sensor is the core of the hardware. (B. Ramkumar et al. 2010). However, most of the commercially available alcohol detectors can only display the results of the wine test, which is limited help to the user. Based on this consideration, this research attempt to develop an app for users to experience a low-cost personal alcohol analysis sensor that is easy to carry. It includes the functions such as making calls to find taxis or valet driving and displaying driving safety videos, to effectively avoid the occurrence of drunk driving cases and achieve the effect of vigilance.

For the software part, to make the system more convenient to operate and maintain, Arduino (A. A. Galadima et al. 2014) and MIT App Inventor 2 (F. Turbak et al. 2014) are chosen for software development. The reason is that the Arduino template is

*Corresponding Author: paggy01272005@gmail.com

relatively cheap, can be used in most operating systems, has a simple programming environment, and has software and hardware. Besides, it is open-source and extensible. On the other hand, MIT App Inventor 2 does not need to learn the complicated Java syntax; it uses a graphical interface. If you use the puzzle mode to combine the programs, you can complete the Android device application. This research is used as a drinking and driving application. The development of the program makes good use of the layered program logic of MIT App Inventor 2 and the low-cost alcohol concentration sensor MQ-3 to measure the alcohol concentration value in the air. The signal of the sensor is then read into Arduino, using the application in this research to calculate and judge the breath alcohol concentration. The data is then transmitted to the Android mobile phone through Bluetooth (C. Bisdikian et al. 2001). This can accurately grasp the value of alcohol concentration, and adopt a more rigorous numerical calculation method. App in the mobile phone displays the measurement results and judges whether you need to rest and watch drinking and driving promotional videos based on the results. In addition, you can understand the need for drinking and create a one-click taxi service. You can directly dial taxis or chauffeurs on the App. The App includes the suggested rest time after drinking and related drunk driving videos to achieve the effect of reminder and vigilance, hoping to avoid the behavior of drunk driving effectively.

2 ARCHITECTURE AND METHODOLOGY

This research is designed with a portable and inexpensive personal alcohol concentration detector, which combines software development and hardware production (as shown in Figure 1); the software is completed in the MIT App Inventor2; Arduino designed a program structure and MIT App Inventor 2 developed the program. In part of this research, the alcohol concentration tools hardware was developed in 4 parts:

1) MQ-3 Gas Sensor: The MQ-3 gas sensor is the concentration detector of this research through analyzing the relevant data.
2) Powerbox: 18650 lithium battery is the power supply for the hardware, and USB provides lithium battery charging, which is convenient.
3) Bluetooth module: Bluetooth function is used to transmit the data read by the MQ-3 gas sensor to the mobile device equipped with the application developed in this research, as the transmission medium of software and hardware.
4) Arduino panel: From the system construction process, it is used to undertake the programming of the hardware, and to provide the integrity of the hardware.

In this research, the introduction of the production principle and flow chart is given as follows. First, open the application, and the application will read whether the alcohol concentration detector is connected. The

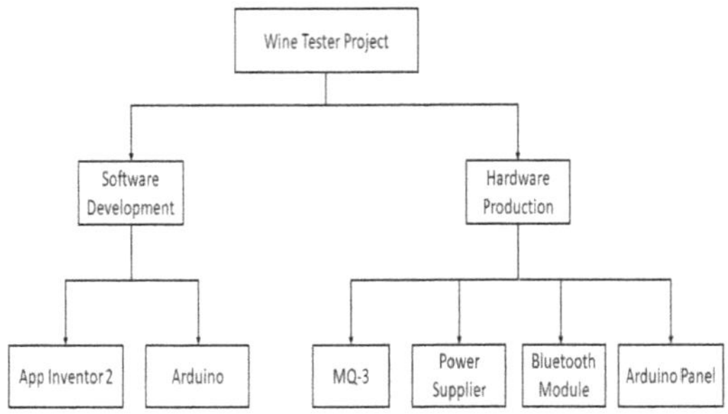

Figure 1. Architecture diagram.

screen will show up "please connect" if the Bluetooth cannot be connected, and must be reconnected. If the connection is successful, the screen will show that the device is connected. The next step is to start the measurement. In this procedure, the user blows into the tube for ten seconds. The MQ-3 gas sensor will convert the alcohol value in the air to the user's body alcohol concentration. The program is based on the alcohol concentration penalty benchmark table issued by the Police Department of the Ministry of the Interior. The App will display the alcohol concentration and recommend the resting time.

If there is no alcohol in the air user blew, return to the home page. If there is an alcohol residual value, the suggested rest time will be displayed, and then will play the driving safety video. After watching the video, the screen will enter the next interface where you can select the ride-hailing choices. It is the end of the procedure. The flowchart is shown in Figure 2.

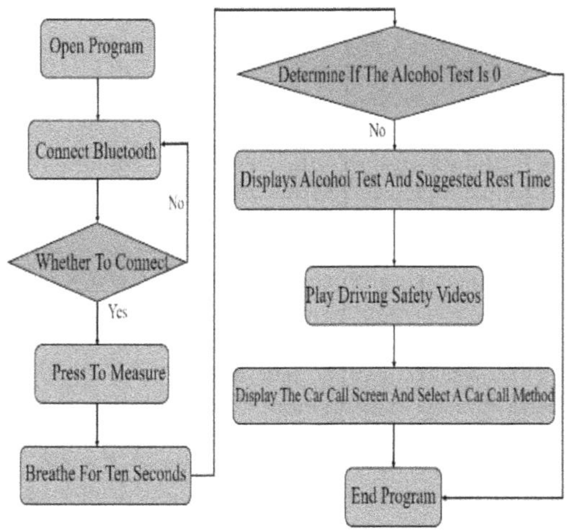

Figure 2. Flowchart.

3 RESULTS AND COMPARISON

Considering the composition of the alcohol concentration detector in this research (as shown in Figure 3), the user's exhaled gas is converted into a corresponding output signal through the MQ-3 gas concentration sensor. The change of its conductivity is sent to the

Arduino template to assist in calculating the wine test value and send the calculated results to the application created by MIT App Inventor 2 in the Android mobile phone through the Bluetooth module.

On the main page of the program (as shown in Figure 4), the connection button will show up and ask the user to connect. After pressing the button, the alcohol analyzer will be connected to the Bluetooth of the mobile phone. If the connection is not successful, the screen will continue to display "Please connect." If the connection is successful, the screen will show that the device is connected (as shown in Figure 5). After the connection is completed, the user could press the measurement button on the screen, and the program will remind the user to wait until the blue light is on which means the measurement can be started. After clicking on the measure button, the screen will display the words "Please start the measurement, countdown: 10 seconds".

During the countdown, the user needs to blow into the equipment. The alcohol concentration value will be sent to the mobile App via Bluetooth. The current alcohol concentration in the maximum value will be displayed on the screen. The unit of alcohol concentration is ppm (as shown in Figure 6). After 10 seconds, the program will determine whether the alcohol measurement value is 0. If it is 0, the screen will tell that the alcohol test is normal and do not need to watch the video (as shown in Figure 7). If it is not 0, the screen will show that the alcohol concentration value has exceeded the standard "Please watch the video and take a break." and will display the recommended rest time: x hours. Afterward, the driving safety video will be played (as shown in Figure 8). After watching the video, provides services such as a dedicated driver line and taxi for users to choose from (as shown in Figure 9). The system will display a car-calling screen and users can choose relevant options. It is suggested to choose a ride-hailing option.

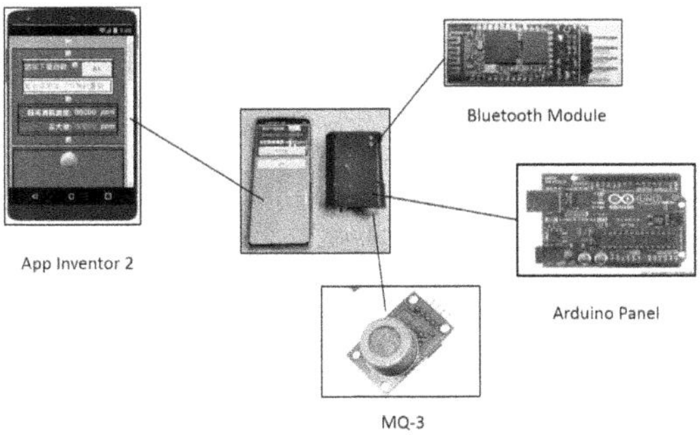

Figure 3. Composition of the alcohol concentration detector.

This research has focussed on building a piece of low-cost equipment, using a low-priced alcohol sensor (MQ3) in the market. The low-priced alcohol detector (AH-395) (as shown in Figure 10) is about NT$350 to 400. It is a mouthpiece-type wine tester and needs to be installed with two AA batteries. It is easy to carry

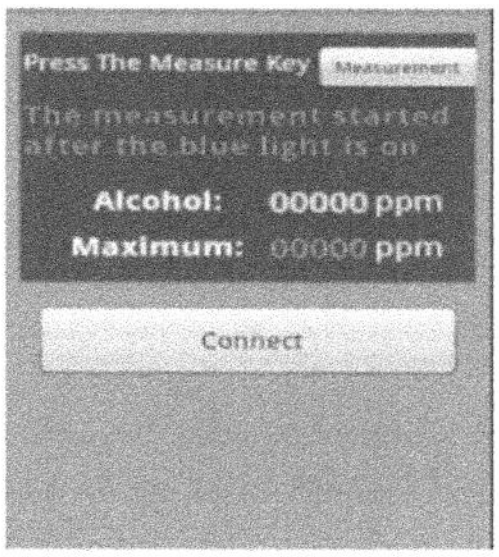

Figure 4. Main page.

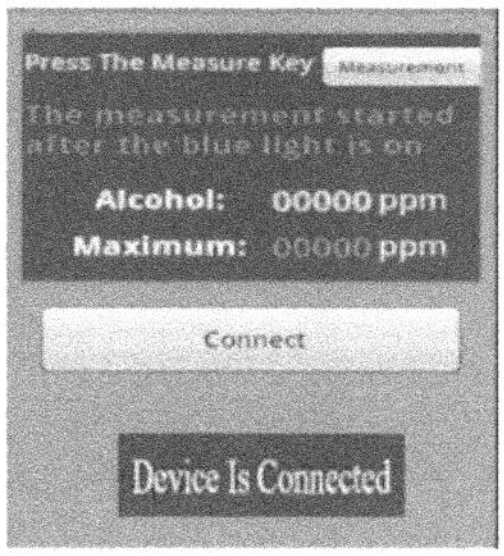

Figure 5. Connection successful.

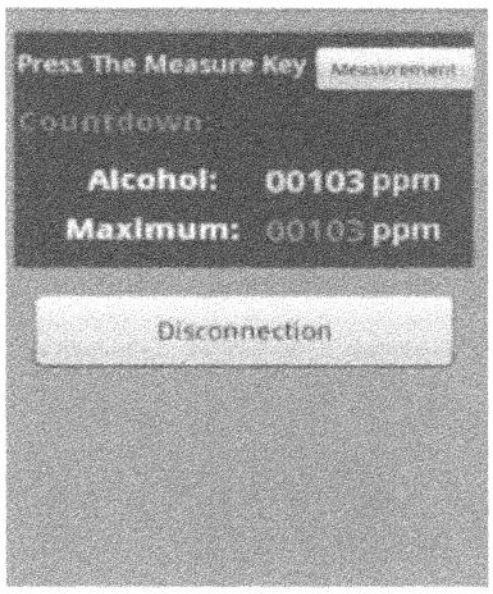

Figure 6. Current alcohol concentration.

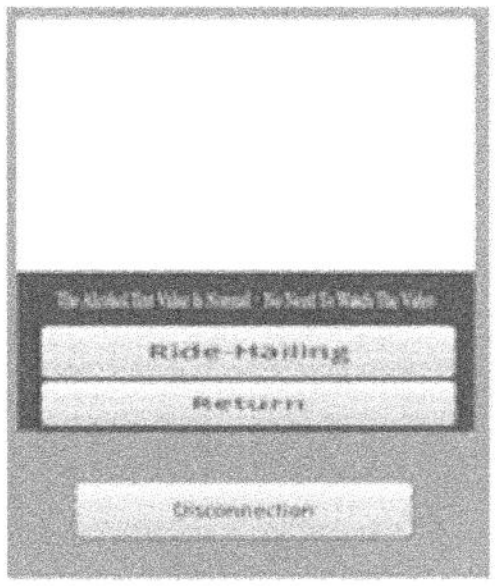

Figure 7. Alcohol concentration value 0.

Figure 8. Suggested rest time & driving safety video.

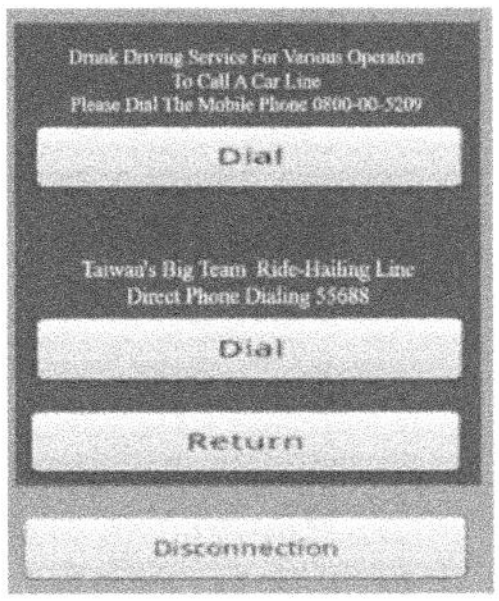

Figure 9. Ride-hailing service.

and its functions are simple and clear. The test results are directly displayed on the instrument. According to the exhalation results, there are three warning functions: danger, attention, and safety. However, the actual measurement found that the gas exhaled from the previous time would remain in the mouthpiece, resulting in errors in the detection value. On the contrary, we can also research and discuss the high-priced wine tester; the price range of the wine tester (CareExpert) (as shown in Figure 11) is about NT$2,000 to 3,000, and the shape of the instrument is more. It is lightweight, has a dedicated app, and has various functions, but the product is expensive. There are two polarized wine testers in the market. This study balances these two testers to develop a low-cost wine tester that has the functions of an app and a ride-hailing service.

Figure 10. AH-395.

Figure 11. Care expert.

We can use commercially available alcohol concentration detectors to test the alcohol concentration in the body after drinking and compare it with the unit conversion table (see Table 1). The unit of concentration measured by this research is the concentration in parts per million (ppm). After the actual measurement, the value of AH-395 (mg/L) is converted into the same unit as this research. From Table 2, we can obtain the wine test developed by this research. The detection results of alcohol concentration are comparable to those of commercial products. Taiwan's alcohol test standard is 0.15mg/L or blood alcohol concentration of 0.03% (subject to the information published by the Ministry of the Interior).

Table 1. Conversion table Of AH-395 Value (mg/L) Conversion Value (ppm).

mg/L	ppm
0.05	25
0.10	50
0.15	75
0.20	100
0.25	125
0.30	150

This study can also be used to replace most commercially available alcohol concentration detectors, such as CareExpert present in the App Store for iOS.

Table 2. Comparison table of measured alcohol concentration.

	This Reaserch	AH-395
800c.c.BEER (5%)	126 ppm	119 ppm
330c.c.WINE (12%)	129 ppm	125 ppm
100c.c.WHISKY (40%)	132 ppm	127 ppm
70c.c.KAOLIANG (58%)	144 ppm	135 ppm

Compared with the alcohol test assistant (CareExpert) and AH-395 in this study, these three equipment have functions such as alcohol value display and alcohol test accuracy. Although CareExpert has the storing value function, it is relatively ineffective for the current needs of users. This study shows the drinking and driving videos and suggests rest time that can prevent drinkers from the behavior of drinking and driving. And also highlight the function of assessing the time required for alcohol metabolism to guide users to choose a safer way to go home. The difference in after-drinking service between this study and CareExpert is that this study can choose a designated driving service. People and cars to be delivered to their homes safely, CareExpert only has a taxi service and does not have the function of a designated driver. In this study, video propaganda is used to deter the occurrence of drunk driving, so that people can be vigilant.

4 CONCLUSION

This study uses a low-cost alcohol concentration sensor (MQ-3) and a free APP development software (MIT App Inventor 2) to form a personal portable alcohol concentration sensor. After performing tests on the app's functionality and hardware devices, a new alcohol concentration residual value app was proposed and compared with commercially available alcohol concentration detectors. The detection accuracy is the same as that of commercially available alcohol detection products, and it can also provide suggested rest times under different conditions so that users can judge how long they need to rest before driving. Watching drunk driving videos can remind people to avoid possible accidents caused by drunk driving, and use the car-hailing service in the app to choose the best mode of transportation to return home, which can greatly reduce the occurrence of drunk driving.

REFERENCES

A. A. Galadima, 2014, "Arduino as a learning tool," *2014 11th International Conference on Electronics, Computer and Computation (ICECCO)*, pp. 1–4.

C. Bisdikian, 2001, "An overview of the Bluetooth wireless technology" *IEEE Communications Magazine*, vol. 39, Issue: 12, pp. 86–49.

B. Ramkumar, H.M. Kittur, and P. M. Kannan, 2010, "ASIC implementation of modified faster carry save adder," *Eur. J. Sci. Res.*, vol. 42, no. 1, pp. 53–58.

P. Sahul, S. Dixitl, S. Mishral, S. Srivastava, 2017, "Alcohol Detection based Engine Locking System using MQ-3 Sensor" *International Research Journal of Engineering and Technology*, vol. 4, Issue 4.

E. Schechtman, D. Shinar, 2011, "An analysis of alcohol breath tests results with portable and desktop breath testers as surrogates of blood alcohol levels" *Accident Analysis & Prevention.*, vol. 43, issue. 6, pp. 2188–2194.

F. Turbak, D. Wolber, P. Medlock-Walton, 2014, "The design of naming features in App Inventor 2" *2014 IEEE Symposium on Visual Languages and Human-Centric Computing*, pp. 129–132.

H. Wilson, S. R. Stoyanov, S. Gandabhai, A. Baldwin, 2016, "The Quality and Accuracy of Mobile Apps to Prevent Driving After Drinking Alcohol" *JMIR mHealth and uHealth.*, vol. 4, No. 3, e5961.

System Innovation for a Troubled World – Tak Lam et al. (Eds)

Removal of Pb^{2+} heavy metals from aqueous solution by novel eco-hydroxyapatite prepared from marble sludge

Kae-Long Lin*, Sheng-Yuan Peng & Yen-Yu Lin
Department of Environmental Engineering, National Ilan University, Yilan city, Taiwan, P.R. China

Ya-Wen Lin & Wei-Hao Lee
Institute of Mineral Resources Engineering, National Taipei University of Technology, Taipei City, Taiwan, P.R. China

ABSTRACT: Eco-hydroxyapatite (eco-HAp) have been synthesized from marble sludge via a simple and hydrothermal process and used the eco-HAp as an adsorbent to remove Pb^{2+} ions from aqueous solutions. X-ray powder diffraction (XRD) and scanning electron microscopy (SEM) characterizations revealed that the increase in the Ca/P molar ratios and the increase in hydrothermal temperature contribute to the increase in crystallinity and peak intensity, while the influence of hydrothermal temperature is more obvious. As the molar ratio of calcium to phosphorus increases, the crystallinity, peak intensity and crystal length are all found to have an increasing trend. When the hydrothermal temperature was 120°C and the Ca/P molar ratio was 1, eco-HAP, the Pb^{2+} removal rate increased from 42.90% to 99.99% because the availability of binding sites that interact with Pb^{2+} at higher doses was higher, which was conducive to excellent removal performance. Remarkably, after 30 minutes of reaction time, the adsorption capacity gradually stabilized and reached adsorption saturation, showing good adsorption efficiency. The adsorption kinetics can be well described by a pseudo-second order kinetic model. These experimental results demonstrated that eco-HAp are promising adsorbents for the efficient removal of Pb^{2+} from wastewater.

1 INTRODUCTION

Adsorption has good performance, cost-effectiveness, and easy handling. It can efficiently remove heavy metals in wastewater even at low pollutant concentrations and is considered to be one of the most practical technologies for removing heavy metals in wastewater. Therefore, adsorbents such as activated carbon, zeolite, biochar, and composite materials have been widely used to remove Pb^{2+} from polluted environments (Zendehdel et al. 2016). Currently, many researchers are focused mainly on the exploration of efficient and economical or novel adsorbents, including zeolite (Shi et al. 2013), clay minerals, and hydroxyapatite ($Ca_{10}(PO_4)_6OH_2$, abbreviated as HAp) (Uddin 2017). Hydroxyapatite is the main inorganic component of human hard tissues and has attracted much attention for its high biocompatibility, biological activity, and unique physical and chemical properties (Q Shi et al. 2019). In the past 20 years, industrial applications of HAp, such as bone and tooth implants, drug/gene delivery, and catalysis, have been reported. Synthetic HAp is considered to be a promising material that can remove heavy metal ions from wastewater or groundwater because of its cost-effectiveness, excellent storage capacity, and biological safety (Hua et al. 2012). Many studies have shown that HAp can effectively adsorb metal ions through the main mechanism of divalent Pb ion exchange in HAp (Nzihou & Sharrock 2010).

In addition, marble sludge is one of the stone waste sludge, which is a waste of the stone processing industry, mainly from the stone chips generated during the cutting process of natural stone, because a large amount of cooling water is used in the cutting process. The cutting saw blade is cooled, resulting in a large amount of muddy waste mixed with wastewater and stone chips after processing, which is called marble sludge. The composition of marble sludge is mainly CaO, about 60%, which is mainly reused as filling material, natural aggregate, asphalt mixture, cement replacement material, lightweight brick material, etc., and the material used to synthesize HAp as adsorbent was not used. In view of this, this study mainly uses marble sludge as calcium source, and hydrothermal synthesis method to synthesize eco-hydroxyapatite (eco-HAp) as an adsorbent. The correlation between changing Ca/P molar ratio, hydrothermal temperature, and the crystallinity and microstructure of synthesized eco-HAp was discussed. In addition, in the process of

*Corresponding Author: klllin@niu.edu.tw

DOI 10.1201/9781003377399-20

synthesizing eco-HAp, waste is recycled and used as an adsorbent to remove heavy metals in an aqueous solution.

2 EXPERIMENTAL

In this study, marble sludge was dried and crushed into powder. Then, 25 g of that powder was dissolved in 100 mL of 3 M HNO_3 and filtered. The initial $(NH_4)_2HPO_4$ concentration was varied to obtain Ca/P ratios of 1, 1.67, 2.33, and 3. NaOH solution was added to keep the solution pH at 10. The mixtures were sealed in polytetrafluoroethylene (Teflon)-lined stainless steel autoclaves, and hydrothermal transformations took place at 120–180°C for 8 h in independent experiments. The autoclaves were allowed to cool to room temperature naturally. The resulting products were powders; they were collected, rinsed with deionized water, and dried at 90°C for 24 h prior to examination.

3 RESULTS AND DISCUSSION

3.1 *XRD patterns of the eco-HAP*

Figure 1(a) shows that when the hydrothermal temperature is 120°C and the Ca/P molar ratio is 1, eco-HAP (MA3) is compared with the JCPDS (09-0432) standard chart. The eco-HAP at 25.90°, 31.74°, 32.54°, 40.19°, 45.30°, 49.46°, and 53.12°2θ is the diffraction peak of eco-HAP; the 2θ of β-TCP is the 34.37° diffraction peak. In the hydrothermal process, calcium ions (Ca^{2+}) react with phosphate functional groups to form nucleation sites, and calcium phosphate is formed after the nucleation sites react. When the hydrothermal temperature is 120°C and the Ca/P molar ratio is 1.67, 2.33, and 3, eco-HAP (MA5, MA7, and MA9) shows the same phenomenon. When the Ca/P molar ratio increases to 2.33, the diffraction peaks with 2θ of 25.90° and 31.74° tend to increase because under high temperature and high-pressure conditions. Water will generate a large number of hydroxyl radicals (OH^-), which will react with β-TCP produced by Ca^{2+} and PO_4^{3-} to gradually form eco-HAP (Lugo et al. 2017), resulting in the diffraction peak intensity of eco-HAP increasing with the increase in the Ca/P molar ratio.

Figure 2 shows the adsorption kinetics of Pb^{2+} by eco-HAP. When the hydrothermal temperature is 120°C and the Ca/P molar ratio is 1, the eco-HAP (MA3), with the increase of the reaction time, under the interaction of the eco-HAP and Pb^{2+}, the adsorption capacity of the eco-HAP and Pb^{2+} ions gradually increased. After 30 minutes of reaction time, the adsorption capacity gradually stabilized and reached adsorption saturation, showing good adsorption efficiency. When the hydrothermal temperature was 120°C and the Ca/P molar ratio was 1.67, 2.33, and 3, eco-HAP (MA5, MA7, and MA9) reached adsorption equilibrium after reaction time for 20 minutes. It is possible that there are different three-dimensional microstructures and more adsorption sites, which can reach adsorption equilibrium in a short time (Webler 2018).

According to Shi et al. (2018), the Langmuir adsorption model assumes that adsorption occurs on an adsorbent with a uniform structure, and the ions of the adsorbate form a monolayer on the outer surface of the adsorbent (Shi et al. 2019). The Freundlich model assumes that the surface of the adsorbent is heterogeneous, multilayer adsorption occurs, the adsorption heat is varied, and the affinity to the heterogeneous surface is irregular. Figure 3 shows the Langmuir model of eco-HAP with different Ca/P molar ratios at a hydrothermal temperature of 120°C. The R^2 of eco-HAP adsorbing Pb^{2+} ions is 0.98-0.99. Figure 4 shows the Freundlich model of eco-HAP with different Ca/P molar ratios at a hydrothermal temperature of 120°C. The R^2 of eco-HAP adsorbing Pb ions is 0.91-0.95.

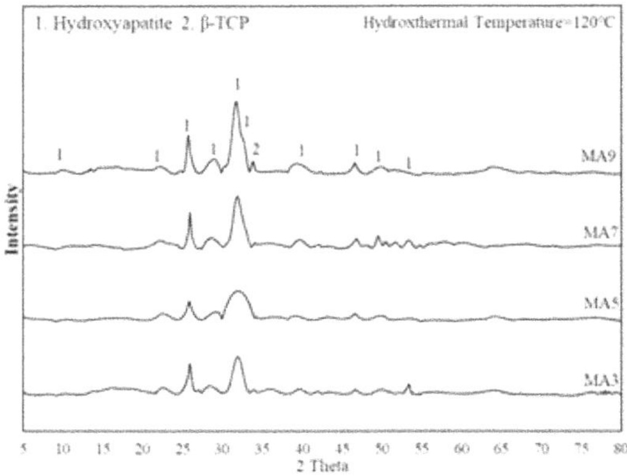

Figure 1. XRD patterns of eco-HAP.

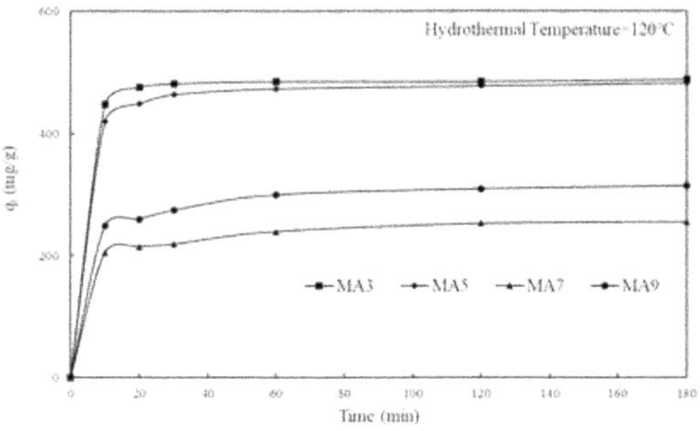

Figure 2. Adsorption kinetics of Pb2+ by eco-HAP.

4 CONCLUSION

In this paper, a higher hydrothermal temperature is beneficial to obtain eco-HAP with higher crystallinity. When the hydrothermal temperature is 120°C and the Ca/P molar ratio is 1, the eco-HAP (MA3), the adsorption capacity gradually stabilized and reached adsorption saturation, showing good adsorption efficiency.

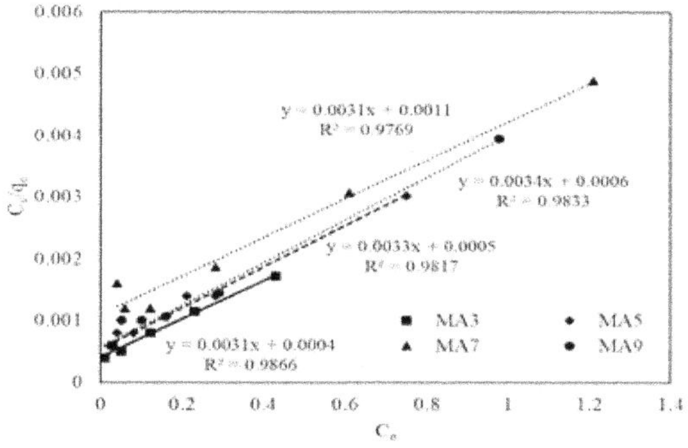

Figure 3. Langmuir diagram of Pb adsorption by eco-HAP.

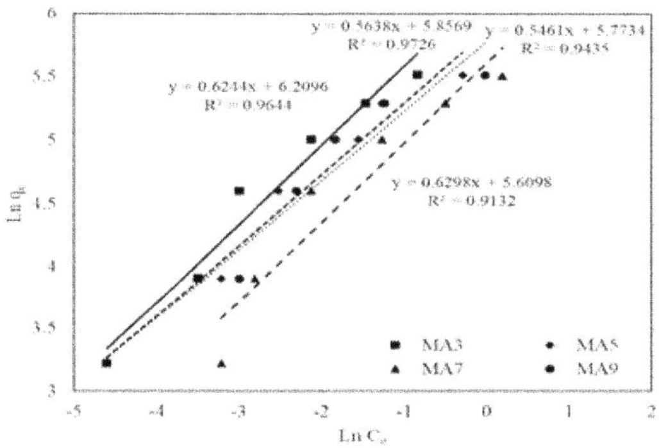

Figure 4. Freundlich diagram of Pb adsorption by eco-HAP.

When the hydrothermal temperature was 120°C and the Ca/P molar ratio was 1.67, 2.33, and 3, eco-HAP reached an adsorption equilibrium time for 20 minutes. It is possible there are different three-dimensional microstructures and more adsorption sites, which can reach adsorption equilibrium in a short time. The adsorption kinetics can be described by the pseudo-second-order kinetic model. The correlation R^2 value fitted by Langmuir is 0.98–0.99, indicating that eco-HAP is more inclined to the Langmuir adsorption model, and the ions of the adsorption form a monolayer on the outer surface of the eco-HAP.

REFERENCES

M. Hua, S. Zhang, B. Pan, W. Zhang, L. Lv, Q. Zhang, 2012. *J. Hazard. Mater.* 211, 317–331.

M.O. Lugo, J.C. Silva, M.L. Muñoz, E. Palacios, G.P. Ramirez, E.R. Cruz, A. Montoya, A.R. Pacheco, 2017. *J Archaeol Sci Rep* 15, 213–218.

A. Nzihou, P. Sharrock, 2010. *Waste Biomass Valorization* 1(1), 163–174.

W.Y. Shi, H. Li, S. Du, K.B. Wang, H.B. Shao, 2013. *Appl Clay Sci.* 85, 103–108.

Q. Shi, A. Terracciano, Y. Zhao, C. Wei, C. Christodoulatos, X. Meng, 2019. *Sci. Total Environ.* 648, 176–183.

M.K. Uddin, 2017. Chemical Engineering Journal, 308, 438–462.

G.D. Webler, W.C. Rodrigues, A.E.S. Silva, A.O.S. Silva, E.J.S. Fonseca, M.F.S. Degenhardt, C.L.P. Oliveira, L. Otubo, D.A.B. Filho, *Appl. Surf. Sci.* 436, 141–151.

M. Zendehdel, B. Shoshtari-Yeganeh, G. Cruciani, 2016. *J. Iran. Chem. Soc.* 13(10), 1915–1930.

System Innovation for a Troubled World – Tak Lam et al. (Eds)
© 2023 The Author(s), ISBN: 978-1-032-45525-9

Factors impacting user innovation project sustainability in crowd intelligence co-innovation communities

Liwen Shi
Department of Mechanical Engineering, Tianjin Ren'ai College, Tianjin, China

Lei Wang
Tianjin Key Laboratory of Equipment Design and Manufacturing Technology, Tianjin University, Tianjin, Tianjin, China

Jinjian Liu & Jing Zhang*
Department of Mechanical Engineering, Tianjin Ren'ai College, Tianjin, China

ABSTRACT: In crowd intelligence co-innovation communities, the health of the project's survival status can predict the possibility of project success. Based on the information adoption model and the diffusion of innovation theory, a conceptual model is proposed to explain how healthy project duration is affected by the characteristics of users and projects from the perspective of quantity and quality. The results demonstrate that improving characteristics of projects and users on the quantitative dimension have a significant positive effect on the long-term health of the project, while it has almost no effect on the quality dimension. This result has significant reference value to maintain the sustainability of project innovation.

Keywords: Crowd intelligence design, Sustainability of project innovation, Information adoption model, Diffusion of innovation theory.

1 INTRODUCTION

In crowd intelligence co-innovation communities, more and more enterprises begin to introduce external innovation and user opinion content, drive product iteration by aggregating information and generating user community, excavate large amount of data, and analyze user needs to ensure that their products can be competitive in the market. Most of the communities gather information, knowledge, user intelligence and capabilities through the form of "Community User Review", and form the evolution of the role attributes of "User-Fan-Designer" to enhance the user's sense of participation and honor. (J. Zhang et al. 2018) This way can not only reduce the costs of innovation, but also improve the efficiency of innovation. Many well-known enterprises at home and abroad have adopted this innovative model, such as Dell's Ideastorm, Haier's Hope, Xiaomi Community, InnoCentive and Local Motors. Its innovation modes can often be summarized as: gathering a wide range of group intelligence resources, solving innovation problems through group collaboration, forming a large number of innovative programs to help guide product design. Thus, this emerging group intelligence is important for complex system decision-making tasks in an open environment. (W. Li et al. 2017) To increase the likelihood and sustainability of their emergence, a healthy and stable project environment is required. This paper mainly evaluates and filters the status of individual projects in the community, and from the characteristics of users and projects, studies the influencing factors of long-term healthy survival of crowd intelligence co-innovation projects, and provides corresponding measures for the continuous emergence and development of Crowd Intelligence. The wide application and popularization of Crowd Intelligence have changed the traditional supply mode of innovative labor force, and has led to a subversive change and development of the entire industry.

2 RELATE WORK

A *Crowd intelligence design*

The crowd intelligence design pattern is a new pattern of innovative design based on social network relationship, which is for aggregating information, user intelligence and ability. It can encourage users to share and contribute knowledge, participate in product innovation design, enable the users, information and other resources to generate value, and promote product innovation. It has the characteristics of interactivity, real-time, dynamic and so on. (J. Zhang et al. 2018) Community members can interact freely and voluntarily, contribute and share innovative ideas, offer

*Corresponding Author: jingzhang@tju.edu.cnd
Foundation items: GTianjin Research Innovation Project for Postgraduate Students (No.2021KJ082)

DOI 10.1201/9781003377399-21

possible solutions, and for an exchange of innovative knowledge.

B *Information adoption model*

The Information Adoption Model proposed for online information sharing can be used to explain how online interactive content affects people's behavior. (S. W. Sussman and W. S. Siegal 2003) In the crowd intelligence co-innovation communities, the interaction of information is random and free. In the process of accepting information, the quality of the information itself and the credibility of information sources are very important. Information from higher quality and more reliable sources will be more authentic and trustworthy, and will be accepted by more community users and influence the participation behavior of other users. Information interaction between users is mostly in the form of comments, so the collection and analysis of users' comments will become a measure of information reliability.

C *Diffusion of innovation theory*

Innovation diffusion theory is used to study the process of people accepting and using new ideas, new technologies and new products. Rogers summarizes the basic law, influencing factors and diffusion process of innovation diffusion in innovation diffusion theory. (E. M. Rogers et al. 2003) In the traditional process of innovation diffusion, the main factors that affect people's decision-making include: the characteristics of innovation, diffusion path, time stage, social system, and promotion strength. The problem studied in this paper is the diffusion of a large number of innovative projects among a large number of potential users in the open community, and the innovation of the project should be completed by the knowledge ability of the users themselves, which is very different from the traditional innovation diffusion. Therefore, with the participation of a large number of users, only time stage and promotion intensity are the potential factors that affect the project diffusion and dissemination.

The above theoretical basis can provide support and help for us to study the influencing factors of the long-term healthy survival of the crowd intelligence co-innovation projects and to explore the guaranteed mechanism of the emergence of Crowd Intelligence.

3 RESEARCH MODEL AND HYPOTHESES

This design takes 572 individual projects on LaunchForth platform (launchforth.io) owned by Local Motors, a well-known foreign transportation tool design and manufacturing company, as examples of continuous innovation design. LaunchForth is a crowd intelligence co-innovation community in the field of automotive product design and manufacturing. (M. Moritz & T. Redlich 2019) There are many projects on the platform and user participation is different. The personal project selected in this study is the initial data of the platform development, but because compared with the platform-led project, this stage of project is closer to the creative stage of brainstorming, there is less external interference, the characteristics of user's spontaneous participation behavior are more obvious, and the spontaneous emergence of crowd intelligence is more pure. Impact factors of project innovation sustainability under user and project characteristics can be better studied.

Combing the project completion first, 141 out of 572 projects were known to be in an incomplete state when data was collected, only two of which were ultimately successful innovations, NO.27 "77 Ford F-150 Restoration" and NO.504 "The CubeRinth MAXIMUS Project / TwoBitCircus STEAM Carinval" So the research focus is 431 completed projects, and then based on the activity of users and the number of innovative programs provided by users, a simple judgment is made on the health of 431 projects based on the RFI improvement model, and 260 health survival projects are screened out.

Based on the above analysis, the sustainability of project innovation is studied from the perspective of team innovation behavior. Combining with the classic innovation diffusion model. It is concluded that time is an important factor influencing user's acceptance of innovation. (I. Arieli et al. 2020) The time span of a project is employed as an indicator of the sustainability of innovation. With the continuous progress of the innovation diffusion process, the number of information receivers and the speed of innovation diffusion are changing. At the beginning of the project, only a small number of users have an interest and involved. As time is there and the influence of the project increases, more users join the project construction. The longer the time span of a project, the more users will participate in the team and contribute their knowledge and ability, which will enhance the possibility of successful project innovation and maintain the continuous output of Crowd Intelligence. Suppose the model is shown in Figure 1.

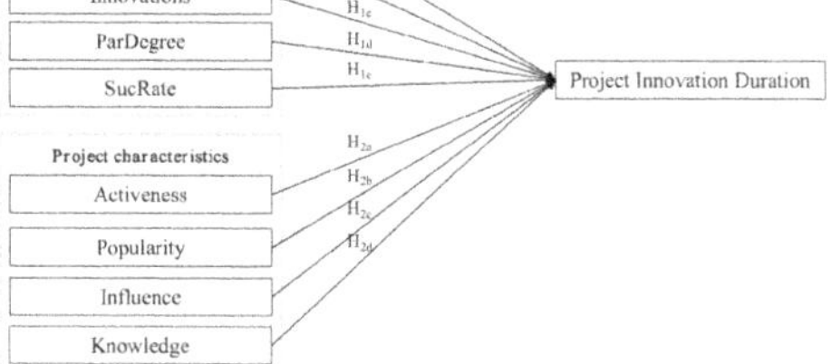

Figure 1. Research model.

A *User innovation behavior*

In the crowd intelligence co-innovation communities, according to the theory of innovation diffusion, the promotion of innovation agents has a positive and constructive impact on the diffusion of innovation. (E. M. Rogers et al. 2003) In the communities, users participate in the design process by submitting ideas, information on the needs, design issues, detailed solutions, and so on. The more members involved, the wider the scope of knowledge dissemination. Users who have participated as agents of innovation, through their own participation behavior, continuously affect other users, whose influence is mainly reflected in two aspects: the number of users participating and the degree of participation. The more users you participate in, the more ideas and solutions you submit to drive your project forward. The more users participate, the more content they submit, reflecting their participation in the project and encouraging users to contribute more innovative content. Therefore, the following assumptions are proposed:

H_{1a}: The number of users participating can positively affect the project's sustained innovation capability.

H_{1b}: The number of user actions can positively affect the project's continuous innovation capability.

H_{1c}: The number of ideas and proposals presented by users can positively affect the project's sustained innovation capability.

H_{1d}: User engagement can positively affect the sustained innovation capability of the project.

In addition, the success rate of a project refers to the proportion of successes that have been established in the community in the past. Many aspects of information about the project creator can be summarized from the project success rate. Creators with high success rates learn how to make better use of their own efforts and contributions from other users, who have strong design capabilities themselves, or the ability to engage and contribute from other users. In addition, projects they publish are more operational and more relevant to the interests of community users. From the perspective of the information adoption model, projects created by users with high project success rates are of higher quality and reliability and are more acceptable to other users. Therefore, the following assumptions are proposed:

H_{1e}: Users' previous project success rates can positively affect their ability to sustain innovation in newly released projects.

B *Project characteristics*

The existing research results show that the information activity, popularity and influence of the project are related to the innovation diffusion ability of the project. Projects with high activity, user popularity and influence are also more likely to be adopted by enterprises. (M. Li et al. 2016) Projects with higher activity, popularity and influence have stronger perceived usefulness to users, (Y. Wang et al. 2019) and their ability to attract customers' continuous participation is also stronger. (J. Lee & I. B. Hong 2021) Therefore, there is a positive relationship between the activity, popularity and influence of the project and the continuous innovation of the project:

H_{2a}: the activity of the project can positively affect the sustainable innovation ability of the project.

H_{2b}: the popularity of the project can positively affect the continuous innovation ability of the project.

H_{2c}: the influence of the project can positively affect the sustainable innovation ability of the project.

Based on the information adoption model, users need to perceive and understand the information, so in crowd intelligence co-innovation, whether the knowledge output in the project can be quickly obtained and perceived by other users will affect the smooth development of the innovation activities and the innovation efficiency of the whole team. Therefore, the following hypotheses are put forward:

H_{2d}: the explicit knowledge output of the project can positively affect the sustainable innovation ability of the project.

4 DATA ANALYSIS AND RESULTS

This data comes from the personal design project block on LaunchForth platform. 260 completed health projects were screened. Based on 12305 records of behavior data generated by 466 users on topics related to project innovation, this paper explores the impact of various factors on the sustainability of project innovation during the emergence of Crowd Intelligence. Based on the research model shown in Figure 1, measurements of independent and dependent variables and explanations for each variable are as follows:

Independent variables include user innovation behavior and project characteristics. User innovation behavior is measured by the number of users participating. Number of user actions; Number of user (creator) creative and schema submissions; User engagement is the average number of words a user participates in at a time. And the user's historical project success rate, which is the proportion of the number of successful projects that the project creator had before to the total number of projects in the same community.

On the level of project characteristics, measures include: project activity, that is, the ratio of the number of user behaviors to the total number of behaviors in the project; The popularity of the item, which is the sum of the number of times users view the item; The impact of the project, that is, the sum of the number of times users who have participated in the project have published new posts; The explicit knowledge output of a project, that is, the ratio of the number of comment keywords to the number of project comments.

The dependent variable is project sustained innovation.

Capability, which is measured by the duration of the project sustained innovation. Since all 260 individual projects in the study were accomplished health projects, it can be concluded that the longer the project lasts, the stronger the project's sustained innovation capability.

Based on the above variable measures, relevant data for all projects are obtained, and the statistical characteristics of the analysis data are shown in Table 1:

Table 1. Description of variables.

Variable	N	Mean	SD	Min	Max
Users	260	6.34	5.01	2.00	55.00
Activities	260	33.10	62.74	3.00	631.00
Innovations	260	11.49	21.15	0.00	155.00
ParDegree	260	46.02	34.71	0.50	235.33
SucRate	260	0.15	0.33	0.00	1.00
Activeness	260	7.47	19.16	0.00	155.15
Popularity	260	3.63	7.28	0.00	62.00
Influence	260	13.62	38.58	0.00	436.00
Knowledge	260	2.61	3.53	0.00	51.00

This paper uses SPSS 25.0 to analyze the data and test the hypothesis model.

1. After unifying dimension of data, the table of correlation coefficient of variables is obtained, as shown in Table 2. The correlation analysis was used to test the correlation between the factors in the emergence of crowd intelligence on the duration of innovation. According to the results, the correlation between variables is acceptable and will not affect the analysis results.

Table 2. Correlations of variables.

Variable	(1)	(2)	(3)	(4)	(5)	(6)	(7)	(8)	(9)	(10)
(1) DUR	1									
(2) Users	0.505**	1								
(3) Activities	0.443**	0.798**	1							
(4) Innovations	0.345**	0.306**	0.307**	1						
(5) ParDegree	0.126*	−0.007	0.097	0.083	1					
(6) SucRate	0.001	0.181**	0.204**	0.459**	0.030	1				
(7) Activeness	−0.236**	−0.170**	−0.115	−0.096	−0.098	−0.053	1			
(8) Popularity	0.148*	0.170**	0.128*	0.183**	0.060	0.143*	0.263**	1		
(9) Influence	0.342**	0.820**	0.914**	0.279**	0.005	0.214**	−0.085	0.103	1	
(10) Knowledge	0.007	−0.053	0.010	0.001	0.735**	−0.014	−0.027	0.076	−0.021	1

Notes: *Correlation is significant at the 0.05 level (2-tailed); **correlation is significant at the 0.01 level (2-tailed).

2. Comparative analysis of the regression model. Depending on the research needs, this paper uses the hierarchical regression analysis method to test the causal relationship between the project continuous innovation time and the variables of user continuous innovation behavior and project characteristics in the process of swarm intelligence emergence. The results of three regression models are shown in Table 3.

Table 3. Results of hierarchical regression analysis (N = 260).

Independent Variable	M1 β	M1 t	M2 β	M2 t	M3 β	M3 t
Constant	−1.359E-16	0.000	−6.757E-17	0.000	5.734E-18	0.000
Users	0.381	4.360***	0.401	4.694***	0.486	5.403***
Activities	0.076	0.870	0.068	0.790	0.484	3.823***
Innovations	0.205	3.696***	0.295	4.981***	0.261	4.581***
ParDegree			0.104	2.025*	0.105	1.410
SucRate			−0.224	−3.923***	−0.211	−3.855***
Activeness					−0.143	−2.730**
Popularity					0.079	1.492
Influence					−0.548	−4.139***
Knowledge					−0.073	−1.011
Model Statistics						
R^2	0.297		0.348		0.420	
Adjusted R^2	0.289		0.335		0.399	
F	36.015***		27.065***		20.117***	
ΔR^2	0.297		0.051		0.072	
ΔF	36.015***		9.889***		7.806***	
f^2	0.297		0.073		0.110	

Notes: *Correlation is significant at the 0.05 level (2-tailed);
**correlation is significant at the 0.01 level (2-tailed);
***correlation is significant at the 0.001 level (2-tailed).

The analysis results are shown in Table 3. The explanatory power of the research model for the duration of project innovation reaches 39.9% (Adjusted R^2). All the hypotheses proposed are strongly supported except for H_{1d}, H_{2b} and H_{2d}.

In check to see the correctness of the results of hierarchical regression analysis, this paper makes two verification analysis. The results of multicollinearity analysis show that the tolerance values of 9 independent variables are greater than 0.1, and the variance expansion factor (VIF) is less than 10 (as showed in Table 4), so there is no multicollinearity problem among variables. The results of the Durbin Watson test show that the value of d statistic is 1.858, which is close to 2, so there is no autocorrelation problem in variables, and the statistical results are significant.

Table 4. Results of multicollinearity test.

Variable	t	Significance	Tolerance	VIF
(Constant)	0.000	1.000		
Users	5.403	0.000	0.287	3.481
Activities	3.823	0.000	0.145	6.902
Innovations	4.581	0.000	0.715	1.398
ParDegree	1.410	0.160	0.421	2.373
SucRate	−3.855	0.000	0.773	1.294
Activeness	−2.730	0.007	0.844	1.184
Popularity	1.492	0.137	0.832	1.202
Influence	−4.139	0.000	0.132	7.566
Knowledge	−1.011	0.313	0.444	2.251

Note: Dependent variable: DUR.

Model 1 shows that the three independent variables account for 29.7% of the total variance, indicating that the parameters in the quantitative level of user innovation behavior have a significant impact on the duration of project innovation, in which the number of users: Users ($\beta = 0.381$, $p < 0.001$), number of user's ideas and proposals submitted: Innovations ($\beta = 0.205$, $p < 0.001$). In model 2, the R^2 value increased from 0.297 to 0.348, with a change of 0.051 ($p < 0.001$), indicating that the newly added influencing factors also have a significant impact on the duration of project innovation: ParDegree ($\beta = 0.104$, $p < 0.05$), which had a significant positive impact on the duration of project innovation: SucRate ($\beta = -0.224$, $p < 0.001$). Model 3 added four parameters of project characteristics, R^2 increased from 0.348 to 0.420, changed by 0.072 ($p < 0.001$). The data showed that the regression model was further optimized, and model 3 was the final model of this paper.

The analysis results of model 3 are presented in Figure 2. According to the hypothesis set out in the present paper, the characteristics of user innovation behavior can positively affect the sustainable innovation ability of projects from both quantitative and qualitative aspects. However, the analysis results show that only H_{1a}, H_{1b} and H_{1c} on the quantitative level of user innovation behavior are valid, and all of them have a significant positive impact on the duration of project innovation. Among them, Users assumes that H_{1a} $\beta = 0.486$, p<0.001; Activities assume that H_{1b} $\beta = 0.484$, p<0.001 and Innovations assume that H_{1c} $\beta = 0.261$, $p < 0.001$. That just says, by increasing the number of users, the number of users' behaviors and the number of users' creativity and scheme submission, the duration of project innovation will be increased by 48.6%, 48.4% and 26.1% respectively.

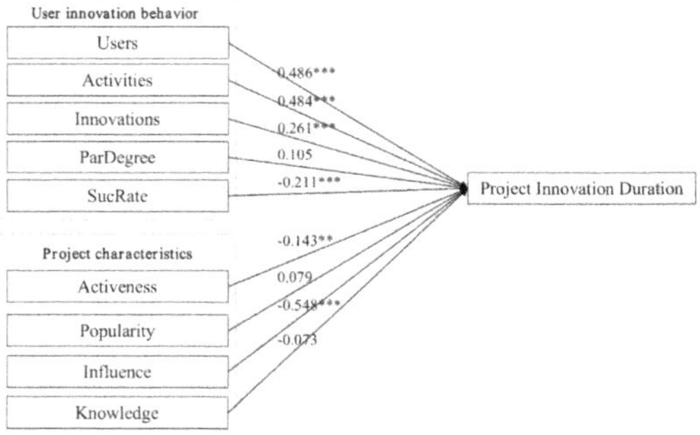

Figure 2. Research results.

On the quality level of user innovation behavior, H_{1d} does not hold, and H_{1e} has the opposite effect. Among them, the degree of user participation is assumed to be H_{1d} $\beta = 0.105$, $p > 0.01$, so the hypothesis is not tenable, that is, the degree of user participation will not affect the sustainable innovation ability of the project. The previous project success rate of users assumes that H_{1e} $\beta = -0.211, p < 0.001$, the hypothesis is true, but there will be a significant adverse impact. That is to say. The duration of project innovation will decrease by 21.1% by increasing the user participation of one unit.

For the project characteristics level, assuming H_{2a} and H_{2c}, the project activity and project influence can significantly affect the duration of project innovation. From the analysis results, the two hypotheses are verified, but the influence direction is opposite. Among them, the activity of the project is assumed to be H_{2a} $\beta = -0.143$, $p < 0.001$. The impact of the project is assumed to be H_{2c} $\beta = -0.548$, $p < 0.001$. That is to say, the duration of project innovation decreased by 14.3% and 54.8% respectively by increasing the activity and influence of a unit. However, the hypothesis that H_{2b} and H_{2d} think that the popularity of the project and the explicit knowledge output of the project can significantly affect the duration of project innovation is not tenable. Among them, the popularity of the project is higher than that of H_{2b} $\beta = 0.079$, $p > 0.01$. The explicit knowledge output hypothesis of the project is H_{2d} $\beta = -0.073, p > 0.01$. That is to popularity and explicit knowledge output of the project will not affect the sustainable innovation ability of the project.

5 DISCUSSION

Depending on the above analysis results, the quantity and quality of user innovation behavior, project characteristics are the key factors affecting the sustainable innovation ability of the project.

The existing research results show that in crowd intelligence co-innovation community, users can create projects at will, and there will be plenty of imaginative and unrealistic innovation projects. The impact of user crowd intelligence behavior flow on continuous innovation ability is greater than behavior quality. And the success rate of users before will also increase the trust of other users, and actively participate in the new projects they create. In addition, because most of the platform users participate in the project out of interests, non-professionals, and participation time is relatively free. Some users only participate in shallow level projects and constantly comment on them. They don't put forward innovative creative ideas or promote projects with other users. Therefore, the degree of user participation does not affect the sustainable innovation ability of the project.

In terms of the characteristics of the project itself, the research results verify that the project activity and influence are the crucial factors for the project to be accepted and promoted, which is consistent with the previous research results. However, project popularity and explicit knowledge output cannot affect the continuous innovation ability in the process of crowd intelligence emergence, which is distinct from previous studies.

6 CONCLUSION AND FUTURE WORK

According to the statistical results of successful projects, (Q. Zheng et al. 2018) among the 260 projects with healthy existence, 88 projects were successful in innovation, accounting for 33.85%. Of 171 projects with unhealthy existence, 43 were successful in innovation, accounting for 25.15%. Considering only two health factors, we can see that the success rate of the project has increased by 8.7%. It shows that the research has certain research value and needs further development. However, this paper only selects local motors as a platform for empirical research, which has certain limitations. There are many existing crowd intelligence co-innovation communities. In the future, confirmatory research will be conducted on several similar open innovation platforms to make the research conclusions more convincing. In addition, the measurement index system of the influencing factors of the project's sustainable innovation ability in the process of crowd intelligence emergence is not perfect. In the future, more influencing factors will be explored, and a more reasonable measurement index system will be put forward to deepen the research on the project's sustainable innovation ability.

REFERENCES

Arieli, I., Babichenko, Y., Peretz, R. and Young, H.P. 2020. *Econometrica.* 88(2), 569–594.

Lee J. and Hong, I.B. 2021. *J. Theor. Appl. El. Comm.* 16(5), 1517–1536.

Li, M., Kankanhalli, A. and Kim, S.H. 2016. *Decis. Support. Syst.* 84 (Apr), 28–40.

Li, W., Wu, W.J., Wang, H.M., et al. 2017. *Front. Inform. Tech. El.* 18, 15–43.

Moritz, M. and Redlich, T. 2019. *Tobias Redlich & Manuel Moritz & Jens P. Wulfsberg (ed.)*, Co-Creation: Springer. 9–20.

Rogers, E.M., Simon, and Schuster, 2003. Diffusion of Innovations, 5th Edition. Greenwood Press.

Sussman, S.W. and Siegal, W.S. 2003. *Inform. Syst. Res.* 14(1), 47–65.

Wang, Y., Wang, J., and Tang, Y. 2019. *Electron. Commer. Res.* 1–28.

Zheng, Q., Guo, W., An, W., Wang, L. and Liang, R. 2018. *Kybernetes.* 47(4), 656–671.

Zhang, J., Wang, L., Shi, L., An, W. and Wei, W. 2018. *2018 IEEE International Conference on Applied System Innovation (ICASI)*. 774–777.

System Innovation for a Troubled World – Tak Lam et al. (Eds)
© 2023 The Author(s), ISBN: 978-1-032-45525-9

A CNN-based model for early discrimination of Alzheimer's disease

Yu-Chieh Wang & Shie-Jue Lee*
Department of Electrical Engineering, National Sun Yat-sen University, Kaohsiung, Taiwan

Jia-Ru Chen
Department of Medical Imaging and Radiology, Kaohsiung Medical University, Kaohsiung, Taiwan

ABSTRACT: Alzheimer's disease (AD) is the main form of dementia. It causes a patient not only a decline in memory but also a considerable deterioration on mental and physical abilities. Alzheimer's disease is irreversible in the sense that it cannot be effectively cured by today's medical technology. All that can be done is to slow down the rate of deterioration with proper care and treatments for the patients in the early stages of the disease. Therefore, early diagnosis of the disease is essential. In this paper, we focus on the problem of how functional magnetic resonance imaging (fMRI) can help physicians diagnose the disease. The goal is to distinguish the difference between a normal person and a patient based on a $61 \times 73 \times 61 \times 140$ pixel size brain image. We propose a CNN-based model, consisting of 151 convolutional layers, 1 fully connected layers and an SVM classification layers, to solve this problem. ReLU is used as the activation function and Avg Pooling is used to reduce the dimensionality of the input images. To improve the recognition ability of the model, normalization is adopted to enhance the resolution of the images in the data preprocessing stage. From experiments, we have found that the proposed model can achieve a good recognition rate in discriminating the Alzheimer's disease.

1 INTRODUCTION

Alzheimer's disease is irreversible because it cannot be effectively cured by today's medical technology. Doctors usually perform clinical diagnosis by examining blood, related cognitive tests, and functional MRI. Because of the irreversibility of the disease, early diagnosis is recommended to facilitate taking preventive actions and appropriate health care.

In recent years, many machine learning techniques have been applied to analyze fMRI images of the brain, such as the use of CNN to detect AD, DNN to detect AD, and RNN to detect AD (Abed et al. 2020; Sarraf et al. 2019; Velazquez et al. 2019). The fMRI images in this study are 4-dimensional images, containing time series which can reveal the change of blood oxygen concentration in the brain over time. As long as the MRI images are available, they can be appropriately handled by a deep neural network to solve the diagnosis problem.

In this paper, the data set is the brain image samples of each patient, and the size of each sample is 61*73*61*140 pixels. There are two types of patients, which are classified as normal (NC) and patient (AD), and the goal is to distinguish them using the image data. In fact, doctors are currently having difficulties in diagnosing the severity of the patient directly from the brain images alone. For example, Figure 1 is a normal person (NC) and Figure 2 is a patient (AD). Looking for the difference between the two images solely by eyes is not an easy task, and seeking the help of machine learning is a tendency in modern medical hospitals.

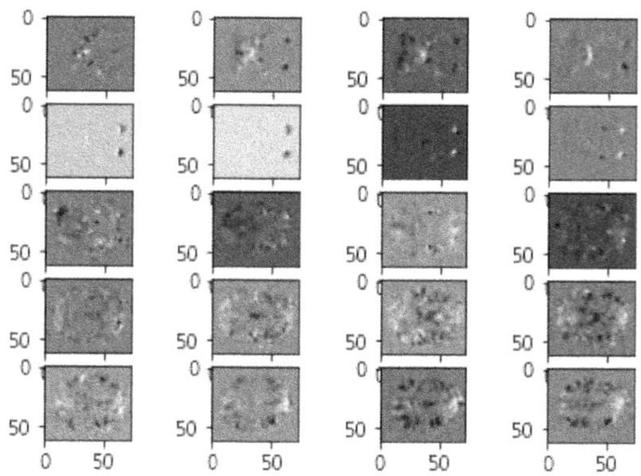

Figure 1. fMRI image of a normal human brain.

In medical imaging, we often encounter unbalanced data sets or insufficient information. Our data are obtained in cooperation with the Kaohsiung Medical University (KMU). However, due to the epidemic in recent years, there is a serious shortage of information and the total number of subjects is only 63, of which 21 are normal persons and 42 are AD patients, causing a serious dataset problem.

*Corresponding Author: leesj@mail.ee.nsysu.edu.tw

DOI 10.1201/9781003377399-22

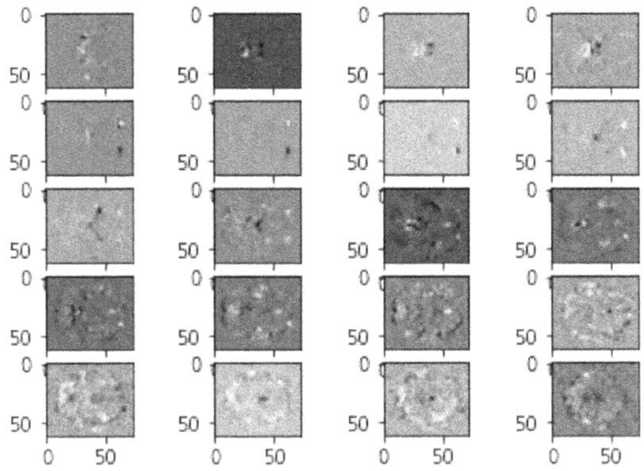

Figure 2. fMRI image of AD patients.

Images usually need to use convolutional layers in deep learning to help find the features associated with disease, and CNN-based model architectures include AlexNet, GoogleNet, VGG16, Resnet, InceptionV3, and so on. These architectures use convolution layers to find deeper features to help detect differences.

In this paper, we propose a method to solve the problem of detecting AD in a faster time. We will use the migration learning approach, and the network is structured with the deep convolutional architecture of Resnet152. We extract the features of this network after training and input them into the SVM classifier for classification. The accuracy of the classification is good.

The rest of the paper is organized as follows, Section 2 describes the dataset we used, Section 3 describes the methodology and framework we used, and Section 4 presents the experimental results. Finally, Section 5 gives a conclusion that describes how the whole method scores.

2 DATASET

The dataset used in the study was provided by Prof. C.-J. Chen of Kaohsiung Medical University. The details of normal subjects and AD patients are shown in Table 1.

Table 1. Dataset.

Number	Class	Count	sample size	Description
0	Normal person	21	61*73*61*140	Normal people aged 0–80 years
1	AD patient	42	61*73*61*140	Patients aged 0–80 years with AD05 and AD10

As shown in Table 1, this data set contains samples of 63 subjects, of which 21 are for normal people and 42 are for AD patients, with a sample size of 61*73*61*140 pixels. Since each person's brain size is different, the original data went through some pre-processing operations, such as Reslice, Coregister: Estimate, Normalism: Estimate & Write, and Smooth, so that the images of all the subjects can be compared together.

The data of medical imaging usually have serious data imbalance and data shortage problems, we will solve this problem when the dataset is used in the learning and testing of the recognition system.

3 THE PROPOSED METHODOLOGY AND MODEL

The structure of this model looks like the one shown in Figure 3, and the parameters involved in the model are shown in Table 2. The original image size was 112*112 pixels. To be adapted to the system, it was changed to a size of 61*73 pixels, and the third dimension was treated as the input of 61 channels.

Table 2. Model.

Layer name	Output size	152-layer		Param
Conv1	61*73	7*7*64 stride 2		35200
Conv2_x	31*37	3*3maxpool stride2		
		$1 \times 1, 64$	$\times 3$	16384
		$3 \times 3, 64$		36864
		$1 \times 1, 256$		16384
Conv3_x	16*19	$1 \times 1, 128$	$\times 8$	65536
		$3 \times 3, 128$		147456
		$1 \times 1, 512$		65536
Conv4_x	8*10	$1 \times 1, 256$	$\times 36$	262144
		$3 \times 3, 256$		589824
		$1 \times 1, 1024$		262144
Conv5_x	4*5	$1 \times 1, 512$	$\times 3$	1048576
		$3 \times 3, 512$		2359296
		$1 \times 1, 2048$		1048576
Avgpool	1*1	1*1 avgpool stride2		
Linear	1000	1*1000		2049000
SVM	2			
Total params		60220602		

The model used in our system is adapted from Resnet152, as shown in Figures 3 and 4, and can be divided into several parts. The first layer, conv1, is a layer associated with a 7*7 convolution kernel, followed by four blocks, conv2_x to conv5_x, which are associated with two kinds of convolution kernels 1*1, and 3*3, 1*1). In these blocks, 3 layers, 8 layers, 36 layers, and 3 layers, respectively, are stacked in each block. The following part is composed of pooling plus a fully connected layer (softmax). We changed the connected layer into a linear structure to extract the characteristic values, and then put them into an SVM classifier for classification, as shown in Figure 4.

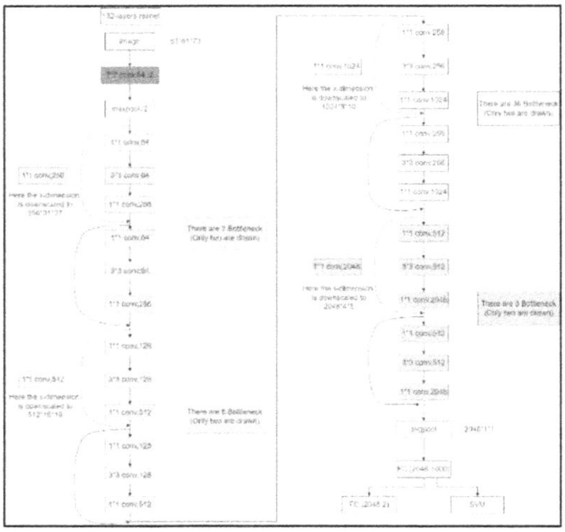

Figure 3. The schematic diagram of Resnet152 architecture.

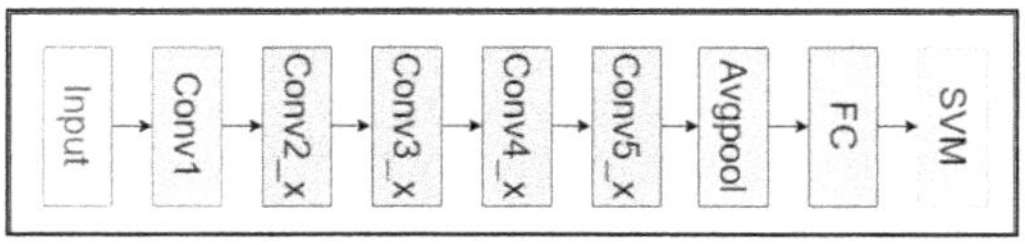

Figure 4. The architecture of our model.

Resnet152 is used because the model has a connection method called "shortcut connection", like the curved arc in Figure 5. This is the so-called "shortcut connection", and Resnet provides two versions, i.e., identity mapping (Figure 5 of X) and residual mapping (Figure 5 of F(X)). In Figure 3, two connections are distinguished by the real line and the virtual line. The real line denotes the same number of channels, using the calculation method: y = F(X) + X; the virtual line denotes a different number of channels, using the calculation method: y = F(X) + WX. This approach can reduce the number of parameters in complex networks, and improve the accuracy of the model.

We use the technique of transfer learning in our model training, which is the fine-tuning of a pre-trained CNN model from a natural image dataset to a medical image task, adapted from the open-source PyTorch. The trained weights were used to train our model, reducing the training time significantly. (Shin et al. 2016).

In addition, we adopted a 5-fold hierarchical cross-validation method for testing the model. The dataset is divided into 5 equal parts, and the proportion of categories in each part is the same. In each run, four parts are used for training and the rest part is used for testing, as shown in Figure 6. This can effectively reduce the effect brought by unbalanced data and produce a more stable classification accuracy.

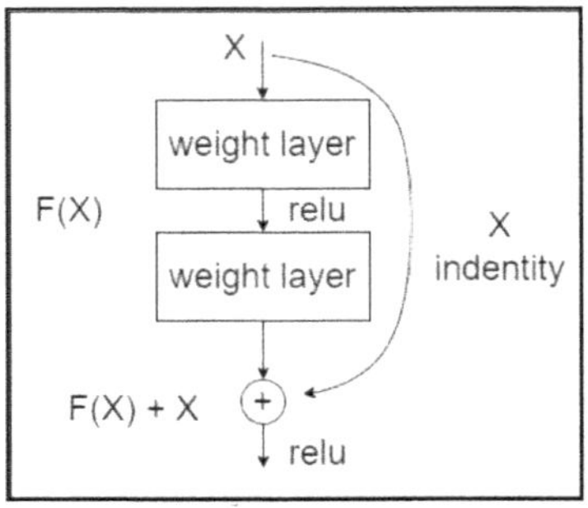

Figure 5. Resnet link structure.

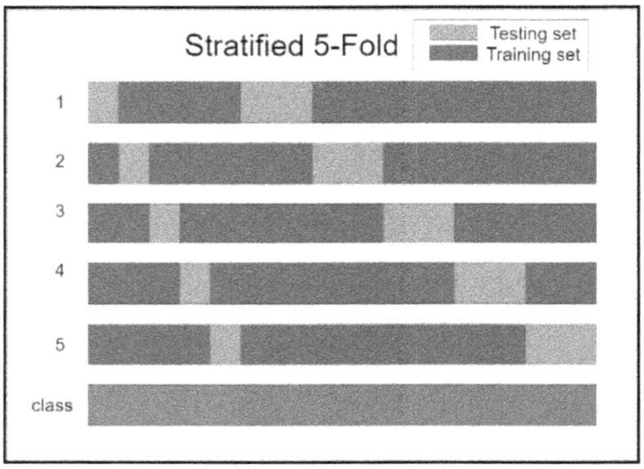

Figure 6. Stratified 5-fold.

In the model training, the loss function adopted is binary cross entropy loss (BCELoss):

$$L = -\frac{1}{N}\sum_{i=1}^{N}[y_i \log(p_i) + (1 - y_i)\log(1 - p_i)]$$

where N is the total number of training samples, y_i is the category of the ith sample, and p_i is the predicted value of the ith sample, which is generally a probability value. To obtain this probability value, we use the sigmoid function whose output is between 0 and 1.

In the model training, the activation function used is ReLU, as shown in Figure 7. ReLU helps prevent the gradient disappearance problem and reduce the computation time. The method used to train the model is SGD with momentum, which makes the training faster in the same direction and slower in different directions, allowing it to accelerate the convergence in the relevant direction and also to avoid being trapped in local minima.

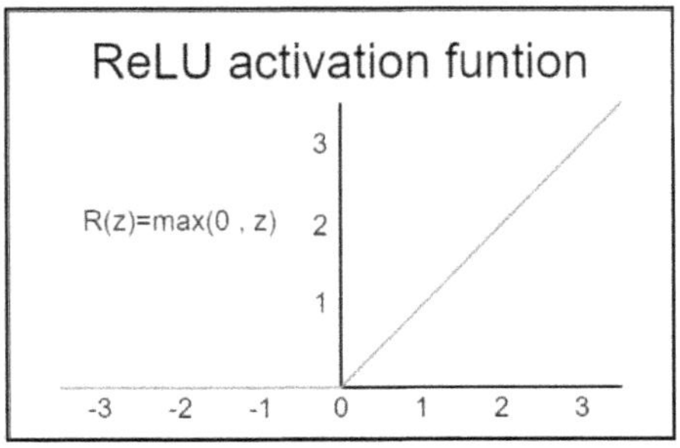

Figure 7. ReLU activation function.

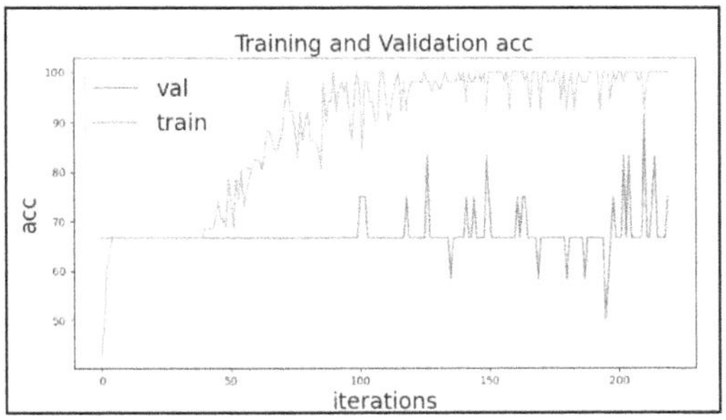

Figure 8. Resnet152 accuracy curves.

The parameters of this model are set as shown in Table 3. An early stop is set to prevent overfitting by stopping the training early as soon as the desired training accuracy is achieved at 8 consecutive iterations.

Table 3. Parameter settings for model training.

Parameter name	Value
image_size	61*73
Cross_vaildation	Stratified 5-fold
Batch_size	4
Epoch	300
Learning_rate	0.00001
Early_stop_hit	8
SGD_momentum	0.9

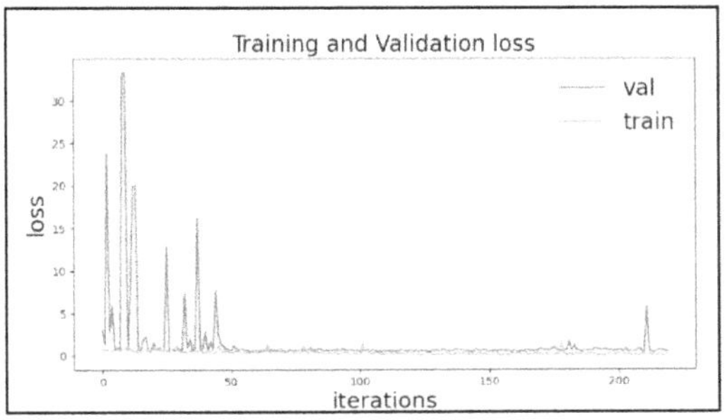

Figure 9. Resnet152 loss curves.

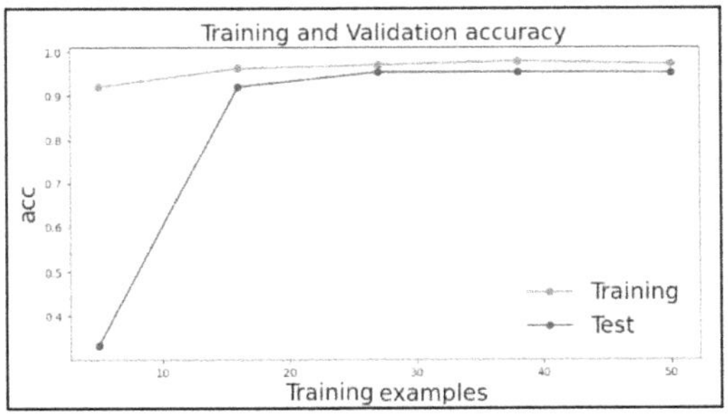

Figure 10. SVM learning curves.

In both the Resnet model and the SVM classifier, the data set is cross-validated using a hierarchical 5-fold approach. The Resnet model uses an early stop, and the average number of training sessions for this 5-fold model is 180, and one of the best-trained models is used to extract the feature values which are used as the input to the SVM classifier.

Our model is trained in open-source PyTorch and the hardware is NVIDIA GPU RTX-3070ti with cuDNN. GPU and NVIDIA Tensor Cores were used to accelerate the training.

4 EXPERIMENTAL RESULTS

The results obtained from the SVM classifier are shown in Table 4, and it can be seen that the average accuracy is 96.8%. The training accuracy is shown in Figure 8, and the training loss is shown in Figure 9. From the curves shown in Figure 10, it can be seen that there is no overfitting with the SVM classifier.

Table 4 shows the confusion table obtained from the proposed classifier. From this table, the values of precision, recall, and f1-score of each category are calculated. In addition, the accuracy, macro-averages, weighted-averages, and micro-averages are also shown in Table 5. Note that 96.8% can be achieved in accuracy.

Table 4. Confusion table.

	Actual label 0	Actual label 1
Predicted label 0	19	0
Predicted label 1	2	42

Table 5. Performance of the proposed classifier.

	precision	recall	f1-score
Class	1.000	0.900	0.950
Class 1	0.950	1.000	0.974
Macro avg	0.975	0.950	0.962
Micro avg	0.968	0.968	0.968
Weighted avg	0.967	0.967	0.966
Accuracy	0.968		

5 CONCLUSION

The model proposed in this paper has several advantages. First, it uses a transfer learning approach without retraining the weights from the scratch, which significantly reduces the training time. The model uses 151 convolutional layers of Resnet152 and one fully connected layer to extract features and then feeds the extracted features into the SVM classifier to do the discrimination. Second, it uses a small dataset without overfitting and achieves a high degree of accuracy. We hope this study will enable doctors to diagnose AD patients early and take appropriate actions to help them.

ACKNOWLEDGMENT

This work was supported by the grants MOST-108-2221-E-110-046-MY2 and MOST-110-2221-E-110-049, Ministry of Science and Technology, the NSYSU-KMU Joint Research Project (#NSYSUKMU 110-I001), and the "Intelligent Electronic Commerce Research Center" from the Featured Areas Research Center Program within the framework of the Higher Education Sprout Project by the Ministry of Education in Taiwan. Ministry of Education.

REFERENCES

Abed, Mahjabeen Tamanna, et al. Alzheimer's Disease Prediction Using Convolutional Neural Network Models Leveraging Pre-existing Architecture and Transfer Learning. In: *2020 Joint 9th International Conference on Informatics, Electronics & Vision (ICIEV) and 2020 4th International Conference on Imaging, Vision & Pattern Recognition (icIVPR)*. IEEE, 2020, pp. 1–6, doi:10.1109/ICIEVicIVPR48672.2020.9606649

Sarraf, Saman, et al. MCADNNet: recognizing stages of cognitive impairment through efficient convolutional fMRI and MRI neural network topology models. *IEEE Access*, 2019, vol. 7, pp. 155584-155600, doi: 10.1109/ACCESS.2019.2949577

Shin, Hoo-Chang, et al. Deep convolutional neural networks for computer-aided detection: CNN architectures, dataset characteristics and transfer learning. *IEEE transactions on medical imaging*, 2016, vol. 35, no. 5, pp. 1285-1298, doi: 10.1109/TMI.2016.2528162

Velazquez, Matthew, et al. RNN-based Alzheimer's disease prediction from prodromal stage using diffusion tensor imaging. In: *2019 IEEE International Conference on Bioinformatics and Biomedicine (BIBM)*. IEEE, 2019, pp. 1665-1672, doi: 10.1109/BIBM47256.2019.8983391

System Innovation for a Troubled World – Tak Lam et al. (Eds)
 ISBN: 978-1-032-45525-9

University lecturers' perspectives on the online learning environment during the COVID-19 pandemic: The Taiwanese context

Siu-Tsen Shen*
Department of Multi-media Design, National Formosa University, Hu-Wei, Yunlin, Taiwan

Stephen D. Prior
Aeronautics, Astronautics and Computational Engineering Design, The University of Southampton, Hampshire, UK

ABSTRACT: This study provides an overview of the COVID-19 (C-19) pandemic and its impact on the challenges faced by University lecturers in Taiwan. Related research was extensively reviewed, which led to an online survey being formulated, based on empirical evidence of the effects of the C-19 guidelines on the lecturer's experience and observations. The results showed that almost three-quarters of the teachers (74%) preferred the traditional face-to-face teaching when compared to distance learning. Further, nearly two-thirds (64%) preferred a mixed learning environment during and after the pandemic. Future work will focus on the empirical evaluation of how to support students more efficiently and effectively who may have limited access to online platforms such as unexpected network technical issues and lack of essential equipment to facilitate efficient learning. In the post-pandemic era, University teachers will need to rethink and adapt new strategies on how to encourage individualistic student online performance and engagement to maintain student learning quality and overall standards.

1 INTRODUCTION

In many ways, Taiwan has been an exemplar on how to fight the C-19 pandemic. It was one of the first countries to shut its borders, and in doing so, prevented a wave of infections and deaths. However, having contained the virus for almost two years it was inevitable that at some point this containment strategy would fail. This situation prompted the authors to conduct a survey on higher education teaching under the first national school lockdown of the C-19 pandemic in Taiwan during May 19 to June 28 of 2021. The objective of this research was to examine the online teaching strategies and challenges, such as network technical issues and the learning process in the implementation of online teaching classes. The authors analysed the immediate effects of the emergency situation on University education together with teachers' reactions, reflections and emotions, with regards to online teaching and learning. Bearing in mind that Taiwan had survived the pandemic for more than a year, since the original outbreak of C-19 in February 2020, thanks to its closed borders, the authors designed an online questionnaire that was completed by 263 University lecturers based in Taiwan. The questions refer to teachers' concerns and their feedbacks towards online teaching. This article offers positive and negative viewpoints of the application of emergency online teaching to address the unprecedented educational issues brought about by the pandemic.

The present survey was conducted online using Google documents, through Line and Facebook with 263 participants over a period of one week (1–7 June 2021). The survey was conducted in three parts. Part I included demographics of the subjects, which consisted of seven questions that captured a general overview of the participants and their background (gender, age, ethnicity, working status, education, social media app use, etc.). Part II consisted of 16 questions related to the participant's health and well-being. Part III from teachers' perspectives concerning the factors influencing their intention to use e-learning system within the Taiwanese universities context, which consisted of eight main questions relating to their online experience of education, whilst being in lockdown during the pandemic.

2 LTERATURE REVIEW

The United Nations Children's Fund (UNICEF) (2020) issued a guide book to provide the instructions for safety measures through the prevention, early detection and control of C-19 in schools and other educational facilities. The guidance, while specific to countries that have already confirmed the transmission of C-19, is still relevant in all other contexts. Education can encourage students to become advocates

*Corresponding Author: stshen@gs.nfu.edu.tw

DOI 10.1201/9781003377399-23

for disease prevention and control at home, in school, and in their community by talking to others about how to prevent the spread of viruses. Maintaining safe school operations or reopening schools after a closure requires many considerations but, if done well, can promote public health (Bender 2020).

Toquero (2020) described how higher education was affected and recommended to educational institutions to produce studies to proliferate and document the impact of the pandemic to the educational system in the context of Philippine. She stated there was a greater need for educational institutions to strengthen the practices in the curriculum and make it more responsive to the learning needs of the students even beyond the conventional classrooms (Toquero 2020).

Rashid and Yadav (2020) stated the use of e-Learning tools and platforms for effective student engagement which may have limitations of accessibility and affordability for many students in India. The need for offering more training of educators in digital technology was important to adapt to the rapidly changing education climate of the world. The use of e-Learning and virtual education may become an integral part of the higher education system in the post-pandemic. The higher education institutions and universities need to plan the post-pandemic education and research strategies to ensure student learning outcomes and standards of educational quality (Rashid & Yadav 2020).

Jena (2020) suggested that virtual education is the most preferred mode of education at this time of crisis due to the outbreak of C-19. The post C-19 education seems to be an education with widely accepted online/virtual education which may perhaps be a parallel system of education. However, the researcher has not covered any statistical analysis on impact of C-19 on higher education and further in-depth study with statistical research may also be undertaken (Jena & Research 2020).

Arora and Srinivasan (2020) conducted two experiments. One set of respondents were those who adopted virtual classrooms and another set that did not adopt virtual classrooms. Descriptive statistics and t – test were performed to analyze the data collected through primary sources. Among those who adopted virtual mode, the mean of actual benefits was significantly less than the mean of expected benefits. Network issues, lack of training, and lack of awareness were stated to be the major challenges faced by them. Lack of awareness was stated to be the most important reason by those who did not adopt virtual classrooms followed by lack of interest and doubts, regarding the usefulness of virtual classes. Less attendance, lack of personal touch, and lack of interaction due to connectivity issues were found to be the significant drawbacks of virtual classes. The study also suggested the ways to overcome the above challenges, drawbacks, and reasons behind non – adoption of virtual classes (Arora & Srinivasan 2020).

Several researches had discussed the impact of C-19 on higher education, its implications and challenges. Furthermore, few researches have been done in relation to C-19 in the field of medicine and other fields related to health sciences (Abdulamir & Hafidh 2020; Gondauri et al. 2020; Holshue et al. 2020; Meng et al. 2020; Usak 2020). However, very little research is done in relation to the field of education on how COVID-19 affected the educational system (Bao 2020; Sintema & Education 2020; Yan & Technologies 2020).

3 SURVEY AND ANALYSIS OF RESULTS

3.1 *Part I – General Overview Review*

In terms of gender, the questionnaire participants were slightly more female dominated (57.4%) as opposed to male (42.6%). There was a broad range of ages represented, with the mean age being 45.5 years old (*SD = 10.6*). The survey being focused on Taiwan, the vast majority of the participants were Taiwanese (95.1%), with other nationalities being British (1.5%), and a range of other nationalities, including Swedish, Norwegian, Greek, Serbian, with smaller percentages (<0.5%). Full-time teachers represented the major grouping (81.4%), followed by those self-employed (4.9%), then Employers (3.8%), Retired (3%) and Unemployed (2.3%). The remainder were made up of smaller percentages of trainee teachers, etc.

3.2 *Part II – Personal Emotional Well-Being*

When asked whether the participant's mental health had been adversely impacted by C-19, it was a surprise to find that on a five-point Likert scale, the highest response was (3) 'Neutral' (28.1%) with the remainder fairly evenly separated between higher and lower scores (1) 16.7% 'Not at all', (2) 17.1%, (4) 23.6%, and (5) 14.4% 'A great deal'. To some extent this is understandable given the difference in perspectives from School teachers to University Lecturers and the relatively low impact to date of C-19 in Taiwan (Note: the recent spike in cases and deaths, starting in mid-May 2021, which resulted in a national lockdown).

The following question asked about how well they adapted to isolation and social distancing. Given that Taiwan has not been under a prolonged lockdown and therefore few people have had to isolate for long periods, it was no surprise that the only 1.5% of the participants answered 'Not at all', the answers then increased (2) 8.7%, (3) 19%, (4) 41.8%, and then (5) 'Very well' 28.9%.

The following question asked how well do you feel you have been able to follow the government rules. Again, there was no surprise to find that most people (67.3%) were able to follow these with no problem, at the other end of the scale, 0% found this very difficult. In terms of worries of developing C-19, (Q.4) was answered Yes (82.5%). When asked about the worry of friends and family developing C-19 this increased to 92% 'Yes'. This clearly shows the level of fear of C-19 within Taiwan.

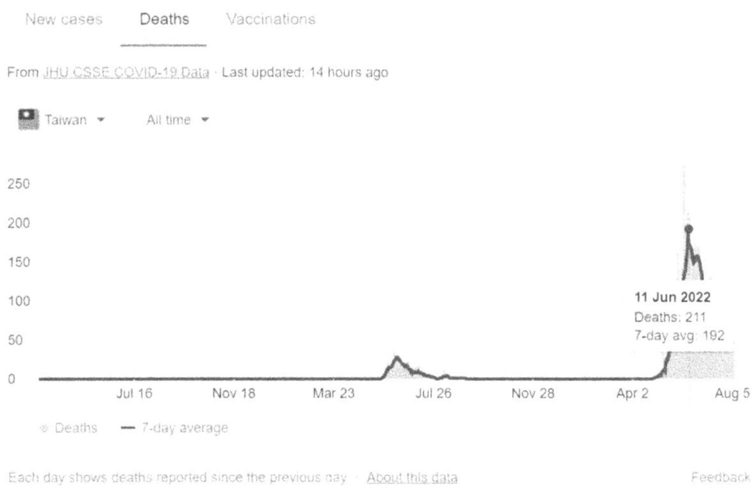

Figure 1. COVID-19 Deaths timeline in Taiwan (Engineering 2022).

When asked whether the participant had ever had a C-19 test, only 6.1% said 'Yes'. Again, Taiwan has experienced a fairly low level of C-19 infection, with only 11,298 cases and 260 deaths (as of 7 June 2021) (CDC 2021). Of these 16 participants, only 4.4% (2) stated that the result was positive. Interestingly, only 3.7% isolated for the required 14 days. Taiwan recently began its vaccination program, so it was no surprise to find that only 2.7% of the participants had had one dose and 2.3% had two doses of the vaccination. When asked if those that had not been vaccinated would accept a vaccination, 82.5% said that they would. Of those that wouldn't accept a vaccination if offered, most cited Side Effects (41), Trust Issues (35), Not Needed (4), Fear of Needles (4) and a small number citing a Religious Belief (Note: participants could select more than one answer).

When asked if they knew of anyone that had contracted C-19, 9.9% said 'Yes'. Of this total 50% were admitted to hospital, and sadly 38.9% of those died.

When asked about their sleep quality, 53.6% had on average 8 hr/night. Just under one quarter said that their sleep pattern was good, with 22.8% stating that they had poor sleep quality. Of those that had poor sleep quality, 46.9% had between 4-6 hr/night, 45.8% experiencing 6-8 hr/night. Rather surprisingly, 7.3% were surviving on 2-4 hr/night.

When asked about what strategies people use to alleviate symptoms of poor sleep quality, there was a range of answers, ranging from Trying to Stay Positive (158), Exercise (118), Online Talking (98), Entertainment (95), and Cooking (49). A very small number were actively seeking medical/professional counselling. (Note, participants could answer more than one category hence the total will not add up to 263).

When asked if they feel anxious, 72.2% said 'Sometimes' and 12.9% answered 'Often'. When asked about depression, 65% answered 'Sometimes', and 11% 'Often'. Clearly, there is a large percentage of the teaching profession with underlying fears and mental issues that they would not normally discuss.

When asked if people sometimes felt lonely, 56.7% stated that they sometimes did, with 3% stating that they often did. In terms of the participants' general health/fitness level, just over half stated that they thought they were about average, with 8% Poor and 1.1% Very Poor. When asked if they thought their general health/fitness had got worse in the last year, 11.8% thought that it had, with 40.3% stating 'Maybe' (a level of uncertainty here, underlying how people perceive this, if maybe they don't regularly exercise).

When asked about whether they thought they were suffering from long COVID, a surprisingly low 4.6% stated that they were.

When asked how long their long COVID had existed, there was a range of answers. The highest percentage being for the 1-3 months' category (53.8%), with 30.8% in the last month, and 15.4% for the 6-9 month's category. Somewhat surprisingly, 38.1% of the participants that thought they had long COVID stated that they were getting medical help with the condition. Therefore, it would appear that nearly two-thirds of unreported and undiagnosed people suffering in silence, and dealing with this challenging condition on their own.

3.3 *Part III – Personal Learning Questions*

When asked if their educational establishment had provided access to a remote online learning platform, 94.7% stated 'Yes'. In terms of which platform was provided, there was a range of answers, Google Meet (195), Microsoft Teams (37), Zoom (20), Discord (12), Skype (5), Vimeo (3) and Goto Meeting (2). There was a multitude of other software packages listed in the 'Others' category such as Adobe Connect, etc.

In terms of the e-classroom interaction, again there were high levels of satisfaction (45.2%) (4 & 5). Likewise, in terms of the e-classroom look and feel, we observed high satisfaction levels (40.1%). In terms of the students' performance, we observed an overall neutral response (43.8%).

In terms of the platform used, it would appear that overall the participants are satisfied with the level of functionality with (49%) in Category 4 (Satisfied) & 5 (Very Satisfied). In terms of the quality of the video/images, there appeared to be no issues, with high levels of satisfaction (52.4%). Finally, in terms of the overall platform performance, we have high levels of satisfaction (53.6%).

From these results, it would appear that most of the available virtual learning platforms are delivering on their expectations in the time of C-19.

When asked whether the use of online distance learning was better than being at school in person, a small majority stated that it was 'Not' (58.6%), with 'Sometimes' (37.3%), and 'Yes' (4.2%). When asked whether they found the use of online distance learning effective for learning, 52.5% said 'No', 38.4% said 'Sometimes' and only 9.1% 'Yes'.

The majority of the participants (59.7%), thought that the digital platform provided was both not efficient and not easy to use!

When asked whether they thought that gaining knowledge online was the same as in person, most thought that it was not, by a margin of (71.9%) to ((28.1%). This result was somewhat surprising. When asked whether the participants preferred the traditional face to face teaching style when compared to distance learning, a very high proportion (74.1%) answered 'Yes, with 22.1% saying 'Sometimes', and only 3.8% answering 'No'.

The final question asked whether the participants would like to keep using the online distance learning tools even after the pandemic was over. Nearly two-thirds (64.3%) stated that only if this was part of a blended learning environment, with 23.2% answering 'Yes', and only 12.5% 'No'.

General Comments

"The Most annoying part of online learning is students not switching on their video or microphones to interact. They prefer chat function. Talking to a screen of initials is not very motivating for the presenter. Equally, students presenting online found it difficult to engage with their audience."

4 CONCLUSION

Due to its early travel ban, extensive use of masks (before C-19) and generally healthy population, Taiwan is an interesting test case for the effects of C-19 on the general population and hence education.

To date the government has still not fully opened its borders, unlike most other countries, despite the population desperately hoping to get back to as normal life as possible. The main reason for this decision is the relatively high numbers of new infections (20,000) and deaths (50) August 2022. The adage of first to close, last to open is very apt. If a country doesn't use the time bought by an early closure to get their populace fully vaccinated, then they will eventually suffer excess deaths, as has been the case with Taiwan.

During the first two years of the C-19 pandemic, the vaccination rate remained very low, with less than 1% having even a single dose (Shen & Prior 2022). This began to change in May 2021 as the infection rate began to increase, and has now reached almost 85% fully vaccinated by August, 2022. The big challenge here being how to encourage vaccinations when the infection rate is relatively low.

This island government, dominated by the Democratic Progressive Party (DPP), who would like to win the Nine-in-One (mayoral) elections that will be held on Saturday, 26 November 2022, to elect all

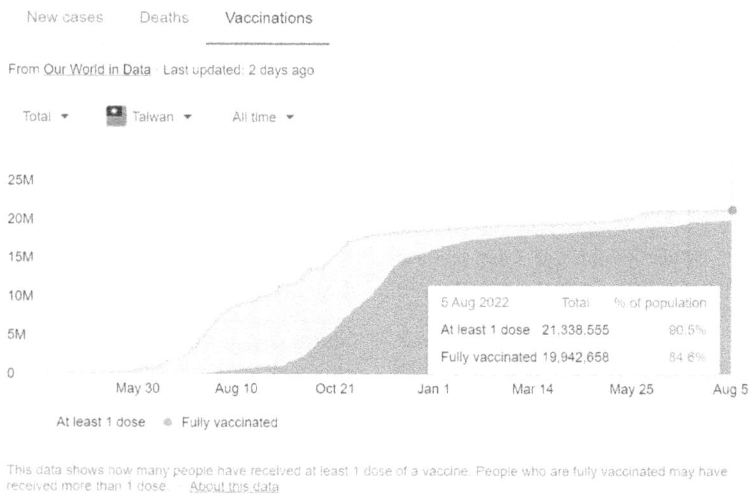

Figure 2. COVID-19 Vaccinations timeline in Taiwan (Engineering 2022).

municipality mayors, county magistrates, city mayors, municipal, county, and city councilmen, township mayors, chiefs of village, and aboriginal district chiefs in Taiwan also plays a part.

The authors presented empirical evidence on the effects of the C-19 pandemic on university teachers. This research was not intended to make a comparison between face-to-face and online teaching styles, but to look into the lessons of what we have learned by this experience and the potential implications for higher educational institutions.

The results pointed to some interesting issues on teaching and learning in higher education, such as how to support the students from behind their screens in terms of technical issues, and how to enhance their learning during the class. Future work will embark on how to improve the online class attendance and knowledge sharing interaction consequences for this epidemic should become routine as part of preparedness efforts worldwide.

Fundamentally, there needs to be a willingness to learn and a capacity for learning from the student. More highly motivated students will succeed despite their circumstances. The C-19 pandemic has highlighted and exaggerated the differences between students who really want to learn and those that are merely 'going through the motions'. In pre-Covid times, weaker students would be carried to some extent by stronger students in their peer groups, C-19 has broken this link and exposed their lack of ability.

Furthermore, the lack of social engagement and interaction with the peer group, has isolated students and caused higher levels of mental illness, causing a strain on the already overloaded systems of support. One other issue worth mentioning are the technical inhibitors, such as the lack of fast and stable internet connections, together with the lack of powerful computing equipment and facilities usually provided by the educational institution. This is probably not the case for Taiwan, however, there are many countries around the world where this would be a much bigger issue.

REFERENCES

Abdulamir, A. and R. Hafidh (2020). "The Possible Immunological Pathways for the Variable Immunopathogenesis of COVID—19 Infections among Healthy Adults, Elderly and Children." *Electronic Journal of General Medicine* **17**(em202).

Arora, A. K. and R. J. P. I. j. o. m. Srinivasan (2020). "Impact of pandemic COVID-19 on the teaching–learning process: A study of higher education teachers." *Prabandhan: Indian journal of management* **13**(4): 43–56.

Bao, W. J. H. b. e. t. (2020). "COVID-19 and online teaching in higher education: A case study of Peking University." *Human Behavior and Emerging Technologies* **2**(2): 113–115.

Bender, L. (2020). Key Messages and Actions for COVID-19 Prevention and Control in Schools. New York UNICEF New York.

CDC, T. (2021). "COVID-19 Data and Statistics." Retrieved May 12, 2021, from https://www.cdc.gov.tw/En.

Engineering, J. H. U. C. f. S. S. a. (2022). "*JHU CSSE COVID-19 Data Taiwan New Cases and Deaths.*" Johns Hopkins University Center for Systems Science and Engineering (JHU CSSE) COVID-19 Data Retrieved August 6, 2022, from https://g.co/kgs/cyrupK.

Gondauri, D., E. Mikautadze and M. Batiashvili (2020). Research on COVID-19 Virus Spreading Statistics based on the Examples of the Cases from Different Countries. *Electron J Gen Med.* 2020; 17(4): em209.

Holshue, M. L., C. DeBolt, S. Lindquist, K. H. Lofy, J. Wiesman, H. Bruce, C. Spitters, K. Ericson, S. Wilkerson and A.J.N.E.j.o.m. Tural (2020). "First case of 2019 novel coronavirus in the United States." *New England Journal of Medicine*.

Jena, P. K. J. I. J. o. A. E. and Research (2020). "Impact of Covid-19 on higher education in India." *International Journal of Advanced Education Research* **5**.

Meng, L., F. Hua and Z. Bian (2020). "*Coronavirus Disease 2019 (COVID-19): Emerging and Future Challenges for Dental and Oral Medicine.*" **99**(5): 481–487.

Rashid, S. and S. S. J. I. J. o. H. D. Yadav (2020). "Impact of Covid-19 pandemic on higher education and research." *Indian Journal of Human Development* **14**(2): 340–343.

Shen, S.-T. and S. D. Prior (2022). The impact of COVID-19 on the well-being of youth and the effectiveness of online learning in higher education—a Taiwanese perspective. *System Innovation in a Post-Pandemic World*, CRC Press: 42–47.

Sintema, E. J. J. E. J. o. M., Science and T. Education (2020). "Effect of COVID-19 on the performance of grade 12 students: Implications for STEM education." *Eurasia Journal of Mathematics, Science Technology Education* **16**(7): em1851.

Toquero, C. M. J. P. R. (2020). "Challenges and opportunities for higher education amid the COVID-19 pandemic: The Philippine context." *Pedagogical Research* **5**(4).

Usak, M., A. R. Masalimova, E. I. Cherdymova and A. R. J. J. o. B. S. E. Shaidullina (2020). "New playmaker in science education: Covid-19." *Journal of Baltic Science Education* **19**(2): 180.

Yan, Z. J. H. B. and E. Technologies (2020). "Unprecedented pandemic, unprecedented shift, and unprecedented opportunity." *Human Behavior and Emerging Technologies*.

System Innovation for a Troubled World – Tak Lam et al. (Eds)
© 2023 The Author(s), ISBN: 978-1-032-45525-9

The golden years of tea: Transition, attenuation, and transformation of tea industry – Case study of tea factories in Taoyuan, Hsinchu, and Miaoli

Po-Jo Lai* & Shyh-Huei Hwang
Graduate School of Design, National Yunlin University of Science and Technology, Yunlin, Taiwan

ABSTRACT: Towns and villages around Taiwan Romantic Route 3 road were once the world's factory active in the global trade market, and they are also a common memory for many Hakka people. This research focuses on the representative "tea industry" among the Hakka Mountain Forest Economy industries. Based on literature research, interviews are conducted on the operation of three tea factories in Taoyuan, Hsinchu and Miaoli, including exploring the origin and transition, attenuation factors, modern transformation opportunities and paths of tea industry development in Hakka Villages along Taiwan Romantic Route 3 road. After a comprehensive analysis, the following points are found: (1) Historical transition of Taiwan's oolong tea: Taiwan's oolong tea was introduced from Fujian at the end of the 18th century. During the Qing Dynasty, that is, in the 1860s, British merchants began to export it. During the Japanese occupation, industrialization continued. In the 1970s, its development was interrupted by the energy crisis, but it turned to develop the domestic market; (2) Historical transition of Taiwan's black tea: Taiwan's black tea originated from the colonial period during the Japanese occupation, and began to be exported to Russia in 1906, and it was not gradually stopped until the Russia's October Revolution in 1917. In 1925, the large-leaf species of Assam black tea was re-introduced, entering the golden age of exporting Taiwan's black tea until 1941, when the Second World War caused export interruption. After the war, National Government of the Republic of China promoted green tea to be the mainstream in 1960s, while black tea gradually declined to the bottom. It was not until the rise of bubble tea in the 1990s and TTES No. 18 (Ruby) after the disaster in 1999 that it was revived. (3) Transformation of tea industry: With rising awareness of creative industries, promotion of country and government policies, and popular development of experience economy, tea factories have started cooperating with corporate brands, creating their own creative tea brands, establishing tea culture experience stores, and building tourism and leisure parks, etc., to gradually transform and upgrade the entire tea industry.

1 INTRODUCTION

Industry constitutes a solid foundation for a country's economic development; however, the amount of resources, terrain, climate, and environmental factors vary from country to country and region, so industries that represent their own characteristics and the most advantageous will be developed. The Hakka group economy is the composition of the migration and movement of the Hakka people, participating in and shaping the regional ethnic relations, the development of industrial resources, and the exchange of commodities, and shaping the lifestyle with ethnic characteristics (Chang & Feng 2020).

It used to be a forest farm with precious alpine wood and a playground for picking up tung tree seeds. On the hills, you can smell the smoky smoke of roasted camphor and smelted lemongrass, silkworm breeding that is well-protected in the slack season, and the sound of folk songs sung everywhere in tea gardens in the valleys ...;" Taiwan Romantic Route 3" retains the wisdom of old Hakka ancestors to reclaim the mountains and forests, and is a historical trajectory created by the Hakka people.

"National-level Taiwan Third-tier Hakka Village Romantic Route" (referred to as "Taiwan Romantic Route 3") is the promotion plan of the Hakka Committee to implement President Tsai Ing-wen's Hakka political views, that is, to create a diverse and rich cultural, ecological, and industrial environment on "Taiwan Romantic Route 3", and to start Hakka Renaissance. The implementation policy of "Taiwan Romantic Route 3" is based on the cultural industry as the driving force for regional development, which has the significance of shaping the identity of the Hakka group. In its policy, it is mentioned that: "Taiwan Romantic Route 3" is an important place of origin of Taiwan tea, so it is planned to be rebuilt as tea garden

*Corresponding Author: trolls5438@gmail.com

DOI 10.1201/9781003377399-24

base and also as an iconic tea fragrance corridor (Liu 2017). From this, it can be seen that the promotion of tea culture and tea industry is a Hakka cultural policy that the state attaches great importance to, and it is regarded as the vision and goal of Hakka culture to inherit, continue and innovate.

This research focuses on the representative "tea" among the Hakka Mountain Forest Economy industries and selects three towns in "Taoyuan, Hsinchu and Miaoli" where Hakka people mainly live along "Taiwan Romantic Route 3" in northern Taiwan as the research area. A tea factory operator in each of these three places is selected as the research object, and an individual interview is conducted based on the literature review to explore the process of each tea factory's development in its own tea industry in each individual region in the past. The specific research purposes of this study are: (1) Discuss the origin and transition of tea industry development in Hakka Villages (Taoyuan, Hsinchu, and Miaoli) along "Taiwan Romantic Route 3". (2) Explore attenuation factors of tea industry in Hakka Villages (Taoyuan, Hsinchu, and Miaoli) along "Taiwan Romantic Route 3". (3) Analyze modern transformation opportunities and paths of tea industry in Hakka Villages (Taoyuan, Hsinchu, and Miaoli) along "Taiwan Romantic Route 3".

1.1 *Introduction to case study*

The research objects are three tea factories in Taoyuan, Hsinchu, and Miaoli, which are named "Fu Yuan Tea Manufactory", "Fuxing Tea Culture Center", and "Yiming Tea Garden" in order from north to south.

1.1.1 *Fu Yuan tea manufactory*

The century-old Fu Yuan Tea Manufactory, located in Longtan District, Taoyuan City, was founded in 1903 by the first generation Huang Caisheng in "Rugu Mountain" and cultivated tea here. The secondgeneration Huang Weihe believed that business had more room for development than farming, so he began to buy the popular Panzhuang Oolong from tea farmers. Pulled by ox carts and transported by manpower carts, tea leaves are transported along the Dahan River to Daxi, and then to foreign firms in Bangka. In 1949, in response to the government's promulgation of management rules for the tea industry, and introduction of dry-fried green tea production technology and foreign tourists from Shanghai, Hangzhou, and other places, Huang Wei-he decided to seize the opportunity and officially built a tea factory in that year, expanding machinery and equipment to improve production capacity. The production of black tea was sold to Japan, and dry-fried green tea was sold to North Africa. The period from 1951 to 1961 can be said to be the golden age of Fu Yuan Tea Manufactory. From 1971 to 1984, the third-generation Huang Wanjun produced green tea, pearl tea, and black tea, and other tea products. Since 1971, the fourth generation Huang Wenliang and Fan Baozhu, Husband and Wife, have produced tea products such as Baozhong tea, Panzhuang Oolong, dry-fried green tea, and black tea, etc. The current stage is toward diversification, producing exquisite handmade teas such as Oriental Beauty, Ruby, and Tsuan-Kan Tea. At present, the fifth generation Huang Zhengwei and Huang Lizhen, young Brother and Sister, have returned to their hometowns to learn tea-making and online business, and inherit good tea-making skills spanning five generations.

Figure 1. Interview with Huang Wenliang, the fourth-generation head of Fu Yuan Tea Manufactory.

1.1.2 *Fuxing tea culture center*

Fuxing Tea Culture Center is located in Fuxing Village, Emei Township, Hsinchu County. The Zeng family started to cultivate the Fuxing area in 1883, and successively established industries such as oil shop, rice shop, sugar factory, and tea factory. In the early 1920s, "Chaliao" was used as a raw tea manufacturing site, and tea was made by hand. The 22nd generation Zeng Xinhu built the Fuxing Tea Factory in 1928, which was the first tea factory in Fuxing Village to introduce tea-making by machinery. In 1935, the "Hsinchu-Taichung earthquake" destroyed the construction of tea factory, but it also transformed the tea factory into mechanized production. After the disaster, it was rebuilt into a modern two-story factory. In 1944, it began to use heavy oil-burning power machinery to make popular black tea at that time. It is also one of three major tea factories in Emei Tea District. After 1980, the traditional tea industry was replaced by technology and foreign trade industry and was regarded as a sunset industry, and Fuxing Tea Factory was closed for this reason. It was not until 2006 that a turnaround occurred. The old tea factory was designated as a historic building by Hsinchu County Government, and after two years of restoration, it was put into trial operation in 2007. In 2011, it was officially named "Fuxing Tea Culture Center", to promote tea industry knowledge and education by means of tea culture tours and tea sensory experience.

1.1.3 *Yiming tea garden*

Yiming Tea Garden is located in Toufen Town, Miaoli County. A family of three generations has been working hard in the tea industry since the Japanese colonial

Figure 2. Interview with Zeng Zhengzhang, the third generation head of Fuxing tea culture center, and Zeng Xianghong, the fourth-generation head of Fuxing tea culture center.

period. The first-generation Lin Shimei used the word "Yi" as the first word of the family-run business name. Under the background of agricultural history as a tenant farmer at that time, it was engaged in the business of growing tea and selling fresh tea leaves. The second generation Lin Mingzong cultivated tea gardens alone, hired tea pickers to pick tea, and sold tea raw materials. In 1996, the thirdgeneration Lin Yuping took over the business and started to learn how to produce and sell alone as a small independent farmer in line with the government's then-implemented agricultural policy of "Production-Marketing Integration". In 2004, Lin Yuping combined the word "Yi" in her grandfather's name with the word "Ming" in her father's name, and officially named the tea garden "Yiming Tea Garden". Before that, there was no brand concept in the operation of tea gardens, and tea merchants called them "the tea seller on Laoqi Mountain". Lin Yuping has currently managed the overall operation of tea garden, tea factory, and landscape restaurant, and continued to operate tea garden and inherit tea industry passed down from previous generations with a grateful mindset.

Figure 3. Interview with Lin Yuping, the third generation head of Yiming tea garden.

2 LITERATURE REVIEW

2.1 *History of tea industry*

In 1866, British tea merchant John Dodd developed tea planting business in Taiwan. He purchased tea seeds and seedlings from Anxi, Fujian, and persuaded tea farmers to plant them (Hsu 2011). Most of early tea production in Taiwan was oolong tea, John Dodd hired a tea master from Fuzhou to come to Taiwan to make oolong tea, and as the sole agent of "Yihe Foreign Firm", he exported Taiwan's oolong tea as "Formosa Oolong" to the world (Tang 2019). Later, around 1873, due to the stagnation of sales, the price plummeted, and foreign businessmen terminated the acquisition (Chang 1993).

During the Japanese occupation period, the rise of black tea reflected the development of the Japanese capital in Taiwan tea. At that time, Taiwan already had small-leaf tea trees for making oolong tea, so they began to study the use of small-leaf tea trees to make black tea. In 1906, Taiwan's first small-leaf black tea was exported to Russia in the form of "red brick tea" (Tang 2019). In 1910, "Japan Taiwan Tea Co., Ltd." was established, which specialized in the production of black tea. This was the beginning of the production of black tea in Hsinchu (Hsu 1996). In 1917, the October Revolution took place in Russia, and the demand for black tea decreased year by year, and the first stage of exporting Taiwan black tea came to an end quietly. In 1925, large-leaf Assam black tea was introduced from India and successfully planted in Yuchi Township, Nantou. Taiwanese black tea began to rise and opened up a golden age (Tang 2019). When the Pacific War broke out in 1941, Taiwan's foreign shipping was blocked, and Taiwan's tea exports were interrupted due to the sharp reduction in the war. After 1944, Taiwan's external communication was cut off, and tea exports were completely halted (Jiang 1960).

After Taiwan's restoration, the National Government of the Republic of China developed green tea, which was sold well in North Africa and Japan, resulting in a booming export of tea industry in northern Taiwan (Hsu 2011). Green tea became the mainstream in the 1960s. However, the sales of black tea outside Taiwan declined year by year, and finally entered a dark period in the 1970s. The world energy crisis in 1975 interrupted the export of various types of tea in Taiwan, allowing Taiwan oolong tea to return to the domestic market. In the 1990s, with the rise of bubble tea black tea gradually recovered. In 1999, Tea Research and Extension Station, Council of Agriculture, Executive Yuan published "TTES No. 18 (Ruby)", which revived the long-dead Taiwanese black tea (Tang 2019).

To sum up, Taiwanese tea can be roughly divided into four types: black tea, green tea, oolong tea, and Baozhong tea due to the different production processes and varieties. The time of its rise for export is also different. During the Qing Dynasty, oolong tea and Baozhong tea were mainly exported. During the Japanese occupation period, Baozhong tea and black tea were mainly exported. Green tea rose after the war (Wu 2007).

2.2 *Transition in tea industry*

Tea industry has been around for more than a hundred years since the Qing Dynasty, the Japanese occupation, and Taiwan's restoration. After transition and reform of various times, what remains unchanged is that tea is an important economic resource for Taiwan. In the early days of Japanese occupation and Taiwan's restoration, it was the "Golden Age of Tea" for tea industry, created brilliant foreign exchange profits for Taiwan, and was the number one export cash crop. The development history of tea industry is like a history of Taiwan's economic development.

In the 1970s, when the world suffered from an energy crisis, Taiwan's tea sales outside the country declined steadily, and it was no longer a major source of foreign exchange. At this time, Taiwan's economy began to take off, and the social type changed from an agricultural society to an industrial and commercial society. Since 1979, the export of Taiwanese tea has declined sharply, and some tea producers have to grow crops to make a living in order to take into account the economic benefits, resulting in a decrease in the area of tea plantations and tea production. In the 1980s, due to factors such as the appreciation of Taiwan dollar and the high domestic wages, the rural labor force was out-flowing and engaged in industrial and commercial employment, resulting in a shortage of labor for tea production and tea making, high labor costs, increased production costs, and tea industry lost its international competitiveness. Later, due to the advocacy of "Tourism Orchard" in Taiwan, industry players set up tea gardens to gather sightseeing and tea sales, and gradually increase the domestic sales and domestic demand of Taiwanese tea. Oolong tea, which caters to the tastes of Taiwanese consumers, has made a comeback and returned to the domestic market (Chiou 2005). In 1983, government abolished the "Tea Production Management Regulations" restrictive licenses and put forward the slogan "Returning Tea to Farmers" to encourage tea farmers to self-production and self-marketing. This mode of production without government regulation has resulted in a decline in the quality of Taiwanese tea, and the export of tea has shifted to domestic sales (Liao 2011). In the 1990s, due to the rise of handshake bubble tea, the consumption pattern of Taiwan's tea industry changed to tea house culture. In 2001, Taiwan implemented the "five-day work weeks" policy, which led to the domestic tourism trend. Tea factory was transformed from the original traditional tea industry operation mode to a leisure and tourism park through independent decision-making or government guidance.

2.3 *Transformation in tea industry*

The implementation of the "five-day work weeks" policy has improved people's concept of leisure and the tourism industry has grown significantly. Citizens can use their time freely, their willingness to travel abroad has increased, and they have paid more attention to the needs of cultural exploration and tourism, and leisure. In recent years, driven by the bilateral cooperation between the implementation of national policies and relevant government units, the awareness of creative industries has sprung up, and domestic tourism industry and the experience economy have become popular. The concept of caring for the earth, environmental protection and sustainability is advocated, and the health-oriented tea industry has an opportunity for transformation.

Traditional agricultural products are on the road to industrial upgrading. Tea factory operators must learn strategies related to serving consumers and experience tours, such as launching local cultural tours, providing immersive experiences in tea gardens or tea factories, building cultural and creative brands, and digital Marketing, etc. to cope with the trend of transformation.

3 RESEARCH METHOD

The purpose of this study is to explore the origin and transition of tea industry development in Hakka Villages (in Taoyuan, Hsinchu, and Miaoli) along Taiwan Romantic Route 3 road, trace attenuation factors of tea industry, and analyze its transformation opportunities and paths. This research explores the root cause from history, and the source of data comes from the literature search and the empirical discussion provided by the research object, so it adopts the textual analysis of qualitative research.

The scope of this research is based on three towns in "Taoyuan, Hsinchu, and Miaoli" where Hakka people mainly live along "Taiwan Romantic Route 3" in northern Taiwan as the research area. After selecting the family background setting in inline with the history of tea industry spanning the Qing Dynasty, the Japanese occupation, and the era after Taiwan's restoration, a tea factory in each of three towns is chosen as the research object, namely "Fu Yuan Tea Manufactory" in Longtan District, Taoyuan City, "Fuxing Tea Culture Center" in Emei Township, Hsinchu County, and "Yiming Tea Garden" in Toufen Town, Miaoli County. This research uses field interviews to conduct individual interviews with the current operators of three tea factories to explore the operation history of their own tea workers in developing their own tea industry in different regions in the past.

The data collection of the research objects of this research mainly comes from the information obtained from the case interviews of three tea factories. Unstructured interview and semi-structured interviews are used as research tools, and new questions and viewpoints are added promptly or the case study subjects are allowed to freely extend the problem orientation, and individual interview consent forms are signed. In order to explore subjective viewpoints of the research subjects, in-depth interviews were used to collect data. Literature review was also used to review the historical context of Taiwan's tea industry in order to understand the phenomenon of interview data.

4 RESEARCH RESULT ANALYSIS

In 1999, Tea Research and Extension Station, Council of Agriculture, Executive Yuan published "TTES No. 18 (Ruby)", which brought the long-dead Taiwanese black tea back to the market, but it was difficult to reproduce the prosperity of the golden age of Taiwan tea industry. Based on case interviews, this research summarizes attenuation factors, modern transformation opportunities, and paths of tea industry development in Hakka Villages (in Taoyuan, Hsinchu, and Miaoli) along Taiwan Romantic Route 3 road as follows.

4.1 *Attenuation factors*

Tea makers are an invaluable resource for tea factories, and their personal tea-making skills can make tea flavors more unique. Tea making is a "practical" experience accumulated year by year on a ten-year basis. With the accumulation of experience over time, tea makers still have to face the reality of limited physical endurance. The young people of the new generation are unwilling to work in agriculture, and even less willing to spend ten years to learn basic skills, resulting in the long-term shortage of workers (tea makers and tea pickers) in the traditional tea industry. The master tea maker has a whole set of tea-making skills but has nowhere to exercise them, and the skills cannot be continued and inherited, resulting in a "fault" in the inheritance of tea industry.

In the 1970s, Taiwan's economy took off, and tea industry, which was dominated by agriculture, was gradually replaced by industrial and commercial society. In the 1980s, factors such as the appreciation of Taiwan dollar and the high domestic labor costs attracted rural labor to outflow for employment, resulting in a rapid aging of the rural population and a shortage of labor for tea production. The occupational structure has also changed relatively, and the agricultural population engaged in primary industry has declined rapidly. The development of industrialization has prompted tea industry to move toward the mechanized operation. At the same time, because industrial construction requires more land to build houses, the planting area of tea gardens has gradually been expropriated or sold and has dropped sharply.

To sum up, the attenuation factors of Taiwan's tea industry can be divided into few categories, such as aging of tea-making professionals, which has resulted in the withering of technology that cannot be passed on, and the industrial and commercial development has led to long-term labor shortages, which has reduced the area of tea gardens and tea production year by year, resulting in the failure of tea industry to reproduce its prosperity.

4.2 *Modern transformation opportunities and paths*

4.2.1 *"Fu Yuan Tea Manufactory" joins hands with "Chai Li Won" to open a tourist factory*

In the advertisement for "The Good Tea Tour of Chai Li Won's Japanese-style sugar-free green tea", the female main character has a solo trip to explore tea culture. The picture shows the misty and rainy tea garden style and the splendid era of the century-old tea factory, which has aroused the audience's yearning for it. The location of the advertisement is the"Fu Yuan Tea Manufactory" located in Longtan District.

Therefore, "Fu Yuan Tea Manufactory" gained the opportunity to start a cooperative relationship with Uni-President Enterprises Corporation in 2001, and became the tea raw material supplier of "Chai Li Won's Japanese-style sugar-free green tea". With the spread of advertisements, the words "the sweet aftertaste is like freshly brewed" on the physical beverage bottle are naturally imprinted in consumers' minds; and "follow the Chai Li Won's footsteps to taste good tea with a sweet aftertaste" at the end of the advertisement has successfully created and combined Chai Li Won's marketing appeal.

"Fu Yuan Tea Manufactory" took this as an opportunity to combine its own tea industry with tourism, and transformed it into a tourist factory. It features the various century-old tea-making machines that are rare but still in use in Taiwan's tea factories, combined with 5 major tea culture experience camps, including guided tours (exploring century-old tea factories), grinding tea powder(stone mortar grinding tea, fine tea fragrance), tea picking (LOHAS village tea-picking fun), kneading black tea (traditional hand kneading black tea) and kneading green tea (traditional hand kneading green tea), providing consumers with an overview of the daily leisure experience of a century-old tea factory.

Figure 4. The transformation of "Fu Yuan Tea Manufactory" into a tourist factory.

4.2.2 *"Fuxing Tea Culture Center" is registered as a historical building, and Oriental Beauty has cross-domain research and development*

"Fuxing Tea factory" was closed in 1980 due to the decline of Taiwan's tea industry. After being designated as a historic building by Hsinchu County Government in 2006, it took two years to restore it under the subsidy of the Hakka Committee and Hsinchu County Government, and it was put into trial operation in 2007. In 2009, "Fuxing Tea factory" was chosen by Uni-President Enterprises Corporation and became the shooting scene of the advertisement of "Chai Li

Won's Baihao Oolong Tea (Oriental Beauty)", attracting many people to make a pilgrimage. In 2011, it was officially opened to the public as "Fuxing Tea Culture Center", and its operating focus has been transformed into providing tea culture tours and tea sensory experiences.

The fourth-generation Zeng Xianghong actively tried innovative research and development of Oriental Beauty Tea. For example, he cooperated with the catering industry to launch a co-branded cold-brewed tea drink, which was sold in department stores. In the near future, it is scheduled to cooperate with creative cafes across domains. In summer, innovative foods such as Oriental Beauty Tea Ice Cream and Oriental Beauty Tea Popcorn will be launched. Zeng Xianghong also teaches at high school. In the multi-flexible curriculum of the school, he teaches tea culture and tea promotion, as well as tea practical classes.

Figure 5. The transformation of Fuxing Tea Culture Center toward promoting tea culture tours and tea sensory experience.

4.2.3 *"Yiming Tea Garden" Landscape Restaurant "serves dishes with tea", the Zen style is integrated into this creative tea culture*

The global environment is abnormal, the temporality of climate and terroir has lost its original normal law, environmental protection and health-oriented policies have been implemented, and the slogan of "non-toxic, organic" has been emphasized and advocated. After obtaining the certification of organic tea garden and its traceability, and in line with the laws of nature and solar terms, "Yiming Tea Garden" adopts the "natural agriculture" to cultivate the tea garden, and launches experience activities such as tea taste appreciation, tea culture experience, and tea table aesthetics to create a brand spirit with respect for the soil.

In the early days, the operation mode of tea factory was simply selling fresh tea leaves. Whenever foreign and domestic tea merchants come to tea garden for tea wholesale, they need to receive and greet customers to eat at the restaurant outside. At the same time, they must pay attention to the sun and weather to avoid damage to the quality of the tea. The round-trip reception process takes a lot of time, so the third-generation Lin Yuping has started to sort out the idle old space, combined with tree planting, and built a glass house landscape restaurant featuring natural ecology, which serves diverse dishes with tea to attract visitors.

Lin Yuping has kept pace with the times and integrated the image of "Zen style" into tea products launched on her own. Through combination of life aesthetics and creative packaging, for example: "Tea Fragrance Postcards" are made of Taiwanese fruits and flowers and other crops combined with portable tea cards. "Xi-Cha-Qu" is an environmentally friendly portable bottle developed by combining the craft of rush weaving in Yuanli Town. Among them, the "butterfly dance flowers speak" series launched in 2013 not only passed the "Miaoli County Creative Product Certification" by the Miaoli County Government Culture and Tourism Bureau but also won the Gold Award in the World Green Tea Competition in the same year.

Figure 6. Yiming tea garden tea market sells creative tea products developed year by year.

5 CONCLUSION

This research focuses on the representative "tea industry" among the Hakka Mountain Forest Economy industries. Based on literature research, in-depth interviews are conducted in Hakka Villages (in Taoyuan, Hsinchu and Miaoli) along Taiwan Romantic Route 3 road exploring the historical context of the tea industry. The data were collected based on the experience provided by the research subjects. The conclusions of this study are as follows.

(1) The origin and transition of Taiwan's tea industry
Taiwan's oolong tea was introduced from Fujian at the end of the 18th century. During the Qing Dynasty, that is, in the 1860s, British merchants began to export it. During the Japanese occupation, industrialization continued. In the 1970s, its development was interrupted by the energy crisis, but it turned out to develop the domestic market. Taiwan's black tea originated from the colonial period during the Japanese occupation, and began to be exported to Russia in 1906, and it was not gradually stopped until Russia's October Revolution in 1917. In 1925, the large-leaf species of Assam black tea was re-introduced, entering the golden age of exporting Taiwan's black tea until 1941, when the Second World War caused export interruption. After the war, the National Government of the Republic of China promoted green tea to be the mainstream in the 1960s, while black tea gradually declined to the bottom. It was not until the rise of bubble tea in the 1990s and TTES No. 18 (Ruby) after the disaster in 1999 that it was revived.

(2) Attenuation Factors in Tea Industry
The tea industry used to be called the "green economy industry". As Taiwan went through the

Qing Dynasty, the Japanese occupation and the Taiwan Restoration period, it faced transitions and reforms under different backgrounds. What has not changed is the important economic role and positioning of "tea" in Taiwan's historical development. "Tea" has created foreign exchange profits for the domestic market and is the number one cash crop for export, driving Taiwan's economic development with its every move. Later, it was replaced due to industrial and commercial development, and tea industry faced the dilemma of a sharp decrease in the areas of tea garden and tea production, failure of tea makers' skills passing on to next generations, and a long-term lack of labor in the traditional tea industry, which led to decline of the tea industry.

(3) Modern Transformation Opportunities and Paths in Tea Industry
With the growth of the national economy and the establishment of the "five-day work weeks" system, the rising awareness of the domestic creative industry, the growing popularity of the experience economy, the advocacy of environmental protection and sustainability issues, and the promotion of healthy diet concept, the tea industry has once again received attention and presented an opportunity for transformation. Taoyuan's "Fu Yuan Tea Manufactory" cooperated with Uni-President Enterprises Corporation to set up a tourist factory, launched a diverse tea culture experience camp, and became a stable supplier of green tea raw materials. Hsinchu's "Fuxing Tea Culture Center" was designated as a historical building, and transformed into a local cultural center that provides tea culture tours and tea experience, and launched the newly developed Oriental Beauty Tea Food, actively conducting cross-regional cooperation. Miaoli's "Yiming Tea Garden" initially created a landscape restaurant to meet the needs of tea merchants from all over the world. It launched a healthy and creative tea meal that "serves dishes with tea". Combined with courses related to tea aesthetics, it has launched creative tea products that are also fashionable year by year. Through the implementation of national policies, the cooperative operation of the public sector, and the cooperation of private enterprises, tea factory operators formulate their own strategies and choose an appropriate path for gradual transformation and upgrading.

Taiwan has a Hakka romantic avenue with rich history, cultural customs, and rich industries – the "Taiwan Romantic Route 3" road, also known as the Hakka Roaming Corridor. Thanks to the combined efforts of the government and the public, many people have come to know Hakka because of the "Taiwan Romantic Route 3" road. Recently, the Hakka Committee launched the period drama "Gold Leaf", which recreated the golden years of tea in Taiwan's glorious era, which attracted the attention of Taiwanese people and set off a wave of Hakka culture. In addition to watching the drama, people also visited the Hakka villages to promote economic benefits. Step on the "Taiwan Romantic Route 3" road to experience the Hakka style rich in history and culture, visit tea factory to experience tea's common culture, and feel the "Golden Years of Tea" at that time in Taiwan.

REFERENCES

Chang, M.H. 1993. *The expansion and achievements of Taiwan's tea industry in the past 300 years.* Taipei Historical Documents Quarterly. 103, 21–43. Taipei City Archives. [in Chinese, semantic translation]

Chang, T.Y., Feng, H.Y. 2020. Review and Analysis of Research on Hakka Industrial Economy Research in Tao-Chu-Miao Area. *Journal of National United University.* 17(2), 41–55. National United University. [in Chinese, semantic translation]

Chiou, N.C. 2005. *Tea Industry in Taiwan: Evolution and Predicament* (Unpublished Master's Thesis). National Tsing Hua University, Hsinchu, Taiwan. [in Chinese, semantic translation]

Hsu, Y.S. 1996. *Journal of Tea Research and Extension Station, Council of Agriculture, Executive Yuan. Taoyuan: Tea Research and Extension Station*, Council of Agriculture. [in Chinese, semantic translation]

Hsu, H.C. 2011. *Operational personality of tea industry among northern Hakka of Taiwan after the Second World War* (Unpublished Master's Thesis). National Central University, Taoyuan, Taiwan. [in Chinese, semantic translation]

Hwang, W.L. (2022, March 27). Interview by P.J. Lai. Interviewer [Tape recording]. *The Golden Leaf Era: The Change, Decline and Transformation of the Tea Industry-Taking Taoyuan, Hsinchu and Miaoli Tea Factory as Examples*, Taoyuan.

Jiang, D.Z. 1960. *History of Taiwan Tea Trade.* 20(12), 14–15. China Magazine Publishing. [in Chinese, semantic translation]

Liao, J.H. 2011. *The Transition of Tea Industry in Taiwan-Construction of Potential Indicator for the Tourism of Tea Factory* (Unpublished Master's Thesis). National Kaohsiung University of Hospitality and Tourism, Kaohsiung, Taiwan. [in Chinese, semantic translation]

Lin, Y.P. (2022, March 26). Interview by P.J. Lai. Interviewer [Tape recording]. *The Golden Leaf Era: The Change, Decline and Transformation of the Tea Industry-Taking Taoyuan, Hsinchu and Miaoli Tea Factory as Examples*, Miaoli.

Liu, H.C. 2017. *Perspective on HAKKA ROMANTIC AVENUE: images, gender, and culture governance* (Unpublished doctoral dissertation). National Dong Hwa University, Hualien, Taiwan. [in Chinese, semantic translation]

Tang, J.H. 2019. *The history of Taiwan tea.* C^3*offee*. 19. New Taipei City. [in Chinese, semantic translation]

Wu, S.C. 2007. *The developments and changes of the tea industry in Taiwan after the Second World War* (Unpublished Master's Thesis). National Central University, Taoyuan, Taiwan. [in Chinese, semantic translation]

Zeng, Z.Z., Zeng, H.H. (2022, March 27). Interview by P. J. Lai. Interviewer [Tape recording]. *The Golden Leaf Era: The Change, Decline and Transformation of the Tea Industry-Taking Taoyuan, Hsinchu and Miaoli Tea Factory as Examples*, Hsinchu.

System Innovation for a Troubled World – Tak Lam et al. (Eds)

Dynamic response analysis of inter-array power cables for a semi-submersible floating offshore wind turbine

Shun-Han Yang*, Yu-Jia Chen, Jun-Hui Huang & Po-Yi Lin
Department of Engineering Science and Ocean Engineering, National Taiwan University, Taiwan

ABSTRACT: The objective of the study is to investigate the dynamic responses of power cables for a floating offshore wind turbine (FOWT). The case study FOWT consists of a reference 15MW wind turbine and a reference semi-submersible platform, proposed by the International Energy Agency and University of Maine. Inter-array power cables are critical for designing offshore renewable energy devices while subjected to various limitations and constraints. Our work provides a numerical simulation approach to quickly determine and analyse the dynamic motion behaviour of umbilical power cables with respect to the floating platform and its mooring systems. In this study, two configurations of the inter-array 66kV power cable, including a catenary shape and a lazy wave shape, are designed and compared. The power cable's characteristics are evaluated using a numerical model subjected to various loading conditions consisting of ocean waves, winds, and currents. The numerical model is simulated using a time-domain coupled dynamic analysis, considering the aero-hydro-servo-elastic behaviours of the floating offshore wind turbine. The effect of environmental loading conditions on the dynamic motions and mechanical characteristics of the power cables are investigated. Factors that are important for the structural integrity of the power cables are identified and discussed.

Keywords: umbilical power cable, inter-array power cable, dynamic response analysis of power cable, metocean environment loads.

1 INTRODUCTION

Offshore renewable energy sources have been the top focused area for reaching the carbon net zero goal. Despite existing experiences with the commercially operating onshore wind turbines, their offshore counterparts pose much more engineering and scientific challenges. However, to obtain more energy, one must reach deep into the heart of the ocean. Numerous designs and concepts have been proposed and made into prototypes on offshore wind platforms. One of the mission critical components found in all offshore energy harvesting systems is the umbilical power cable connecting the actual device and the rest of power transmission systems. This is the component that requires the highest possible integrity throughout the entire system to ensure uninterrupted power harvesting.

2 UMBILICAL POWER CABLES

The overarching objective of this paper is to investigate different metocean loads and their impact on different power cable configurations. We have created two power cable configurations: the catenary, also known as free-hanging, and the lazy wave. The cable profile is shown in Figure 1 together with the three major metocean loads simulated: wind, wave, and current. Figure 2 shows six common cable configurations based on experiences from the oil and gas industry, in which cable configurations can also be referred to as riser configurations. An overview and comparison of different variants is presented in Karegar (2013).

Figure 1. Cable profile of two modelled configurations: catenary and lazy wave.

Despite the relatively short history of full-scale development on umbilical power cables, a few literatures have been targeting the dynamic umbilical power

*Corresponding Author: hedyshyang@ntu.edu.tw

DOI 10.1201/9781003377399-25

cables and their simulations throughout past years thanks to its imperative importance and ubiquitous needs in all floating energy devices.

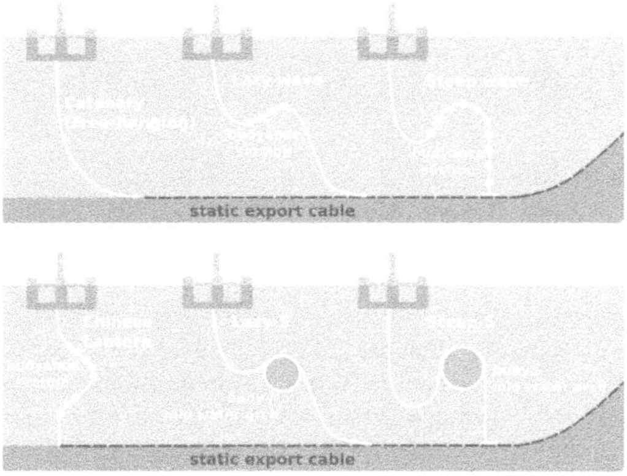

Figure 2. Six cable profiles of common cable configurations.

Rentschler et al. (2019) presented a dynamic analysis on two selected umbilical configurations, catenary and lazy wave, using a genetic algorithm and a hybrid MATLAB-OrcaFlex simulation environment to calculate the fatigue life of the power cables under the environment loads of wave and wind. For their genetic algorithm, they have proposed a fitness scoring function to evaluate the behavioural differences between different cable configurations. The proposed fitness function is as following:

$$\text{fitness} = \frac{\rho_{\max}}{\text{MAC}} + \frac{T_{\max}}{\text{MBL}}$$

where $\rho_{\max}$ refers to the cable maximum curvature, $T_{\max}$ refers to the cable maximum effective tension. These two terms are then normalized by Maximum Allowable Curvature (MAC) and Minimum Breaking Load (MBL) respectively to sum up to a fitness score. While this fitness score is not directly applicable to our evaluation in present paper, it provides comparable metrics to compare our simulation results with their results. Following up this paper, Rentschler et al. (2020) presented another parametric study on varying water depths. In both of their papers' concluding remarks, the need of considering current loads has been mentioned as important future work.

Thies et al. (2012) also presented OrcaFlex based dynamic cable simulations for both wave energy converters (WECs) and offshore floating wind platforms. The authors also compared different cable configurations and analysed different load distributions based on different cable configurations (Thies et al. 2017). It is worth noting that the cable property presented in the aforementioned paper has been followed as a reference cable property by later papers, including present study.

Hall et al. (2021) presented a dynamic cable analysis using an open-source simulation software alternative: MoorDyn. They compared the dynamic results against OrcaFlex based on a lazy wave cable configuration attached to a WEC device.

3 METHODOLOGY

3.1 *Power cable properties*

Table 1 shows basic mechanical properties of the 66kV power cable simulated in present study. Recent study suggested the transition towards 66kV AC power cable as a more cost-effective alternative and it has been adopted in most recently proposed offshore floating wind designs (Ferguson et al., 2012). Our power cable properties are adopted from Theis et al. (2017) and the same cable properties have been used in most of the aforementioned studies.

Table 1. Basic mechanical properties of simulated power cable.

	2-Armor cable	Buoyancy section
Outside diameter [mm]	200	306
Nominal weight in air [N/m]	706	423
Nominal weight in seawater [N/m]	390	−316
Bending stiffness [kNm^2]	10	10
Axial stiffness [MN]	700	700
Torsional stiffness [kN m^2/°]	600	600
Minimum Breaking Load [kN]	100	100
Minimum bending radius [m]	2	2

3.2 *Simulation setups*

In this subsection we describe important details of our simulation setup including cable configurations, floating platform configurations, metocean loads and simulated cases.

Floating wind turbine system and cable configurations

The IEA 15 MW wind turbine (Gaertner et al. 2020) and the UMaine VolturnUS-S semi-submersible platform (Allen et al. 2020) are used together as the case study floating offshore wind turbine (FOWT) system in present study. Because the present study aims to design power cable systems for a site condition suitable for Taiwan strait, a shallower water depth is considered and is set to 70 m. A three-leg catenary configuration, using the R3 studless mooring chain, is defined. Each mooring line is set to be 445.4 m in length and anchored to the seabed with a radius of 490 m. Except of the mooring line length and the anchor radial spacing, the readers are referred to Gaertner et al. (2020) for the properties of the wind turbine and Allen et al. (2020) for the properties of the semi-submersible platform, respectively.

Figure 3 presents the isotropic view of the FOWT system. The origin of the coordinate is defined at the geometric centre of the waterplane of the platform under the still water condition. Irrespective of the type of the cable configuration, the cable is connected to the platform at the position of (0, 0, −20) m and anchored to the seabed at the position of (140, 0, −70) m. The length of the power cable is set to 170 in length for the catenary configuration. For the lazy-wave configuration, the power cable is set to be 69, 69, and 39 m from the segments 1 to 3, where the segment 1 is the one connected to the platform.

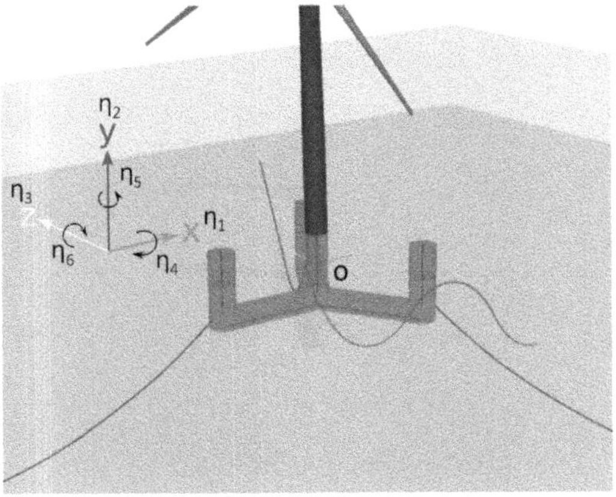

Figure 3. DNV SIMA model of the IEA 15MW semi-submersible platform with mooring system and umbilical power cable (shown in red) in lazy wave configuration.

Metocean loads

Metocean loads, current, wave, and wind, are individually simulated for our study. Case C1 is configured with minimum current speed allowed at 0.001 m/s with no wind or wave load, simulating near absence of environment loads. For the case of wind load, a wave is added as necessary environmental load in conjunction with the wind, the added wave is configured at Tp = 5.5 s and Hs = 1.5 m. The direction of the added wave is configured to be in the same direction of incoming wind. X indicates catenary configuration failed to converge during simulation whereas X indicates lazy wave configuration failed to converge. All the cases are simulated with 1400 seconds simulation time with the first 200-second results are removed from the data analysis to avoid the transient effect.

Table 2. List of cases of the simulated **current** loads.

Case	Current direction	Current speed [m/s]
C1	0°	0.001
C2	0°	0.500 X
C3	0°	1.0 X
C4	0°	1.5 X X
C5	180°	0.5
C6	180°	1.0
C7	180°	1.5

Table 3. List of cases of the simulated **wave** loads.

Case	Wave direction	Hs [m]	Tp [s]
W1	0°	1.5	5.5
W2	0°	3.5	7.5
W3	0°	7.5	1
W4	60°	1.5	5.5
W5	60°	3.5	7.5
W6	60°	7.5	1

Table 4. List of cases of the simulated **wind** + **wave** loads.

Case	Wave (direction/Hs/Tp)	Wind direction	Wind speed [m/s]
WW1	0°/1.5/5.5	0°	3
WW2	0°/1.5/5.5	0°	10
WW3	0°/1.5/5.5	0°	16
WW4	60°/1.5/5.5	60°	3
WW5	60°/1.5/5.5	60°	10
WW6	60°/1.5/5.5	60°	16

4 SIMULATION RESULTS

4.1 *Cable dynamic under current load*

Figure 4 shows two vertical cable profiles under lazy wave configuration, with current speed at 0.5 m/s and 1 m/s, corresponding to case C2 and C3 respectively. While the cable profile and tensile snapshot of C2 (solid blue line in the figure) matches literature simulations using OrcaFlex, the tensile snapshot of C3 illustrates a new loading issue never reported before. The power cable in C3 is completely adrift due to the

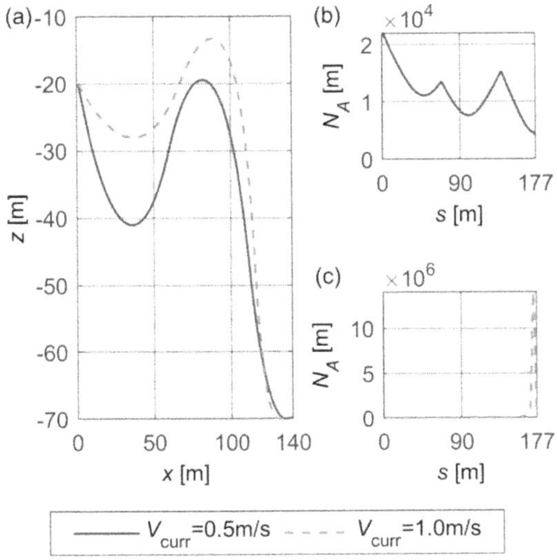

Figure 4. (a) Lazy wave cable profile under two different current speeds corresponding to case C2 and C3. (b)(c) show the tension snapshot over cable length of (a) respectively.

combination of added buoyance section and strong current, leaving extremely biased tensile distribution along the cable length.

Figures 5–7 showcase cable dynamics under three different current loads C1, C3, and C7. In case C3, the catenary configuration is absent due to simulation divergence. We suspect this is due to the non-compressible criteria enforced internally of the simulation tool we used. When comparing case C1 and C7, one can observe that the current direction can have significant impact on dynamic responses of the power cable. With case C1 a near static current speed of 0.001 m/s is the only environment load to the platform but the displacement in η_1 already matches the displacement found in case C7, a strong but opposite direction current load.

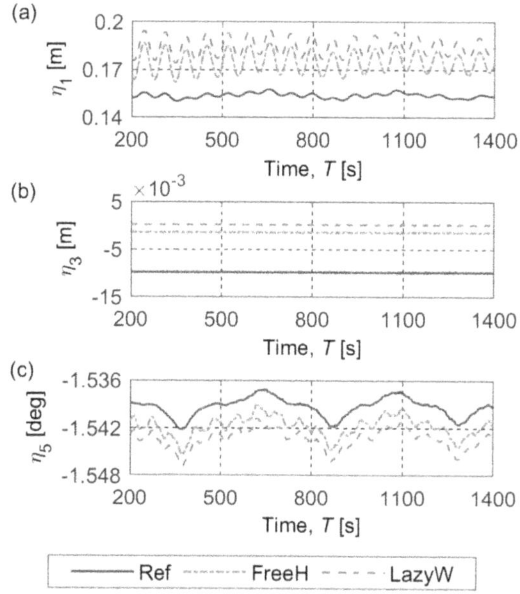

Figure 5. Three cable configurations under current load case C1. (See Figure 3 for corresponding η definitions).

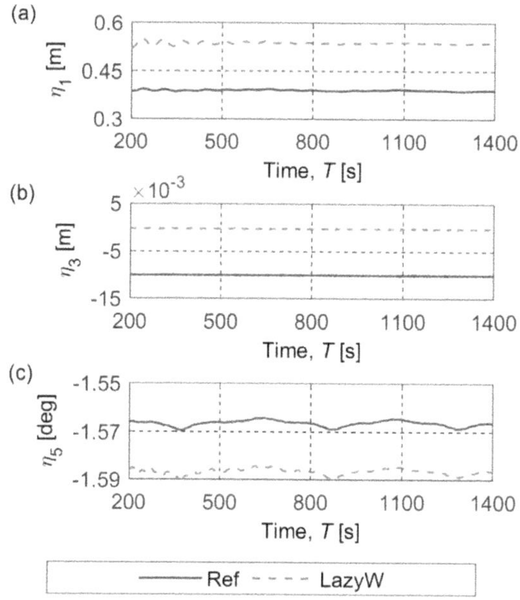

Figure 6. Three cable configurations under current load case C3. (See Figure 3 for corresponding η definitions).

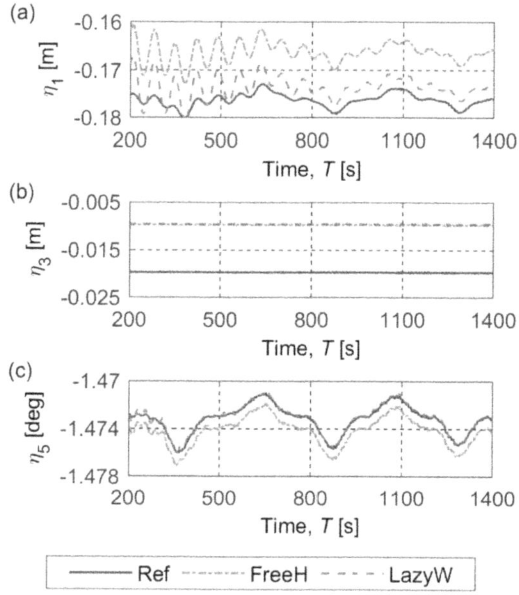

Figure 7. Three cable configurations under current load case C7. (See Figure 3 for corresponding η definitions).

4.2 *Comparing cable dynamic under wave load and wind + wave load*

Figures 8–9 showcase mechanical properties representing cable structural integrity under wave and wind + wave loads. While most of other mechanical properties including curvature (κ_y, κ_z), tensile force (N_A), and total stress (σ_{total}) remains comparable between wave versus wind + wave, the overall displacement Δ_s shows significant increases due to wind + wave load. In our future work we intend to continue investigating the overall structural integrity impact of the displacement and if this would be an important factor to evaluate the fitness of the given cable configuration.

5 SUMMARY AND DISCUSSION

This paper presents a case study on metocean environmental loads on the dynamic responses of the umbilical power cable used in floating offshore wind turbines (FOWTs). To the best of our knowledge, this study presents the first metocean load on power cable with a 15MW floating platform for offshore wind farm. Three different metocean loads are simulated, current, wave, and wind, mimicking the weather condition of our operation site. Two different cable configurations, catenary and lazy wave, are modelled and simulated for 1400 seconds simulation time and the dynamic responses are presented.

The design of dynamic power cable is often considered to bring only small system behavioural changes in comparison to the overall floater (Thies et al. 2019). With our simulated result, we show that when considering current load impact to the overall FOWT platform, different cable configurations can cause platform motion when considering *different current directions*.

The presented study considers site-specific metocean loads, current, wave, and wind, in isolation. Whilst this approach provides a clear overview of different metocean load components causing different motion impacts, our future work aims to simulate combined metocean loads thus being able to provide a realistic metocean simulation environment for further evaluations. On top of a mixed metocean load, our future work also aims to include fatigue life analysis to evaluate the overall structural integrity over the designed operation timeframe of the FOWT.

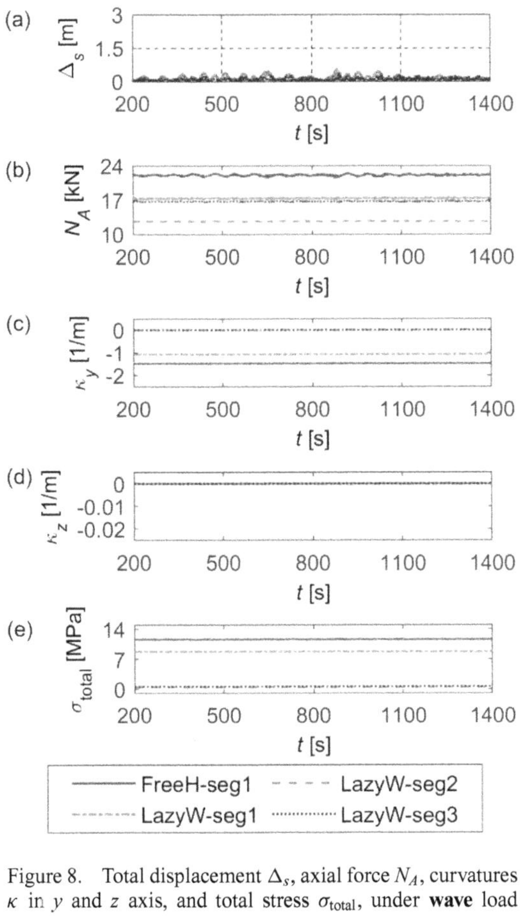

Figure 8. Total displacement Δ_s, axial force N_A, curvatures κ in y and z axis, and total stress σ_{total}, under **wave** load case W1.

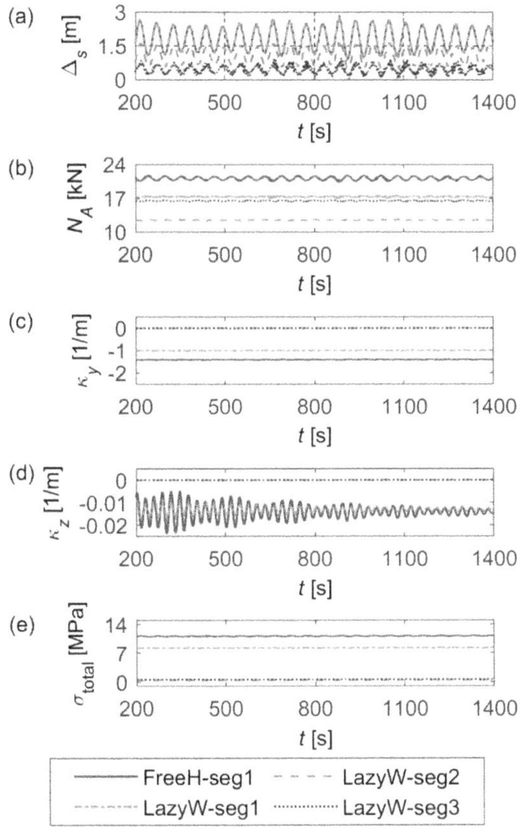

Figure 9. Total displacement Δ_s, axial force N_A, curvatures κ in y and z axis, and total stress σ_{total}, under **wind + wave** load case WW3.

REFERENCES

Allen, C., Viscelli, A., Dagher, H., Goupee, A., Gaertner, E., Abbas, N., Hall, M., & Barter, G. 2020. Definition of the UMaine VolturnUS-S reference platform developed for the IEA Wind 15-megawatt offshore reference wind turbine. No. NREL/TP-5000-76773. National Renewable Energy Lab. (NREL)

Blanco, M.I. 2009. The economics of wind energy. *Renewable and Sustainable Energy Reviews*, 13(6), 1372–1382.

Ferguson, A., de Villiers, P., Fitzgerald, B., & Matthiesen, J. 2012. Benefits in moving the inter-array voltage from 33 kV to 66 kV AC for large offshore wind farms. *EWEA 2012*.

Gaertner, E., Rinker, J., Sethuraman, L., Zahle, F., Anderson, B., Barter, G.E., Abbas, N.J., Meng, F., Bortolotti, P., & Skrzypinski, W. 2020. IEA wind TCP task 37: definition of the IEA 15-megawatt offshore reference wind turbine. No. NREL/TP-5000-75698. National Renewable Energy Lab. (NREL)

Hall, M., Sirnivas, S., & Yu, Y.-H. 2021. *Implementation and Verification of Cable Bending Stiffness in MoorDyn.* Proceedings of the ASME 2021 3rd International Offshore Wind Technical Conference. ASME 2021 3rd International Offshore Wind Technical Conference, Virtual, Online.

Karegar, S. 2013. *Flexible riser global analysis for very shallow water.* University of Stavanger, Norway.

Rentschler, M.U.T., Adam, F., & Chainho, P. 2019. Design optimization of dynamic inter-array cable systems for floating offshore wind turbines. *Renewable and Sustainable Energy Reviews*, 111, 622–635.

Rentschler, M.U.T., Adam, F., Chainho, P., Krügel, K., & Vicente, P.C. 2020. Parametric study of dynamic inter-array cable systems for floating offshore wind turbines. *Marine Systems & Ocean Technology*, 15(1), 16–25.

Sobhaniasl, M., Petrini, F., Karimirad, M., & Bontempi, F. 2020. Fatigue Life Assessment for Power Cables in Floating Offshore Wind Turbines. *Energies, 13*(12).

Thies, P., Harrold, M., Johanning, L., Grivas, K., & Georgallis, G. 2019. *Load and fatigue evaluation for 66kV floating offshore wind submarine dynamic power cable.* Jicable International Conference on Insulated Power Cables, France.

Thies, P.R., Johanning, L., & Dobral, C. 2017. *Parametric Sensitivity Study of Submarine Power Cable Design for Marine Renewable Energy Applications.* Paper presented at the Proceedings of the ASME 2017 36th International Conference on Ocean, Offshore and Arctic Engineering, Norway.

Thies, P.R., Johanning, L., & Smith, G.H. 2011. Assessing mechanical loading regimes and fatigue life of marine power cables in marine energy applications. *Proceedings of the Institution of Mechanical Engineers, Part O: Journal of Risk and Reliability*, 226(1), 18–32.

System Innovation for a Troubled World – Tak Lam et al. (Eds)

A container-based mixed-criticality real-time system platform

Jun Wu* & Sheng-Kai Hsu
Department of Computer Science and Information Engineering, National Pingtung University, Pingtung, Taiwan

ABSTRACT: In this paper, a container-based mixed-criticality real-time system (ComityRT) platform is presented to help developers to build a MCRTS faster and easier. ComityRT supports real-time tasks with different degrees of timeliness and criticality to be deployed and executed on containers running in a multicore environment. It can be configured according to developers' preferences, e.g., a container can be created for a set of tasks with the same timing requirement and/or criticality level in a specific processing core or across multiple cores. Furthermore, task migration among different containers and cores is allowed, and the computing resources (such as CPU and memory) can be managed for each container statically or dynamically. This paper also includes an MCRTS example to show the capabilities and benefits of CommunityRT.

1 INTRODUCTION

Since 2007, Steve Vestal proposed the well-known critical-dependent task model (Vestal, 2007), the research of mixed-criticality real-time systems (MCRTS) has received much attention in many safety critical application domains, such as avionics, automotive, healthcare, and transportation. Similar to real-time systems, the research of MCRTS also focuses on theoretical task scheduling and schedulability analysis in the early stage. In the past decade, many excellent scheduling approaches have been proposed for MCRTS (a complete survey can be found in Burns and Davis (2018)), while comparatively little work is done on system implementation (such as Cinque and Tommasi (2017), Cinque et al (2019), Anderson et al (2009), Kim et al (2014)).

In this paper, we are interested in the implementation of a platform, called a container-based mixed-criticality real-time system (ComityRT), to help developers to build MCRTS. ComityRT is also a testbed for theoretical scheduling approaches so that the correctness and the performance of existing and new scheduling approaches can be evaluated and compared. Based on container technology, ComityRT is capable to isolate the execution of real-time tasks with different degrees of timeliness as well as criticalities by separating them into different containers in a multicore environment. Developers can configure ComityRT according to their preferences and/or requirements such that specified tasks can be assigned to a container (it can be configured to operate in a specific core or across multiple cores) with predefined resource allocations (e.g., number and utilization of cores, amount of memory, etc.). For instance, developers can configure ComityRT to assign tasks with higher and lower criticality levels to containers with more and less allocated resources, respectively, so that the high-criticality tasks might have more chances to meet their timing constraints. Furthermore, task migration among different containers (including the containers operated in different cores) is allowed; and the allocated resources of containers can be adjusted dynamically at the run-time to meet the requirements of unpredictable workload of tasks (e.g., aperiodic tasks).

ComityRT is an ongoing project, and its current progress is reported in this paper, which mainly focuses on the task separation based on Docker containers (while the support of containers with a patched real-time Linux kernel will be the further work). At the end of this paper, an MCRTS example will be presented so that the functionalities and the benefits of ComityRT can be demonstrated for which we have some encouraging results.

2 SYSTEM AND TASK MODELS

We consider an MCRTS running on a multicore processor. The processor $\mathcal{P}$ has M cores $\{core_1, core_2, \ldots, core_M\}$ and the system has L criticality levels $\mathcal{L} = \{l_1, l_2, \ldots, l_{\mathcal{L}}\}$, where $l_1 < l_2 < \ldots < l_{\mathcal{L}}$ (in other words, $\forall l_x > l_y$ if $x > y$). At the run-time, an MCRTS can be operated at (and switched to) one of its criticality levels. Typically, it starts at its lowest criticality level (i.e., l_1), and it switches to a higher level when a predefined condition or event occurs. For ease of presentation, we define the current criticality level of the system as l^*.

*Corresponding Author: junwu@mail.nptu.edu.tw

DOI 10.1201/9781003377399-26

A set of n mixed-criticality real-time tasks $\mathcal{T} = \{\tau_1, \tau_2, \ldots, \tau_n\}$ is considered for an MCRTS. Each task $\tau_i \in \mathcal{T}$ is defined by a tuple $(A_i \vec{T}_i, \vec{C}_i, \vec{D}_i, L_i)$ where A_i, $\vec{T}_i$, $\vec{C}_i$, $\vec{D}_i$ and L_i are the arrival time, period, worst-case computation time, relative deadline, and criticality level of task τ_i respectively. Based on Vestal's work, we assume that a higher-criticality task might has a longer computation time, a longer deadline, and a shorter period (Vestal, 2007, Burns & Davis, 2018). Therefore, $\vec{T}_i$, $\vec{C}_i$ and $\vec{D}_i$ are vectors of values, one per criticality level, with the following constraints:

$$
\begin{aligned}
l_x > l_y \Rightarrow\ & T_i\,(l_x) \leq T_i(l_y) \\
& C_i(l_x) \geq C_i(l_y) \\
& D_i(l_x) \geq D_i(l_y)
\end{aligned}
$$

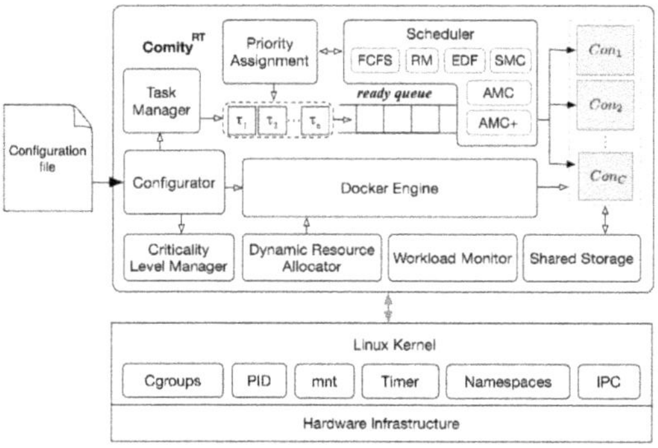

Figure 1. The architecture of ComityRT.

We also consider tasks are well-formed, i.e., $C_i(l_x) \leq D_i(l_x) \leq T_i(l_x)$, $\forall \tau_i \in \mathcal{T}, l_x \in \mathcal{L}$. Each task $\tau_i \in \mathcal{T}$ can be considered as a template of its instance and each instance will be instantiated for every request of the task. The first request of τ_i arrives at time A_i and the reminder requests will arrive regularly for every period $T_i(l^*)$ where l^* is the current criticality level of the system at the run-time. Let $\tau_{i,j}$ and $A_{i,j}$ denote the jth instance of task τ_i and its arrival time (i.e., the instantiate time), respectively. The worst-case execution time of each task instance $\tau_{i,j}$ is no more than $C_i(l^*)$. We say a task instance $\tau_{i,j}$ meets its deadline requirement if it is completed at or before its deadline (i.e., $A_{i,j} + D_i(l^*)$); otherwise, it misses the deadline.

The major objective of the scheduling of mixed-criticality tasks is to assure to criticality level L_i that τ_i won't miss a deadline. In particular, the tasks must meet their deadlines if their criticality levels are no less than the system's current criticality level l^* such that the objective is to obtain. In other words, the execution of a task τ_i can be omitted when l^* is higher than its criticality L_i.

3 A CONTAINER-BASED MCRTS PLATFORM

3.1 *Architecture*

The architecture of ComityRT is shown in Figure 1. At the current stage, ComityRT is built upon Linux (kernel version 3.0 or above) and it works with some kernel components, such as Cgourps, PID, mnt, Timer, Namespaces, and IPC), to accomplish task separation as well as resource (e.g., CPU and memory) management for mixed-criticality real-time tasks. As shown in Figure 1, ComityRT also works with Docker engine 1.3 so that multiple containers can be created to execute tasks separately.

Developers can use ComityRT to build an MCRTS according to their preferences: it only needs to provide a *configuration file*. In particular, the number of containers with their allocated resources (including the processing cores and the amount of memory) can be defined in a configuration file and it can be loaded by the *Configurator* module. According to the configuration file, the *Configurator* module works with *Docker engine* to create a certain number of containers with predefined resource allocations in which a set of mixed-criticality real-time tasks can be executed. Note that the allocated resources of a container can be adjusted dynamically at the run-time by the *Dynamic Resource Allocator* module.

The *Task Manager* module is responsible for loading the definitions of tasks from a configuration file, and it will generate the task instances of the tasks, e.g., generate a task instance of a periodic real-time task regularly. The task set is then scheduled by the *Scheduler* module which supports FCFS, EDF (Liu & Layland, 1973), RM (Liu & Layland, 1973), SMC (Baruah & Vestal, 2008), AMC (Baruah et al., 2011) and AMC+ (Bate et al., 2015) at the current stage. Since most scheduling algorithms are priority-driven, the *Scheduler* module cooperates with the *Priority Assignment* module to assign priorities to tasks. Note that the priorities of tasks can be also adjusted dynamically by the *Priority Assignment* module at the run-time, which is necessary for dynamic priority scheduling algorithms, such as EDF.

The *Shared Storage* module is designed for inter-task communication, which is a memory/disk space that can be accessed by all task instances from all containers. Furthermore, at the run-time, the *Criticality Level Manager* module and the Workload Monitor module are responsible for switching the criticality level of the system and for providing the online status of all tasks.

3.2 *Configuration file*

The *configuration file* is used to define important parameters for a to-be-built MCRTS on ComityRT. The configuration file is a plain text file organized by sections which are SYSTEM, CRITICALITES, CONTAINERS, TASKS and TASK_ASSIGNMENTS sections. Please refer to https://resl.csie.nptu.edu.tw/projects/ComityRT for more details.

The SYSTEM section has 5 parameters: *Num_Cores*, *Used_Cores*, *Preemptible*, *Migratable*, and *Default_Sched* where *Num_Cores* is the number of cores in the system (i.e., M). Note that we also number the cores by 1, 2, ..., M (i.e., the core ID). *Used_Cores* is the number of cores that will be used in the to-be-built MCRTS. Note that *Used_Cores* can't exceed the value of *Num_Cores* (i.e., *Used_Cores* $\leq M$). Also note that *Used_Cores* is optional parameter. When *Used_Cores* is omitted, the default value is set as M. *Preemptible* is Boolean value that indicates the allowance of task preemption in the system. *Preemptible* and *Migratable* indicate that the task preemption and migration are allowed or not, respectively. In particular, tasks are allowed to migrate between containers when the value of *Migratable* is True. The last parameter, i.e., *Default_Sched*, is the default scheduling algorithm of tasks in the system.

The CRITICALITES section defines the available criticality levels in the system. Parameters *Num_Criticality_Levels* and *Init_Criticality_Level* are the number of available criticality levels and the initialized criticality level when the system starts. The details of each criticality level are defined in the subsection LEVEL. Parameters in LEVEL include *CLID*, *Name*, and *Weight*, where *CLID* and *Name* are the ID and name of the criticality level. The *Weight* of a criticality level reflects the criticality (i.e., the urgency or the importance). In particular, a level with a higher weight value is more critical compared to a level with a lower weight.

The CONTAINERS section defines the containers (i.e., Docker containers) in the system. Each container is defined in the CON subsection. A CON subsection has *COID*, *Name*, *Allocated_Core*, *Core_Utilization*, *Allocated_Memory*, and *Sched_Algorithm* parameters, where *COID* and *Name* (optional) are the ID and name of the container. In order to provide more flexibility to developers, ComityRT allows a container to running on one or more cores defined by the *Allocated_Core* parameter (a list of cores represented by their ID's). The utilization ratio of each core is defined in the *Core_Utilization* parameter (a list of ratios, one per core). Furthermore, *Allocated_Memory* and *Sched_Algorithm* are the sizes of memory allocated to the container and the scheduling algorithm for tasks in the container. Note that the default scheduling algorithm is determined by the *Default_Sched* when *Sched_Algorithm* is omitted, the default scheduling algorithm is determined by *Default_Sched* (defined in the SYSTEM section).

The TASKS section contains the definitions of tasks. The detail of a task is defined in a TASK subsection which contains *TID*, *Name*, *Arrival*, *Period*, *WCET*, *Deadline*, and *Criticality_Level*, where *TID* and *Name* are the ID and name of the task. The arrival time, period, worst-case execution time, deadline, and criticality level are defined in parameters *Arrival*, *Period*, *WCET*, *Deadline*, and *Criticality_Level*, respectively. Note that *Period*, *WCET*, *Deadline* are vectors of criticality levels, one per criticality level.

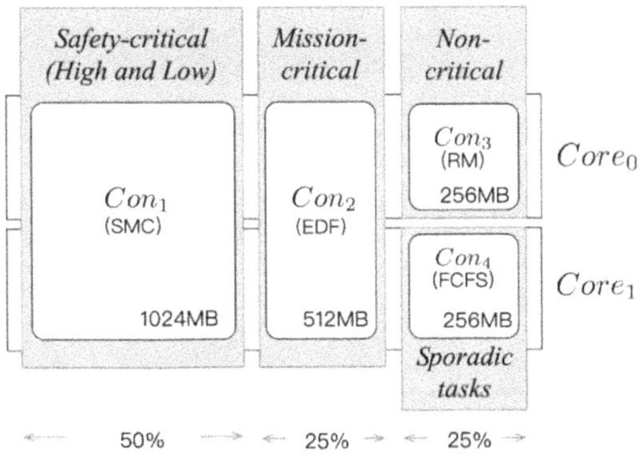

Figure 2. The container layout of the example MCRTS.

Finally, the TASK_ASSIGNMENT section defines the assignments of tasks to containers; each assignment is defined in an ASGN subsection which contains *Assigned_Tasks* and *Designated_Containers*. Where *Assigned_Tasks* are the tasks assigned to be executed on the designated containers (defined in *Designated_Containers*).

3.3 *Implementation*

Currently, the implementation of ComityRT is on an x86 platform (Intel Core i5-7500 quad-core processor with 4GB ram) with CentOS 7 (Linux kernel 3.10) and Docker 1.13. The modules and tools are built in Python and shell script. In particular, a universal timer is used to trigger the arrival of task instances (in the *Task Manager* module), to check the deadline expiration (in the *Task Manager* and *Workload Monitor* modules) and to perform related event handlers. The *Workload Monitor* module works with *proc* and Docker's commands (such as inspect and stats) to obtain the online status of task instances. For more details about the implementation, please refer to https://resl.csie.nptu.edu.tw/projects/ComityRT.

4 DEMONSTRATION

We built an MCRTS on ComityRT with a quad-core processor (Intel Core i5-7500) with 4GB RAM but we only use two cores and 2GB ram for simplification The container layout of the example MCRTS is shown in Figure 2. Specifically, there are 5 criticality levels are considered: *high safety-critical, low safety-critical, mission-critical, non-critical*, and *sporadic tasks* from the highest to the lowest. We build 4 containers for this MCRTS, where Con_1 is allocated on $Core_0$ and $Core_1$ with 1024MB ram, each has 50% utilization, Con_2 is also allocated on $Core_0$ and $Core_1$, with 512MB ram, each has 25% utilization, Con_3 is allocated on $Core_0$ (25% utilization and 256MB ram), and Con_4 is allocated on $Core_1$ (25% utilization and 256MB ram). The most critical tasks, i.e., high and low safety-critical tasks, are assigned to Con_1 and are scheduled by SMC algorithm while mission-critical tasks and noncritical tasks are assigned to Con_2 and Con_3, respectively. Note that they are scheduled by EDF and RM. Finally, the least critical tasks, i.e., sporadic tasks, are assigned to Con_4 (and are scheduled by FCFS). This example shows that ComityRT is capable to build an MCRTS. Moreover, the online monitoring tool provides great help to understand the online usage of cores, containers as well as the status of task instances. Figure 3 shows a screenshot of the online workload monitoring.

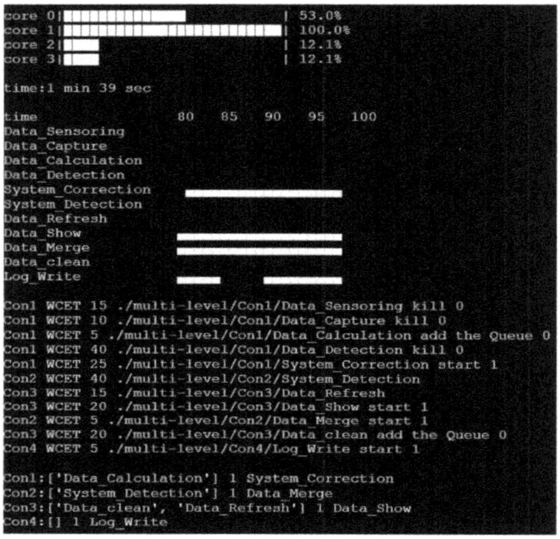

Figure 3. A screenshot of the example MCRTS.

5 CONCLUSION

In this paper, ComityRT is proposed to provide developers with a platform to build or test their own MCRTS systems. ComityRT can be configured based on developers' requirements, such as a container that can be created for a set of tasks with the same timing requirement and/or criticality level in a specific processing core or across multiple cores. ComityRT also supports task migration, and the computing resources (such as CPU utilization and memory) can be managed for each container statically or dynamically. The example proposed in this paper has demonstrated that ComityRT is capable to build and to run such a mixed-criticality real-time system. For future work, we will focus on providing more predefined scheduling algorithms and planning to support task synchronization.

ACKNOWLEDGEMENT

The authors gratefully acknowledge the support of the Ministry of Science and Technology (MOST) of Taiwan under the grants MOST-109-2221-E-153-002 and MOST-110-2221-E-153-001-MY3.

REFERENCES

Vestal, S. (2007). Preemptive scheduling of multi-criticality systems with varying degrees of execution time assurance. *Proceedings of the 28th IEEE International Real-Time Systems Symposium*, 239–243. https://doi.org/10.1109/RTSS.2007.47

Burns, A. & Davis, R. I. (2018), A survey of research into mixed criticality systems. *ACM Transactions on Embedded Computing Systems*, 50(6), Article 82, 1140–1152. https://doi.org/10.1145/3131347

Cinque M. & Tommasi, G. D. (2017), Work in progress: real-time containers for large-scale mixed-criticality systems, Proceedings of the 38th IEEE Real-Time Systems Symposium, https://doi.org/10.1109/RTSS.2017.00046

Cinque, M., Corte, R. D., Eliso, A. & Pechia, A. (2019), RT-CASEs: container-based virtualization for temporally separated mixed-criticality task sets, Proc. of the 31st Euromicro Conference on Real-Time Systems, Article 5, 5:1–5:22.

Anderson, J., Baruah, S. & Brandenburg, B. (2009), Multicore operating-system support for mixed criticality, *Proceedings of the Workshop on Mixed Criticality: Roadmap to Evolving UAV Certification.*

Kim, K., Erickson, J. & Anderson, J. (2014), Mixed-criticality on multicore (MC^2): a status report, *Proceedings of the 10th International Workshop on Operating Systems Platforms for Embedded Real-Time Applications*, 45–50.

Liu, C. L. & Layland, J. W. (1973), Scheduling algorithms for multiprogramming in a hard real-time environment, *Journal of the Association for Computing Machinery*, 20(1), 46–61, https://doi.org/10.1145/321738.321743

Baruah, S. & Vestal, S. (2008), Schedulability analysis of sporadic tasks with multiple criticality specifications, Proc. of the 20th Euromicro Conference on Real-Time Systems, 147–155, https://doi.org/10.1109/ECRTS.2008.26

Baruah, S. K., Burns, S. K. & Davis, R. I. (2011), Response-time analysis for mixed criticality systems, Proceedings of the 32nd IEEE Real-Time Systems Symposium, 34–43, https://doi.org/10.1109/RTSS.2011.12

Bate, I., Burns, A. & Davis, R. I. (2015), A bailout protocol for mixed criticality systems, Proceedings of the 27th Euromicro Conference on Real-Time Systems, 259–268, https://doi.org/10.1109/ECRTS.2015.30

System Innovation for a Troubled World – Tak Lam et al. (Eds)
© 2023 The Author(s), ISBN: 978-1-032-45525-9

Machining time estimation of the EDM process

Wei-Chen Lee* & Chang-Lin Yang
Department of Mechanical Engineering, National Taiwan University of Science and Technology, Taipei, Taiwan

ABSTRACT: Electrical discharge machining (EDM) is widely used in the mold industry because EDM can make complex shapes unlike traditional machining. Much research has been conducted how to increase EDM material removal rate, improve the surface roughness of the workpiece, and reduce the electrode wear in EDM. However, due to the randomness of the EDM processes, the research on the prediction of EDM time is limited. The objective of the research was to build a model to predict the EDM machining time accurately, which can significantly increase the usability of EDM machines through performing computer-aided process planning. In this research, a full factorial experiment of 270 trials was performed and we considered four factors: pulse on time, pulse off time, low-voltage current, and machining depth. The data of the trials were used as training data, which were put into a neural network model. We then prepared 100 sets of test data, and the average error of the neural network model was 8%.

1 INTRODUCTION

Electrical discharge machining (EDM) is a nontraditional machining method that uses electrical energy to generate heat to melt and evaporate workpieces. EDM can process conductive materials with high toughness and hardness, which cannot be easily machined by traditional machining processes such as milling or lathing. However, EDM also has drawbacks such as high cost of the electrode and the time-consuming process. With the development of industry 4.0, a small amount of customized, flexible production seems to be one of the current industry trends. To face the various products, estimating the EDM time and using the result to plan the production schedule and maximize the utility of the production machines has become the key task of the current industry.

Several researchers have studied the related topics of EDM time. Joshi et al. (2009) conducted finite element analysis to simulate the operation of EDM with the four input machining parameters (discharge current, discharge voltage, pulse on time, and duty factor), and the output included the material removal rate (MRR) and the tool wear rate (TWR). The analysis results were used to train an artificial neural network (ANN) so that the ANN model can be used to generate more results without performing finite element analysis. It was found that the MRR initially increased with pulse on time, reached a maximum, and then decreased gradually under the situation that the pulse on time and the pulse off time were the same. Liang et al. (2017) analyzed the effects of current, voltage, pulse on time, and duty factor on the MRR and other responses. The results show that the current and pulse on time have the most significant influence. As the current increased, the MRR is also increased. Besides, at 70% duty factor, the increase in pulse on time resulted in the fluctuating MRR. Straka et al. (2018) also reported similar results: The higher the current, the higher the material removal rate; in addition, the higher the pulse on time, the material removal rate first increased and then slightly decreased. Bédard et al. (2020) explored the influence of 10 machining parameters of EDM on MRR, TWR, and surface roughness (SR). The most significant factors regarding the MRR are current, pulse on time, and voltage. Payal et al. (2020) used Taguchi's L36 orthogonal table to design experiments. By using the analysis of variance (ANOVA), it was found that the current had the most significant contribution to the MRR, about 42%. The electrode lift time was 12.38%, and the electrode material was 3.21%. Dikshit et al. (2019) studied the effects of the three factors, current, pulse on time, and pulse off time, on the MRR and SR in the EDM of Inconel 625. They used regression analysis to obtain a quadratic mathematic model for the MRR based on the three factors, and the predicted error for MRR was 2.19%. Rajeswari et al. (2018) discussed the influence of pulse on time, current, voltage, and duty factor on the MRR. It was found that the discharge voltage waveform was different in each discharge, which illustrated the randomness of the EDM.

In the literature mentioned above, it can be found that current and pulse on time are the most frequently discussed, and found that their influences on material removal rate are significant. In addition, other

*Corresponding Author: wclee@mail.ntust.edu.tw

DOI 10.1201/9781003377399-27

machining parameters have also been addressed in different documents, showing that many factors affect the results of EDM.

Some researchers used different approaches for EDM performance prediction. Mandal et al. (2007) used backward propagation neural network to model the EDM process. Three machining parameters, current, pulse on time, and pulse off time, were used in the experiments to obtain the MRR and TWR. They found that using ten nodes in each hidden layer for two layers can yield the best results. The mean prediction error of such a model was 3.06%. To design experiments, Pradhan et al. (2010) did a similar study by using four parameters, current, pulse on time, duty factor, and voltage. In terms of the MRR, the current was the most significant factor, contributing 87.61%. The experimental results were used to train two neural-fuzzy models and a neural model. For 15 sets of test data, the prediction errors of the three neural networks were around 5%. Belloufi et al. (2020) used a fuzzy logic model based on the pulse on time, current, and voltage, to predict the performance of EDM. The average error of the model prediction was 1.51% for MRR. Vanderauwera et al. (2010) took the flushing and efficiency into consideration when estimating the EDM time better. For a simple prismatic electrode, the average roughing time prediction error is 13.8%, which was better than the mean error of 68.9% estimated by the machine tool manufacturer. Based on the same idea, an EDM time prediction system (Lauwers et al. 2010) was developed. The results of this system were compared with the estimated time provided by the EDM manufacturer, and the prediction error was about 25%. Watanabe (2004) also developed an EDM time prediction software package, EDcam. The software categorized the machining content into deep, standard, and shallow machining. The EDM time prediction error was within 40% of the actual time.

From the previous discussions, most research focuses on the MRR. In addition, most of the research that study the machining time by using the MRR did not mention the machining depth. However, Chian (2001) shows that the MRR changed with the machining depth. To accurately predict the EDM time, the influence of the machining depth should be considered. The objective of this research was to develop a system to be able to accurately estimate the EDM time. In addition to the machining depth, we will take pulse on time, pulse off time, and low-voltage current to explore their influence on machining time through experiments, and then use the experimental data to establish a machining time prediction model based on a neural network.

2 METHODS AND MATERIALS

All trials were conducted using the Sinking EDM machine (Model Number D433CL, CHMER, Taichung, Taiwan), as shown in Figure 1.

Figure 1. The Sinking EDM machine used in this research.

The material of the workpiece used in the experiment was cold work tool steel SKD11, as shown in Figure 2. Its density is about 7.85 g/cm^3, and its hardness is about HRC 20. The electrodes used in the experiment were copper rods with a diameter of 10 mm.

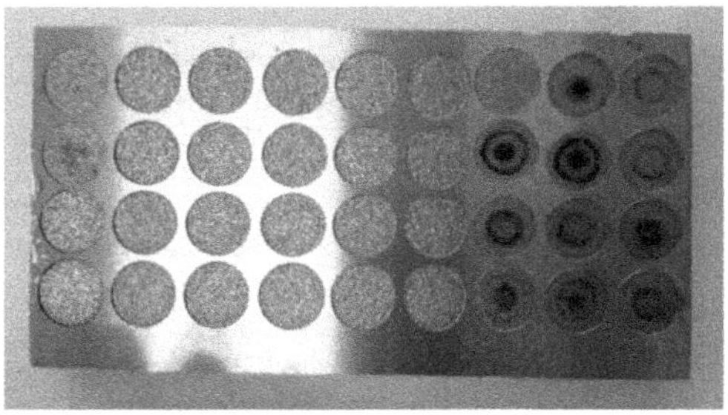

Figure 2. SKD11 tool steel used in this research.

This research aims to predict the EDM time and explore the relationship between machining parameters and time. The experimental design is shown in Table 1. We selected three 3-level factors: pulse on time, pulse off time, low-voltage current, and one 10-level factor machining depth. A full factorial experiment of 270 trials was conducted. Each trial was performed twice, and then an average was taken as the EDM time under the trial's condition. The 270 sets were then used as training data to establish the neural network-based machining time prediction model. In addition, within the range of our experimental plan, 100 sets of trials are conducted as testing data for verifying the machining time prediction model.

Table 1. The design of experiment.

Pulse on time (μs)	Pulse off time (μs)	Low-voltage current (A)	Machining depth (mm)
50	25	12	0.5 to 5.0 for every 0.5
150	75	18	
300	150	24	

3 RESULTS AND DISCUSSION

To study the effect of pulse on time on machining time, we selected 27 sets of data at the machining depth of 0.5 mm, as shown in Figure 3 (Yang, 2021). Figure 3(a), (b), and (c) correspond to the pulse off time of 25, 75, and 150 μs, respectively.

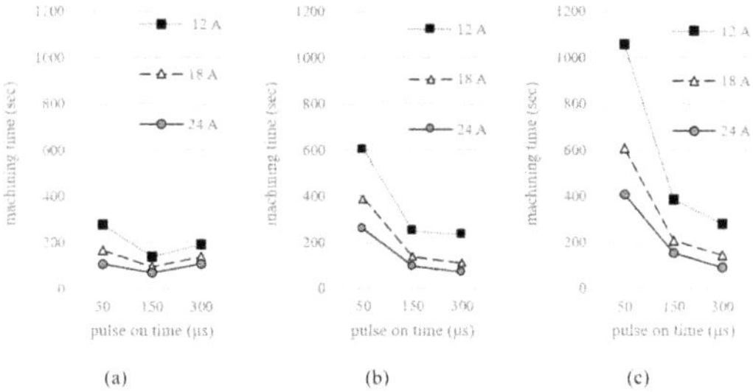

Figure 3. Effect of pulse on time on machining time, pulse off time (a) 25 μs; (b) 75 μs; (c) 150 μs.

Generally, when the pulse on time increased, the machining time decreased. However, there was an exception. From Figure 3(a), we can see that when the pulse on time increased from 150 μs to 300 μs, the machining time increased. The pulse off time was 25 μs for the data shown in Figure 3(a). Because there was an apparent difference between the pulse on time and the pulse off time, the debris generated during the discharge cannot be flushed away during the short pulse off time. The debris accumulated and affected the discharge mechanism. As a result, the machining time is not shortened but increased.

When the pulse on time increased, the reduction of the machining time slowed down. The higher the pulse on time, the higher the material removal per discharge, but the cycle time of a single discharge also increased, resulting in a decreased material removal rate. The trend of this result is the same as the results of previous research (Joshi et al. 2009; Straka et al. 2018).

To study the effect of pulse on time on machining time, we also selected 27 sets of data at the machining depth of 0.5 mm, as shown in Figure 4 (Yang, 2021). Figure 4(a), (b), and (c) correspond to the pulse off time of 50, 150, and 300 μs, respectively.

Generally, when the pulse off time increased, the machining time also increased. However, from Figure 4(c), it can be seen that when the pulse off time increased from 25 μs to 75 μs, the machining time reduced. This may be due to the debris accumulation for the short pulse off time of 25 μs.

To study the effect of low-voltage current on machining time, we selected 27 sets of data at the machining depth of 0.5 mm, as shown in Figure 5 (Yang, 2021). Figure 5(a), (b), and (c) correspond to the pulse on time of 50, 150, and 300 μs, respectively.

An increase in low-voltage current means an increase in discharge energy. The higher the discharge energy, the more material is removed. Therefore, from

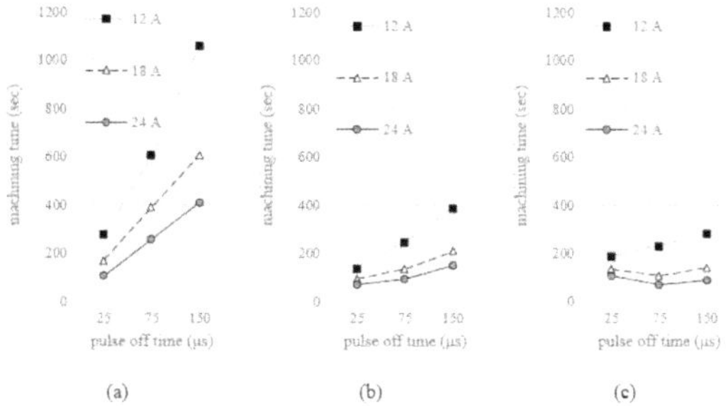

Figure 4. Effect of pulse on time on machining time, pulse on time (a) 50 μs; (b) 150 μs; (c) 300 μs.

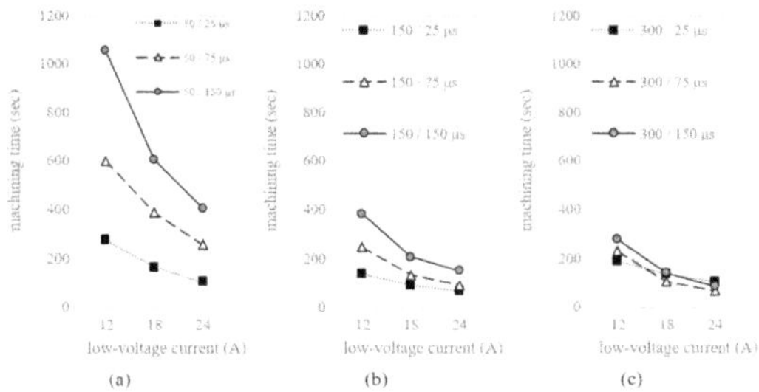

Figure 5. Effect of low-voltage current on machining time, pulse on time (a) 50 μs; (b) 150 μs; (c) 300 μs.

Figure 5, we can see that the higher the low-voltage current, the shorter the machining time.

To study the effect of machining depth on machining time, we selected 270 sets of data at the machining depth from 0 to 5 mm for every 0.5 mm, as shown in Figures 6, 7, and 8 (Yang, 2021). Figure 6(a), (b), and (c) correspond to the pulse on time/pulse off time 50/25, 50/75, and 50/150 μs, respectively. Figure 7(a), (b), and (c) correspond to the pulse on time/pulse off time 150/25, 150/75, and 150/150 μs, respectively. Figure 8 (a), (b), and (c) correspond to the pulse on time/pulse off time of 300/25, 300/75, and 300/150 μs, respectively.

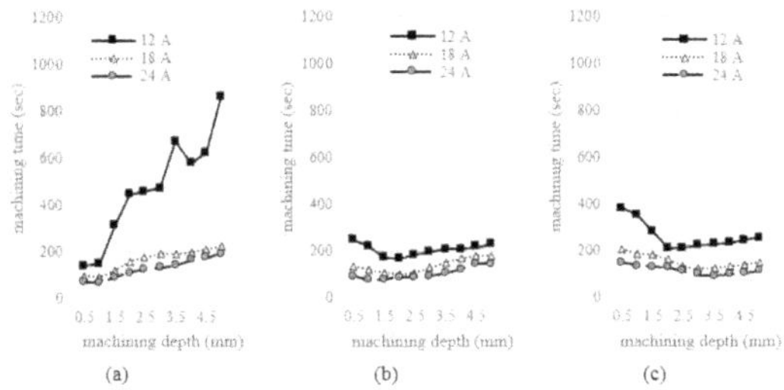

Figure 6. Effect of machining depth on machining time, (a) T_{on}: 50 μs, T_{off}: 25 μs; (b) T_{on}: 50 μs, T_{off}: 75 μs; (c) T_{on}: 50 μs, T_{off}: 150 μs.

In Figures 6 to 8, we can see that the required machining time also changed as the machining depth changed. Figure 8(a) is the result under the pulse on/off time of 300/25 μs. Due to the long pulse on time and short pulse off time, the debris removal effect is poor.

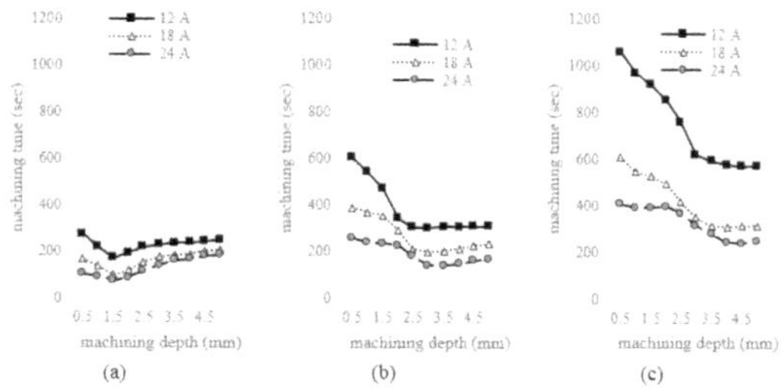

Figure 7. Effect of machining depth on machining time, (a) T_{on}: 150 μs, T_{off}: 25 μs; (b) T_{on}: 150 μs, T_{off}: 75 μs; (c) T_{on}: 150 μs, T_{off}: 150 μs.

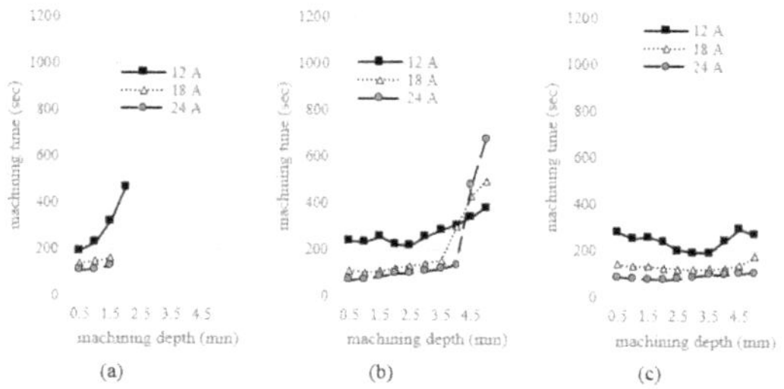

Figure 8. Effect of machining depth on machining time, (a) T_{on}: 300 μs, T_{off}: 25 μs; (b) T_{on}: 300 μs, T_{off}: 75 μs; (c) T_{on}: 300 μs, T_{off}: 150 μs.

Eventually, the EDM could not be carried out to the target depth of 5.0 mm. On the contrary, Figure 6(c) is the result under the pulse on/off time of 50/150 μs. The pulse off time was long enough to remove the debris by flushing. Therefore, as the machining depth increased, the machining time decreased.

4 MACHINING TIME PREDICTION MODEL

The deep learning toolbox of MATLAB was used as a development tool to define the neural network architecture. A total of 270 sets of data are used as training data. Inputs are pulse on time, pulse off time, low-voltage current, machining depth, and the ratio of pulse on time to pulse off time, and the output was machining time. We designed and tested two neural networks with a different number of hidden layers to find the most suitable machining time prediction model.

The neural network architecture with one hidden layer is shown in Figure 9. We tried 20 different nodes from 10 to 200 for every ten nodes. The outcomes of the 20 trials were compared, and the hidden layer with 130 nodes yielded the best result, with an average error rate of 8%. The final hyperparameters we used for the neural network were as follows: the activation function was a rectified linear unit (ReLU), the loss function was a mean square error (MSE), the optimizer was stochastic gradient descent with momentum (SGDM), the learning rate was 0.0004, the epoch was 150, the learning rate decay was 0.0001, and the momentum was 0.9. The training can effectively converge, and the training time was about 30 s.

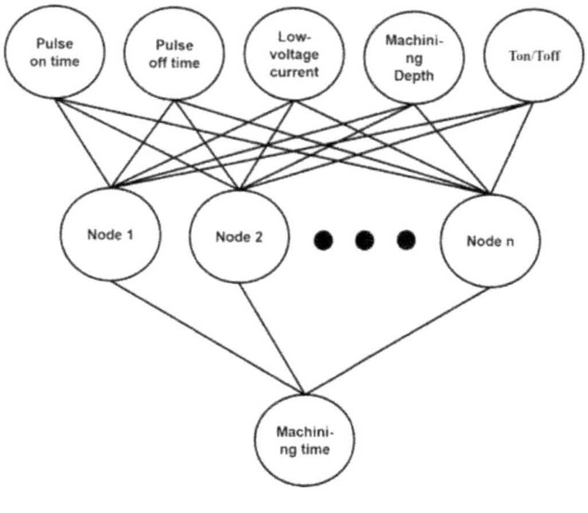

Figure 9. One hidden layer neural network architecture.

Furthermore, we used two hidden layers to construct the neural network model. We tried ten different numbers of nodes, from 10 to 100, for every ten nodes for each layer. As a result, we had 100 different neural network models. The average errors of the 100 neural networks are shown in Figure 10. The best outcome out of these modes was obtained when the number of nodes in the first and second layers was 90 and 10, respectively. Under the condition, the error rate of the test data was still 8%. The hyperparameters were the same as the neural network with one hidden layer, except that the learning rate was 0.00002, and the epoch was 200.

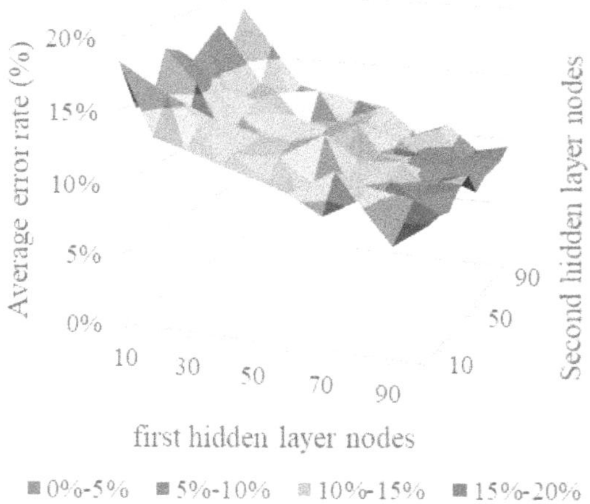

Figure 10. The average error rate of the two hidden layer neural network models.

Comparing the results of the neural networks with one hidden layer and two hidden layers, we can understand that the neural network with one hidden layer can yield the same error rate as that with two hidden layers. Therefore, the simple one hidden layer neural network with 130 nodes model was adopted as our EDM time prediction model.

5 CONCLUSIVE REMARKS

Through 270 sets of full-factorial experiments in this research, the relationship between the machining time and the four machining parameters, pulse on time, pulse off time, low-voltage current, and machining depth was explored. The neural network model we established in this research can yield an average error rate of the testing data of 8%. However, the influence of other factors such as electrode area and electrode shape on the EDM time was not considered. Further research will be conducted for a more extensive study.

REFERENCES

Bédard, F., Jahazi, M. & Songmene, V. 2020. *Int. J. Adv. Manuf. Tech.* 107 333–342.

Belloufi, A., Mezoudj, M., Abdelkrim, M., Rezgui, I. & Chiba, E. 2020. *Int. J. Adv. Manuf. Tech.* 109 2065–2093.

Chian, Y.F. 2001. Master Thesis, Dept. of Mech. Eng. National Taiwan Univ.

Dikshit, M.K., Anand, J., Narayan, D. & Jindal, S. 2019. *J. Braz. Soc. Mech. Sci. Eng.* 41, 302.

Joshi, S.N. & Pande, S.S. 2009. *Int. J. Adv. Manuf. Tech.* 45 300–317.

Lauwers, B., Oosterling, H. & Vanderauwera, W. 2010. *CIRP Ann.* 59 223–226.

Liang, J.F., Liao, Y.S., Kao, J.Y., Huang, C.H. & Hsu, C.Y. 2017. *Int. J. Adv. Manuf. Tech.* 95 1743–1750.

Mandal, D., Pal, S.K. & Saha, P. 2007. *J. Mater. Process. Tech.* 186 154–162.

Payal, H., Bharti, P.S., Maheshwari, S. & Agsrwal, D. 2020. *Teh. Vjesn.* 27 761–772.

Pradhan, M.K. & Biswas, C.K. 2010. *Int. J. Adv. Manuf. Tech.* 50 591–610.

Rajeswari, R. & Shunmugam, M.S. 2018. *Int. J. Adv. Manuf. Tech.* 100 1945–1964.

Straka, Ľ. & Hašová, S. 2018. *Int. J. Adv. Manuf. Tech.* 97 2647–2654.

Vanderauwera, W. & Lauwers, B. 2010. ISEM 41–46.

Watanabe, Y. 2004. *Int. J. Elec. Mach.* 9 59–61.

Yang, C.L. 2021. Master Thesis, Dept. of Mech. Eng. National Taiwan Univ. of Sci. & Tech.

System Innovation for a Troubled World – Tak Lam et al. (Eds)

The research of the value of the influence of Mazu belief on folk culture

Hsu-Yen Lin* & Deng-Chuan Cai
Graduate School of Design, National Yunlin University of Science & Technology, Douliou, Taiwan, PR China

ABSTRACT: The preservation of the intangible cultural heritage is a universal value, and the trend of preservation has become a comprehensive conservation that crosses the ethnics, cultures, and generations from individual conservation. Mazu belief is the specific Intangible Cultural Heritage in Taiwan since it has the largest number of congregations. The goal of this research is to analyze the current condition of registration of Mazu belief in folk culture, and propose the value feature of the Mazu belief Intangible Cultural Heritage.

The research methods include document analysis and comparative analysis. With textual analysis, we analyze the literature and secondary data relating to the intangible cultural heritage, which include the announcement, preservation and conservation plans, and the thesis of registration of Mazu belief in Cultural Heritage.

The research result indicates that Mazu belief culture has ten value features, including ten properties of collectivity, traditionality, variability, totipotency, cohesion, universality, sociality, innovation, industrial applicability, and abundance. The research result can be provided for the competent authority's reference in formulating the policy of preserving and conserving Mazu beliefs, and then they can enhance the preservation and inherent value of the intangible cultural heritage.

Keywords: Intangible Cultural Heritage, Mazu belief, preservation and conservation.

1 INTRODUCTION

In 2003, United Nations Educational, Scientific and Cultural Organization, UNESCO passed "Convention for the Safeguarding of the Intangible Cultural Heritage", which brought the cultures like social practices, concept expression, expression form, and knowledge and skill into protection. Aiming at pursuing cultural diversity, respecting human creativities, and becoming the source of law when many countries are formulating relevant institutions. The preservation of Intangible Cultural Heritage is a universal value, and the trend of preservation has become a comprehensive conservation that crosses the ethnics, cultures, and generations from individual conservation.

In Taiwan, the "Cultural Heritage Preservation Act" was been legislated in 1982, which is aimed at conserving folk and local specific traditional arts, folklore, and some relevant cultural relics. In order to enhance the preservation of each type of Intangible Cultural Heritage in Taiwan, the act has been amended eight times.

Belief is a strong faith that is usually expressed as a stubbornness towards something for which there is insufficient evidence to convince every rational person to trust (Kaufmann 1961). It is also the option and posses of our own lives philosophy, values, worldview, etc.

Mazu belief is one of the specific Intangible Cultural Heritage in Taiwan, which is the representation of immigration and marine culture. Taiwan is an area that owns the most prosperous development and the largest number of congregations of Mazu belief culture now. Mazu belief influents folk culture deeply, and it is present in every lifestyles such as traditional customs, ceremonies, rites, festivals, habits, languages, words, religions, beliefs, moralities, etc.

Mazu belief was selected into United Nations Educational, Scientific, and Cultural Organization (UNESCO) "Masterpieces of the Oral and Intangible Heritage of Humanity" in October 2009. Moreover, it also became one of twelve Potential Intangible Cultural Heritage Sites in Taiwan in 2010.

On 3/23 of the lunar date every year, which is Mazu's birthday, there is grand series of folk pilgrimage activities take place, which is generally called "The Mazu pilgrimage craze in March". And the activity has become a Taiwan special folk culture.

The research purpose is to analyze the situation of registration of Mazu belief in the intangible cultural heritage and then provide the value of Mazu belief.

2 RESEARCH METHOD

The research methods include document analysis and comparative analysis. And with textual analysis, we

*Corresponding Author: ho960316@gmail.com

DOI 10.1201/9781003377399-28

analyze the literature and secondary data relating to the intangible cultural heritage, which include the announcement, preservation and conservation plans, and the thesis of registration of Mazu belief in Cultural Heritage.

The information is from all competent authorities that logged Mazu's belief in Cultural Heritage, including eleven municipalities, competent authority of each city or county, and the Bureau of Cultural Heritage, Ministry of Culture, BOCH. (Taipei City, New Taipei City, Miaoli County, Taichung City, Changhua County, Yunlin County, Chiayi County, Tainan City, Kaohsiung City, Taitung County, Kinmen County, etc). Total twelve departments.

Besides, we compared each national information of Mazu belief in Cultural Heritage through comparative analysis, and explored the distinctiveness and differences between each information with the data we have now.

In the end, we divided the value of Mazu belief through document analysis, then induced, connected, and interpreted the different cultural value features presented by Mazu belief.

3 RESULTS AND DISCUSSION

3.1 *Taiwanese folk intangible cultural heritage believed in Mazu the most, then followed by WangYe worship*

Taiwanese "National Cultural Heritage Network" collected the intangible cultural heritage of the category of Han ethnic group folklore in March 2022, which amounts to 173 items, including 157 items of local folklore and 16 items of national-level folklore.

Of the top five registered congregations by classification, Mazu belief has the most items with 32, 28 of which are local folk Mazu belief and 4 of which are national-level Mazu belief. Moreover, it accounts for 18.5% of the total number of folk registration and is distributed in eleven areas including Taipei City, New Taipei City, Miaoli County, Taichung City, Changhua County, Yunlin County, Chiayi County, Tainan City, Kaohsiung City, Taitung County, and Kinmen County,.

The second most likable site is "WangYe worship"(29 items), which accounts for 16.76% of the total number of folk registration. The third place goes to "Lantern Festival"(15 items), which accounts for 8.67% of the total number of folk registration. The fourth is "Ghost Festival" (13 items), which accounts for 7.51% of the total number of folk registration. The fifth most likable site is "Parade Formation (DinTao)"(12 items), which accounts for 6.94% of the total number of folk registration.

3.2 *The influences of Mazu belief on folk culture*

The followings are the influences of Mazu belief on folk culture analyzed in this research.

3.2.1 *The relationship between Mazu and its congregations is just like family*

Mazu, a Fujianese shamaness, whose name is Lin Mo, and whose nickname is Moniang, is seen as a celestial empress in most of the historical data. The fact on National religion Information Network and Wikipedia indicated that Mazu has many names like Tianshang Shengmu, Tianhou, Tian Hou Niang Niang, Tianfei, Tianfei Niang Niang, Mei Jhou Niang Ma, Mazu po, Guan Yin Da Shih, etc. In the early period, Mazu is only called "Ling Nyu", titled "Fu Ren", and titled "Tianhou"or "Shengmu" in late period (Hsun Chang 1995). People like to call Mazu "Mazu po", "Niang Ma", "Po Zai", "Gu Po Zu". These names with a maternity-elderly connotation reflect that the relationship between Mazu and its congregations is similar to that of family, implying the imagery of "mother."

3.2.2 *Mazu is an almighty goddess:*

Mazu belief originated as a "witch" belief. Witchcraft is a supernatural power that originates from the association, and the ancestors believed that this kind of power could make them avoid disasters (Chih-Ming Cheng 2008). There are many miracle tales and domestications of Mazu, for example, inspiring people to worship, blessing and protecting sailing off, assisting in the war and resisting the enemy, stopping drought, curing disease, and eliminating pandemics, warning and preventing misfortune, prohibiting floods, helping people revive, blessing pregnant women, subduing demons, saving the famine, and boosting the growth of crops, etc (Jui-Long Hsieh 2015).

3.2.3 *Mazu is a local patron saint:*

The Penghu Tianhou Temple is the earliest fortified point when Mazu belief entered Taiwan, later, most places in Taiwan used Mazu Temple as the Dagong Temple in the region. Gradually, some localization of names like "Kai Tai Ma", "Kai Ji Ma", "Bei Gang Ma", "Sin Gang Ma", "Pu Zih Ma", "Lu Gang Ma", "Tai Ping Ma", "Mai Liao Ma", "Bei Dou Ma", "Guan Du Ma", "Da Jia Ma", etc. And these names reflect that there is a deep connection between the thoughts of congregations and the belief in place emotion. Although Mazu didn't have any children before her death, she is still a local patron saint forever (Mei-Ying Huang 1994).

3.2.4 *The Personalization of goddess:*

In Meizhou, Mazu was named "Mei Jhou Ma"; in Tongan, "Yin Tong Ma"; in Anxi, "Ching Si Ma"; in Quanzhou Sam-ip, "Wun Ling Ma"; and in Zhangzhou Zhangpu County, "Wu Shih Ma" (Jin-Lang Lin 2018). Additionally, in Putian, Mazu was named "Da Ma"; in Xianyou, "Er Ma"; and in Hui'an, "San Ma".

There is only one "Mazu", but there are always many josses of Mazu in Mazu temples, and sequence Mazu in "Da Ma", "Er Ma", and "San Ma". To aid in identification, temples provide sculptures of Mazu to congregations; in this way, congregations can idolatrous the goddess and endow Mazu with a name and number, as well as numbering the sculptures in the order they appear (Da MaLiou Ma). Moreover, each

Mazu has a different division of duties because of the difference in numbers (Mao-Hsien Lin 2008).

3.2.5 *Endowing the belief with sociality:*
Mazu belief has cultural elements such as pilgrimage procession and Jin Siang (Pilgrimage)(congregations worship in temples and strengthen friendship). Pilgrimage procession usually refers to patrol areas or settlements, and the congregations or temples in one area have the right and obligation. The biggest social meaning behind is about the cohesion that can connect people. What's more, the pilgrimage procession is also a trip of spiritual healing that represents the Taiwanese spirit of supporting each other all the time (Ya-Lin Chang 2019).

Jin Siang (Pilgrimage) usually means that congregations worship in temples. Mazu Pilgrimage, also known as "Huei Niang Jia" (Hsun Chang 1995), takes Joseph's legacy as its main symbolic meaning. In fact, the purpose of the Mazu Pilgrimage is to confirm the subordinate relationship between ancestral temples and fenxing satellite temples, with the ceremony and core value of supporting ancestral temples, bunrei, and restoring spiritual power. .

3.2.6 *Forming a special culture:*
Mazu belief was selected for the United Nations Educational, Scientific, and Cultural Organization (UNESCO) "Masterpieces of the Oral and Intangible Heritage of Humanity" in October 2009. It has also been selected as one of the twelve potential intangible cultural heritage sites in Taiwan for 2010. Every year, on Mazu's birthday – which is 3/23 of lunar date every year – there is a grand series of folk pilgrimage activities known as "the Mazu pilgrimage craze in March". And the activity has become a Taiwan special folk culture.

3.3 *The value feature of the registration for the intangible heritage of Mazu belief*

This research has collected 32 announcements and reasons for the intangible heritage registration of Mazu belief. First, we unified its cultural features, then induced and interpreted the features with textual analysis, and finally analyzed and proposed ten common cultural values of Mazu belief (Table 1), and the following are the descriptions.

3.3.1 *The collectivity is the core value of the belief (Collectivity):*
An important folk belief that has lasted for more than a hundred years. Presenting the collectivity core value of regional community and life values.

3.3.2 *Having traditional continuity (traditionally)*
Mazu belief has a long history and it also preserves traditional ceremonies. As it continued till date, the public still passes down this culture from generation to generation, and it never changes or fades away as times go.

3.3.3 *Having variability on history (variability)*
With the changing of some external factors such as the living environment, Mazu belief has brought influence on culture. Then they created historical and special folk activities together.

3.3.4 *The totipotency of the miracle tales and domestication of Mazu (totipotency)*
Mazu has turned to the goddess of totipotency from the goddess of the ocean, water, and agriculture. For instance, Mazu inspires people to worship, blesses and protects sailing off, assists in the war and resists the enemy, stops drought, cures disease and eliminates pandemics, warns and prevents misfortune, prohibits flood, helps people revive, blesses pregnant women, subdues demons, saves the famine, boosts the growth of crops, helps with coming-of-age ceremonies, and promotes unity, etc.

3.3.5 *Agglomerating organization and help with each other (cohesion)*
Agglomerating the areas, clan identity, internal cohesion, and participation, this shows the functions of reunification and unity, which even cross the frontier of each area. In other words, Mazu belief is a demonstration of ethnic integration.

3.3.6 *The widespread influence in Taiwan (universality)*
Mazu is a widespread belief that has the longest pilgrimage procession days and the furthest pilgrimage procession route. People automatically make pilgrimages on foot, and the ways of worship along the whole pilgrimage procession are distinctive.

3.3.7 *Promoting social harmony (sociality)*
People have passed down the common values of Mazu such as a compassionate heart, reassuring the public, soothing people's minds, etc. The gods will interact, identify, and then get along well with each other.

3.3.8 *Setting a good example of innovation (innovation)*
Conserving various types of customs, including the performance of Yi-Ge and Parade Formation, Firecrackers at Hu Yeh, etc., especially the Parade Formation, which produces typical symptoms of the "March Mazu pilgrimage craze."

3.3.9 *Boosting the development of local businesses (Industrial Applicability):*
The business opportunities brought by Mazu culture have promoted the development of international tourism and relevant industries indirectly.

3.3.10 *The abundance of cultural expression (Abundance):*
The cultural expressions of Mazu belief are diverse and abundant, for instance, literature, art, dance, proverbs, rituals, cultural relics, etc.

Table 1. The statistics of the ten common culture values of Mazu belief.

NO	Name	Collectivity	Traditionally	Variability	Totipotency	Cohesion	Universality	Sociality	Industrial Innovation	Applicability	Abundance
1	Dajia Jenn Lann Temple (Important Folklore)	■	■	■	■	■	■	■	■	■	■
2	Baishatun Gong Tian Temple (Important Folklore)	■	■	■	■	■	■	■	■	■	■
3	Beigang Chaotian Temple (Important Folklore)	■	■	■	■	■	■	■	■	■	■
4	Yunlin Liufang ma (Important Folklore)	■	■	■	■	■	–	■	■	■	■
5	Guandu Temple	■	■	■	■	■	–	■	–	■	■
6	JinBaoVillage CiHu Temple	–	■	■	■	■	–	■	–	–	–
7	Baishatun Gong Tian Temple	■	■	■	■	■	■	■	■	■	■
8	Jhonggang Cihyu Temple	■	■	■	■	■	–	■	–	■	■
9	Dajia Jenn Lann Temple	■	■	■	■	■	■	■	■	■	■
10	Leh Cherng Temple	■	■	■	■	■	–	■	–	■	■
11	Xinshe Jiuzhuang ma	■	■	■	■	■	–	■	–	■	■
12	Taichun Wanhe Temple	■	■	■	■	■	–	■	–	■	■
13	Wuqi Chaoyuan Temple	■	■	■	■	■	–	■	–	■	–
14	Dajhuang Haotian Temple	■	■	■	■	■	–	■	–	■	–
15	Shetou Fangqiaotou Tianmen Temple	■	■	■	■	■	–	■	–	■	■
16	TongAnLiaoShiErZhuang	■	■	■	■	■	–	■	–	■	■
17	Changhua Nanyao Temple	■	■	■	■	■	–	■	–	■	■
18	Changhua Twenty-Four Zhuang	■	■	■	■	■	–	■	–	■	■
19	ShenGang FuAn Temple	■	■	■	■	■	–	■	–	■	■
20	Beigang Chaotian Temple	■	■	■	■	■	■	■	■	■	■
21	Yunlin Liufang ma	■	■	■	■	■	–	■	■	■	■
22	Beigang Pilgrimage	■	■	■	■	■	■	■	■	■	■
23	Hsin Kang Feng Tian Temple	■	■	■	■	■	■	■	■	■	■
24	Hsin Kang Feng Tian Temple Welcome Mazu	■	■	■	■	■	–	■	–	■	–
25	Tainan Grand Mazu Temple.	■	■	■	■	–	–	■	–	–	■
26	Luermen Tianhou Temple	■	■	■	■	■	–	■	■	■	■
27	Luermen Mazu Temple	■	■	■	■	■	–	■	■	■	■
28	Anping Kaitai Mazu Temple	■	■	■	■	■	–	■	–	■	■
29	Lioujia Heng An Temple	■	■	■	■	■	–	■	■	–	■
30	Linyuan Fong Yun Temple	■	■	■	■	■	–	■	■	■	■
31	Taitung Tianhou Temple	■	■	■	■	■	–	■	–	■	■
32	Fengshang Tianhou Temple	■	■	■	■	■	–	■	–	–	■

4 CONCLUSION

According to the findings of this study, there are ten common cultural values of Mazu belief: 1. collectivity is the core value of the belief; 2. having traditional continuity; 3. having variability on history; 4. the totipotency of the miracle tales and domestication of Mazu; 5. agglomerating organizations and helping each other; 6. the widespread influence in Taiwan; 7. promoting social harmony; 8. setting a good example of innovation; 9. boosting the development of local businesses; 10. the abundance of cultural expression. The research result can be provided for the competent authority's reference in formulating the policy of preserving and conserving Mazu belief, and then they can enhance preservation and inherently preserve the intangible cultural heritage.

REFERENCES

Chih-Ming Cheng. (2008). Wushu Wunhua De Jhesyueh Singsih. Dichijieh Ru Fo Hueitong Ji Wunhua Jhesyueh(67). Cultural Affairs Bureau of the Changhua County.

Graebner, F. (1911). Methode der ethnologie (1). C. Winter.

Hsun Chang. (1995). Nyushen Sinyang Yu Mazu Chongbai De Bijiao Yanjiou. Bulletin of the Institute of Ethnology Academia Sinica, 79, 185–203.

Hsun Chang. (2008). Mazu D Sinyang De Jhueisyun: Jhang Syun Zihsyuan Ji. Bo Young Cultural Enterprise Inc.

Hsun Chang. (2013). Mazu Zaosiang Yu Biaojhunhua Wunti Taolun. Zongjiao Renleisyueh., Social Sciences Academic Press. China, 4, 149–167.

Jin-Lang Lin. (2018). Shenling Taiwan · Diyi Ben Chinjin Shenming De Siao Baike: Jhangbei Mei Jiao, Dan Ni Yidingyao Jhihdao De Shenming Wei Lishih, Baibai Siao Jhihshih. Persimmon Cultural Enterprise Co., Ltd.

Jui-Long Hsieh. (2015). Mazu Sinyang Gushih Yanjiou-Yi Jhongguo Yanhaidichyu, Taiwan Weijhu Yao Kaocha Fanwei. (PhD thesis).Department of Chinese Literature National Chung Cheng University.

Mao-Hsien Lin. (2002). Taiwan Mazu Chuanshuo Jichi Bentuhua Siansiang. Jingyi Renwun Syuehbao, 17, 91–113.

Mao-Hsien Lin. (2008). Taiwan Ren De Mama – Mazu. Traditional Arts, 76, 33–36.

Mao-Hsien Lin. (2014). Wusing Wunhua Zihchan Minsu Lei Denglu Siankuang Yanjiou (I) Cross-Strait Intangible Cultural Heritage Forum in 2014 (Taichung). Headquarters Administration of Cultural Heritage, Taichung.

Mei-Ying Huang. (1994). Taiwan Mazu De Sianghuo Yu Yishih. The deparment of culture and publishing in Independence Evening Post.

Ming-yu Lin. (1988). The legend of Mazu. Dong Men Publisher.

National Cultural Heritage Network(2017).https://nchdb.boch.gov.tw/

National religion Information Network The folk belief, Mazu.https://religion.moi.gov.tw/Knowledge/Content?ci=2&cid=167

Shi-Zhen Wu. (1987). Shenhua Sihwei Bian Si. Shenhua Sin Lun(5). Shanghai Literature & Art Publishing House.

Walter Arnold Kaufmann. (1961). The Faith of a Heretic. Princeton University Press.

Ya-Lin Chang. (2019). Mazu Raojing Yanlu Doushih Gandong, Mao-Hsien Lin: "Buchiou Hueibao De Fuchu, Tsaishih Taiwan Zuei Chunpu Deyangzih". 2019 Weisiao Jikan Chunji Hao-Genjhe Shenming Chyu Lyusing

System Innovation for a Troubled World – Tak Lam et al. (Eds)
© 2023 The Author(s), ISBN: 978-1-032-45525-9

An innovative teaching model for the imagination and practice of digital crafts in the future

Shu-Huei Wang*
Department of Digital Design, MingDan University, Wen-Hua, ChangHua, Taiwan

ABSTRACT: Reasons like sub-replacement fertility, a reduced birth rate, etc. forced many countries to propose the policy of developing a digital economy. For example, Japan's Super Smart Society (Society 5.0) used digital technology to address changes in population structure. Crafts in the cultural and creative industry were served as the object of digitization in this research. As hands were prerequisite in traditional crafts, their production was limited. The focus would be on how to increase the output of crafts by introducing digitization or developing the possibility of digital crafts in the future through curriculum planning of industrial issues. The futuristic imagination of digital crafts would be the theme of the experimental course for students to learn to apply the digital tools. In addition to experiencing crafts and learning about AR (Unity & Vuforia), students may learn about digital tools like 3D printing (123Ddesign) and Premiere Pro in the curriculum. It was expected that students would have more imagination and application when using the futuristic technology in traditional crafts. The radar chart of the questionnaire survey at the end of the course rendered to an average of 4.5, whose perfect score was 5. It implied that most of the students gave positive recognition to the course, which not only helped them with digital learning but also overthrew their previous cognition of digital application.

1 INTRODUCTION

Based on the Research Report on Human Resource Planning and Development, Vol. 20 published by the National Development Council in July 2020, Shu-ting Huang (2020/07) indicated that the workforce aged between 15 and 64 reached to a peak in 2015 and began to decline when the number got to 7.37 million people. It was predicted that the number would decrease to 8.62 million people in 2065, which was about 50% of employees in 2018 (8.49 million people). It was also predicted there would be 7.15 million elderly people in 2065, which increased by 1.1 times compared to the aged people in 2018 (3.72 million). Reasons such as an aging population and sub-replacement fertility forced many countries to propose the policy of developing a digital economy. For instance, the Super Smart Society presented by Japan solved the changes in the population structure with digital technology (Society 5.0). Hence, innovative teaching should combine digital technology and industrial demands to cope with the epochal requirements. In the survey excerpt of Manpower Supply and Demand Investigation and Estimation Result in Key Industries – Cultural and Creative Industry, the National Development Council indicated a gap between theory and application generally existed in the Cultural and Creative Industry. Currently, students were short of creative thinking in their fields and lack executive power and an industrial understanding [Ministry of Culture, (103)][Report on Cultural and Creative Manpower Supply and Demand Gap Survey and Coping Strategies, collected and organized by National Development Council].

*Corresponding Author: wangsh@mdu.edu.tw

2 CURRICULUM DESIGN

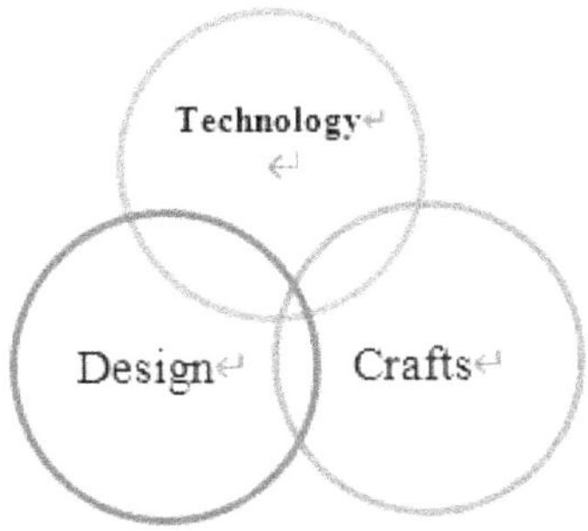

Figure 1. Interdisciplinary drawing.

Crafts in the cultural and creative industry served as the object of digitization for course planning. As hands were a prerequisite in traditional crafts, their production was limited. How to increase the output of crafts by introducing digitization or developing the possibility of digital crafts in the future through curriculum

DOI 10.1201/9781003377399-29

planning of industrial issues would be focused on. The futuristic imagination of digital crafts would be the theme of the experimental course for students to learn to apply digital tools, including AR (Unity & Vuforia), 3D printing (123Ddesign) and Premiere Pro. Students were expected to be able to 1. learn to apply the digital tools for creative design thinking, problem-solving, and creating; 2. overcome interdisciplinary learning problems; 3. keep a learning diary by videotaping, film editing, and recording the learning process of each group; 4. present diversified results by applying various digital tools; and 5. comprehensively learn and record the theory, digital tools, creative design thoughts, and creation via curriculum planning, diversified learning, and project design to solve industrial issues.

Course content: the learning goals would be based on imagination and practice of the future of digital crafts, which could be expressed as innovative design thinking X interaction technology X crafts X videotaping and film editing X digital crafts, shown in Figure 1.

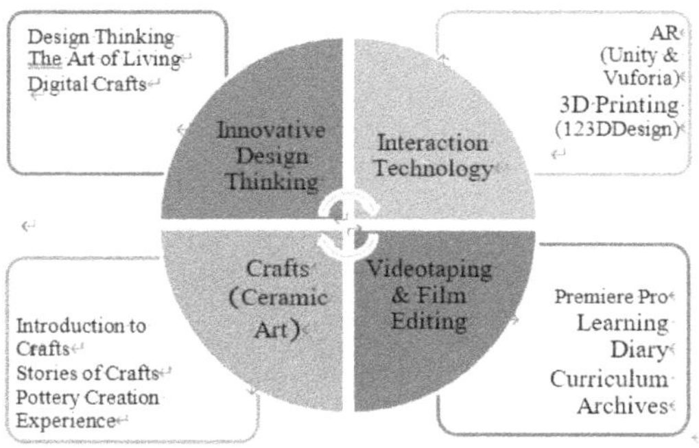

Figure 2. Curriculum module.

The course steps and flow were planned for 18 weeks and explained as follows:

Step 1. Knowledge of craftsmanship and the art of living (discovering problems): greeting, grouping (course instruction), and craftsmanship for Week 1; arts and crafts aesthetics for a new life for Week 2; videotaping practice and film editing for Week 3.
Step 2. Innovative model (introducing interaction technology): innovative design thinking for Week 4; knowledge of interactive media design for Week 5; imagination of virtual reality for Week 6; AR creation experience for Week 7; learning 3D printing for Week 8; 3D printing creation for Week 9.
Step 3. Conceptual design and pottery experience: group report and progress for Week 10; discussion of design concepts for Week 11; a field trip to Tien Chung kiln to experience creation for Weeks 12 and 13.
Step 4. Integrated creations of digital crafts (by groups): group discussion (with proprietors & teachers) for Week 14; integration and design of digital crafts creations for Week 15.
Step 5. Exhibiting design and feedback: exhibition planning, discussion, and production for Weeks 16 and 17; presentation of final accomplishments for Week 18.

Figure 3. Curriculum steps & flow.

3 RESULTS AND DISCUSSION

Students from various departments took the general course on Craftsmanship and the Art of Living, including 6 from the Department of Post-Modern Agriculture, 7 from the Department of Digital Design, 1 from the Department of Chinese Culture and Communication, 2 from the Department of Information Communication, 8 from the Department of Applied Foreign Languages, 2 from the Department of Fashion Imaging, 3 from the Department of Business Administration, 2 from the Department of Marketing and Logistics and 2 from the Department of Hospitality Management. Total of 33 students were taking this course and 8 of them were foreign students from Vietnam and Indonesia, which was really interdisciplinary, international, and interscholastic (teachers and assistants from different schools).

1. ***Invite experts to give speeches***: Experts of digital tools in different fields would lecture on practice and problem-solving ability to motivate students' interest in learning. Digital tools consisted of Premiere Pro, 123Ddesign, and AR (Unity & Vuforia). Not all students liked all of them because some only preferred one or two.
2. ***Give topics with the industry***: The Tien Chung Pottery Workshop gave topics to the assignments of the students, who would videotape, edit films and record learning diaries. Students would spend three weeks on AR practice and cases, two weeks on 3D printing practice and finally pottery practice. It was expected that students would be able to collaborate to complete each group's project using a variety of digital tools. Currently, there were four groups and two of them were successful and the other two needed more adjustment. The attitude of the group leader could affect the progress and learning interests of the whole group members.
3. ***Invite teachers to instruct at school***: The Project Director would invite teachers of various fields (such as AR, editing, 3D printing, and ceramic) to give a course together or conduct collaborative teaching for a certain issue on digital pottery. The progress was quite smooth for the moment.

It was planned students would be grouped to select one teacher on Week 14 and they could take in-depth action to explore the issue they felt interested. Specific results would be exhibited in the final stage. Teachers and students might take the initiative in discovering problems and presenting solutions. Meanwhile, the motive of spontaneous learning of students could be ignited.

4. ***Field trip/experience away from home***: Trips to the Mixed Reality Research Teaching Center (MRRTC) of National Formosa University and Tien Chung Kiln were arranged for students to experience and actually understand the industry in person.
5. ***The industry gave topics for the course***:
   (1) Pottery samples were replaced by 3D printing, which not only could verify customers' requirements, but also save costs and time.
   (2) AR was used as a marketing tool. International and cross cities and country marketing could be made by AR on the digital network.
   (3) Students could review their learning process at any time which was preserved through videotaping and film editing.
   (4) Ceramic experience helped students to feel the temperature of handmade pottery and understand the flow of ceramic production.

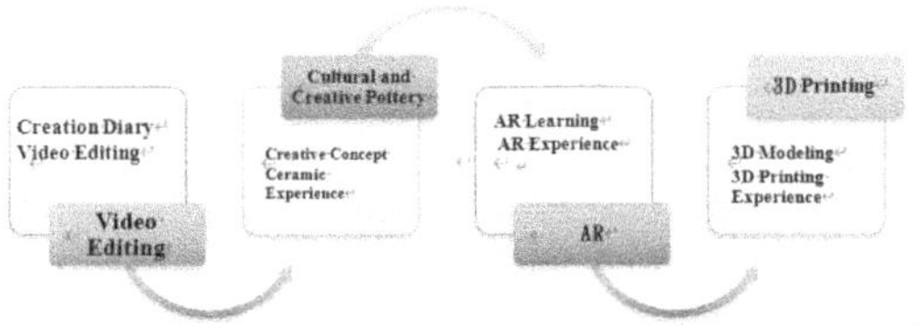

Figure 4. Application of digital tools.

6. ***Apply digital instant messaging for teaching aid***: The app Line was mainly used for real-time communication. In addition to the course group, there were four other team groups established. Besides, an FB web page was set up to record and share all information about the curriculum, explained as follows:
   (1) Online classes for the first four weeks mostly depended on communication through Line and the school announcements on the digital platform.
   (2) Team groups were further established under the course group for discussion respectively.
   (3) Questions from students were mainly responded through the course group on Line; major announcements and assignment uploading were made via the digital platform of the school.

4 CONCLUSION

According to the feedback on the final questionnaire survey, there were six sub-problems for Question 1, "How do I actually feel about this curriculum?" with a mean of 4.5; eight sub-problems for Question 2, "Did this curriculum enhance my ability and knowledge?" with a mean of 4.5. There were two other qualitative essay questions. Generally speaking, diversified learning in the curriculum was recognized. Some students considered that a slower pace would be better. Course suggestions for the future were described as the following:

(1) It would be too ambitious to integrate these four techniques; however, the integration of two to three of them might work. (2) Imagination could only be verified by experiment. (3) Though interdisciplinary, international, and interscholastic factors were accidentally involved; however, the process was challenging. For instance, foreign students had to get out of class on time and couldn't do their homework after class. Thus, the group leader needed to know how to assign tasks and integrate everything during class. Besides, group members needed to explain everything to the foreign students over and over again to make sure they understand since their comprehension in Chinese was not good. (4) Time for pottery production was inadequate. The priority of these four specialties for an 18-week project needed to be adjusted. For example, the ceramic experience came first, followed by video editing, 3D printing, AR, and then pottery production. (5) Field trips should be made twice (one during the first three weeks to understand and experience ceramic and the other during Weeks 12 through 15 for pottery production).

ACKNOWLEDGEMENTS

The sponsor from the MOE Talent Cultivation Project for Digital Humanities was appreciated.

REFERENCES

Archives of MOE Talent Cultivation Project for Digital Humanities. The official website was visited on March 28, 2021.

Shu-ting Huang, 2020/07. "*A Study on Priority Issues and Measures to Solve Population Transition with Digital Technology*" Research Report on Human Resource Planning and Development, Vol. 20. Compiled by Department of Human Resources Development, National Development Council. P:2, ISBN: 978-986-5457-16-7.

Excerpt of the National Development Council, "*Manpower Supply and Demand Investigation and Estimation Result in Key Industries – Cultural and Creative Industry.*" Ministry of Culture (103), "Report on Cultural and Creative Manpower Supply and Demand Gap Survey and Coping Strategies," collected and organized by National Development Council. Date of reference: https://theme.ndc.gov.tw/manpower/cp.aspx?n=3E8CACC405CD9AAE

System Innovation for a Troubled World – Tak Lam et al. (Eds)
© 2023 The Author(s), ISBN: 978-1-032-45525-9

A review on the structural analysis of leaning suspension system of three-wheeled vehicle

Jian-Liang Lin* & Kuo-Hung Hsiao
National Science and Technology Museum, Kaohsiung, Taiwan

ABSTRACT: This paper reviews the design of the leaning suspension system. The historical development of three-wheeler and its leaning suspension mechanism for handling a wider variety of terrain and improving the riding stability and the maneuverability are introduced. According to the structure and the operation of leaning system, the leaning suspension system is categorized into lateral mechanical design, mixed mechanical design, and hydraulic design. The topological analysis of structure and their advantages are presented.

1 INTRODUCTION

Today, people have several land vehicles to transport people or cargo for various terrains by wheels and continuous tracks. The three-wheeled vehicle, also called the three-wheeler, is one kind of available vehicle and develops for a long time, even tracking back to Leonardo da Vinci's automobile between 1475–1500 (E. Payne 2012). The three-wheeler is regarded as a car or a bike since it can be controlled by a steering wheel or a handlebar, such as Toyota i-road and Yamaha Tricity.

The available three-wheelers can be electric vehicles or fuel vehicles, and the fuel three-wheeler almost uses the layout of the front engine. The three-wheelers are categorized into two forms according to the different arrangements of wheels, as shown in Figure 1. The three-wheeler with one front wheel and two rear wheels (1F2R in short) is the delta form or called "trike." The front-wheel controls the steering, and its corresponding mechanism is similar to the motorcycle. The rear two wheels and their mechanism like the one of a car, and the power drives the rear wheels. The advantage of the delta form is the simple design to reduce the cost and the weight of the steering mechanism. Nevertheless, the lateral stability is low, especially for the variation of the motion direction. The three-wheeler with two front wheels and one rear wheel (2F1R in short) is the tadpole form and called "reverse trike." The engine usually drives the rear wheel. The motion direction of the vehicle can be controlled by the rear wheel or the front wheels. The tadpole form has the excellent characteristic of aerodynamics (Engineering a 3 Wheel Vehicle Chassis 2015; J. Sindha 2018; Y. Rajnathn et al. 2020).

*Corresponding Author: jllin927@mail.nstm.gov.tw

The cornering of the three-wheeler is the same of automobiles or multiple-wheeled vehicles. It is hard to turn without some special devices to resolve the lengths of motion paths for the inner and the outer wheels, i.e., the path taken by the outer wheels is larger than the path taken by the inner wheels. The differential is the power distribution device to adjust the speed of the inner and outer wheels to turn smoothly. Some three-wheelers use the differentials for the turning without slipping, like BRP Can-Am Spider and Arcimotor.

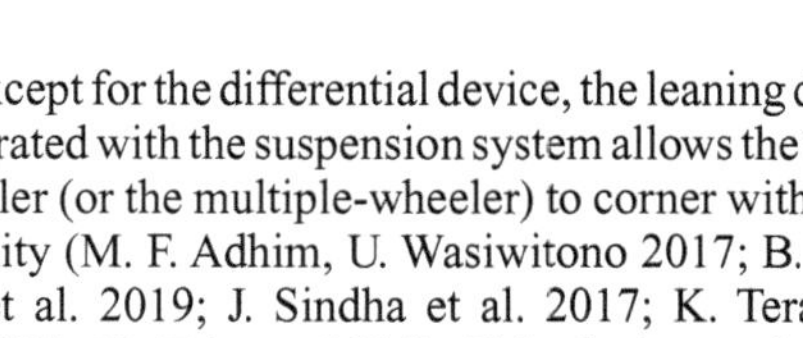

Figure 1. Categorizations of three-wheelers.

Except for the differential device, the leaning design integrated with the suspension system allows the three-wheeler (or the multiple-wheeler) to corner with good stability (M. F. Adhim, U. Wasiwitono 2017; B. Haridas et al. 2019; J. Sindha et al. 2017; K. Terada et al. 2017; K. Takano 2014). This design makes the vehicle can lean during a turn or the process of changing the moving direction through a slight rotation radius. The leaning three-wheeler has developed for various leaning designs. Hence, this paper reviews the development of the leaning three-wheeler. The paper categorizes the available leaning designs and introduces the operation process respectively for these designs to satisfy the function of leaning vehicles. The structural analysis of these available designs is presented.

 DOI 10.1201/9781003377399-30

2 CHARACTERISTICS OF LEANING THREE-WHEELER

The three-wheeler can be regarded as the mixed design of motorcycle and car and develops various designs based on the orientations of vehicles. In the early times, the three-wheeled car, such as cyclecars or microcars, streamlines the design of cars for economic and legal reasons. With the development of a three-wheeled vehicle, the three-wheeled bike that has the structural layouts and the weight similar to a motorcycle but enables to take more cargos is presented. The three-wheeled bike enables the application of the structure of the scooter, like the chassis and the power transmission device.

In reality, the structure of the three-wheeler has good lateral stability when the vehicle is vertically in a stationary situation, moving forward, crossing a frictionless obstacle. The gravity center of the vehicle is in the range of the triangle by three wheels. The wheel track and the extra wheel reduce the maneuverability and the stability when they overturn in corners. The problem is resolved by the leaning device integrates with the suspension system. The electronic control systems with sensors are attached to enhance the efficiency and stability. The leaning design allows the vehicle with the narrow track to corner safely and improve the sport performance of the traditional three-wheeler (M. F. Adhim; U. Wasiwitono 2017; B. Haridas et al. 2019; Y. Rajnathn et al 2020; M. A. Saeedi, K. Takano 2014; R. Kazemi 2013). Their wheel longitudinal movement can advance the roll stability of the vehicle (C. Tang et al. 2018). The electrical or mechanical parking systems are used to lock the vehicle to prevent fall down for the decreasing wheel track.

The controlling methods of leaning three-wheelers are active and passive. The passive leaning methods include the driver skill, the controlling of the motorcycle, and the counter-steering. The active learning methods mean reducing the demand of counter-steering to allow direct steering inputs and to maintain the vehicle in an upright position when stationary. The active action makes the three-wheeler like the controlling of cars (J. Sindha 2018).

The existing designs of leaning three-wheelers are both of the delta form and the tadpole form. The important design concept of the leaning suspension system is to make these wheels maintain contact with the ground, whatever terrains and riding situations. For the shocks from the bumps of the road, the leaning suspension system can eliminate the resulting displacements of the wheels through the compressive or tensile actions of absorbers and the operations of the learning design. Such a leaning design significantly contributes to the bumping of a single wheel. Furthermore, the operation situation is the same as the tilting for cornering. Based on the existing designs, the leaning function can be generated through the motions of links and the strokes of the hydraulic cylinder. To maintain the balance of chassis is the common point for various leaning designs.

3 CATEGORIZATION AND STRUCTURAL ANALYSIS

According to the structure, the absorbers arrangement, and the operating principle, the leaning suspension system is categorized into the lateral mechanical design, mixed mechanical design, and hydraulic design. Most me-chanical designs are simple linkage mechanisms, and few mechanical designs used gears or a combination of chain and sprocket. The hydraulic designs applied actuators, hydraulic valves, and gear pumps. In what follows, these types are introduced in detail.

3.1 *Lateral mechanical design*

The lateral mechanical design of the leaning system is the application of a parallelogram mechanism. The available designs can be roughly divided into two types: a parallelogram linkage and two separate parallelogram linkages.

One parallelogram linkage

One parallelogram linkage is the typical design of the leaning system, such as Tricity and Niken of Yamaha (K. Takano 2014; K. Terada et al. 2016), Piaggio MP3 (S. Bartolozzi et al. 2010), Honda Neowing. The parallelogram places above the absorbers or the forks respectively for the front wheels to form the leaning design. The leaning parallelogram cannot rotate with the steering and usually has a tilting front end. The corresponding generalized mechanical device is shown in Figure 2(a). The topological analysis indicated that the design is a (5, 6) planar mechanism. Owing to the concept of symmetric structure and the parallel rotation axes, the generalized chain is regarded as (4, 4) a generalized chain, as shown in Figure 2(b).

Furthermore, by applying absorbers or the forks, Kawasaki presented an approach for using one parallelogram as the leaning system (M. Hinchliffe). The original design seems to be complex with mono-like BMW telelever arrangement. The normal-looking front forks replaced the telelever suspension in the revolutionary design. The parallelogram that is still in the bottom to connect to the hubs makes the leaning mechanism rotate with steering.

Two separate parallelogram linkages

The combination of two separate parallelogram linkages developed several different designs by the arrangement of absorbers (S. Carabelli et al. 2008; S. Carabelli & A. Tonoli 2007; D. Mercier 2012; R. J. Suhre et al. 2007; A. Tonoli et al. 2007), as shown in Figure 3(a). The corresponding generalized mechanical device is shown in Figure 3(b). When one wheel takes the bump forces, its corresponding parallelogram operates independently, and another parallelogram is stationary. The absorber connects to the chassis and the upper or the lower swing arm. Such a design is the same as the structure of double-wishbone suspension.

For one side mechanism, it is a (6, 7) planar mechanism with 1-DOF, and the generalized chain is shown in Figure 3(c).

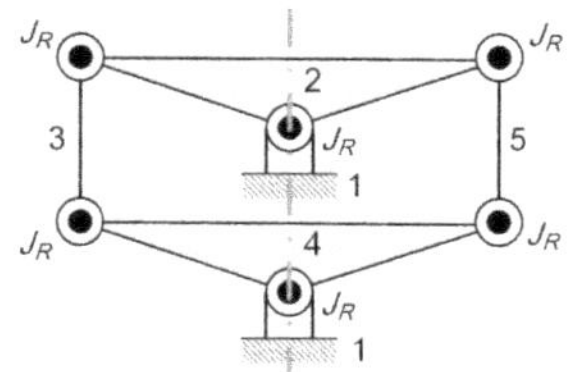

(a) Corresponding generalized mechanical drawing

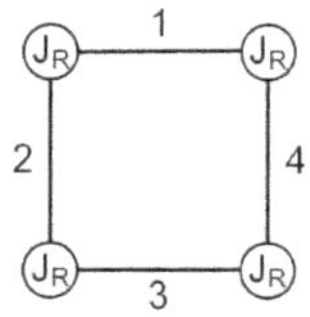

(b) Corresponding generalized chain

Figure 2. Lateral mechanical design_one parallelogram.

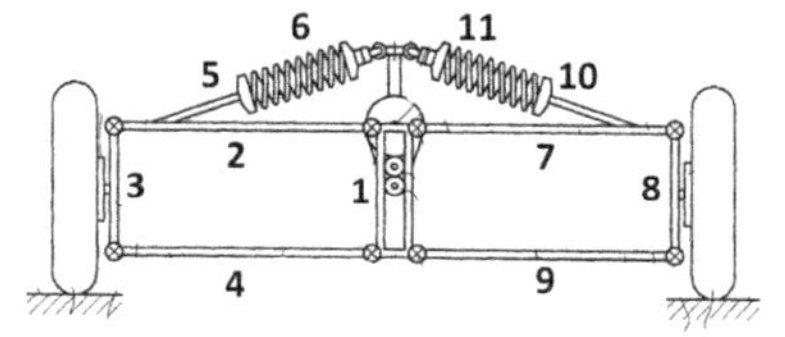

(a) An existing Design of leaning two-separate parallelogram

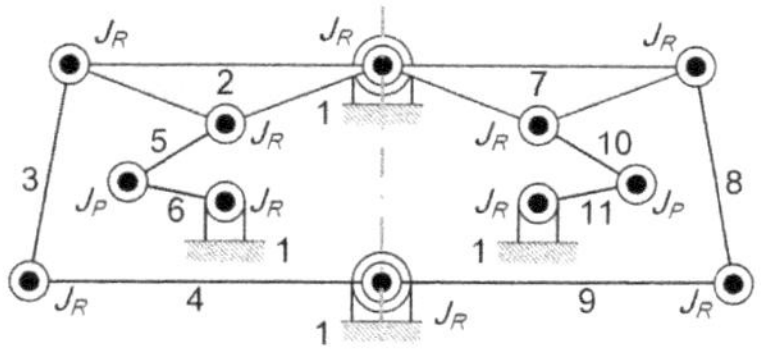

(b) Corresponding generalized mechanical drawing

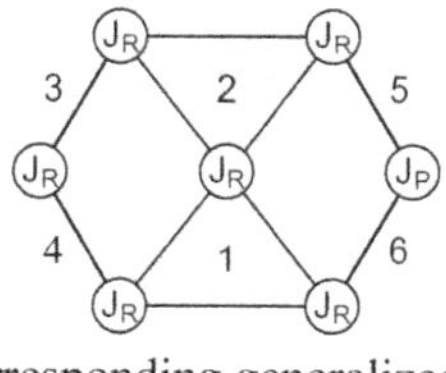

(c) Corresponding generalized chain

Figure 3. Lateral mechanical design_two separate parallelograms with two absorbers.

There is a unique design presented by Peugeot (M. Doveri 2004). It can be called "Dual Tilting Wheels." The design used only one absorber for the front suspension of the three-wheeler. This absorber is installed horizontally to connect the two parallelogram linkages. The design likes a variable parallelogram. It is a (9, 11) planar mechanism with ten revolute joints and one sliding joint and has two degrees of freedom, as shown in Figure 4. The two inputs are the contact forces from the ground for generating the constrained motion of the mechanism.

3.2 *Mixed Mechanical Design*

A particular mechanism design of the leaning system by Honda (B. Gomes 2018). The design mixes the design concept of the lateral parallelogram and the longitudinal parallelogram. The design is complex and seems to be heavy for maneuvering. The function of the lateral parallelogram is the same as the above lateral mechanical design. However, the left and the right side respectively extend a parallelogram. This conceptual design can increase the tilting angles even the adding of the wheel base would reduce the maneuverability. This mixed design is a spatial mechanism. According to the geometrical characteristics, its structural analysis can be divided into three planar mechanisms. The longitudinal design is a (6, 7) mechanism with 1 DOF. The lateral design can be regarded as a (4, 4) mechanism with 1-DOF through the symmetric structure.

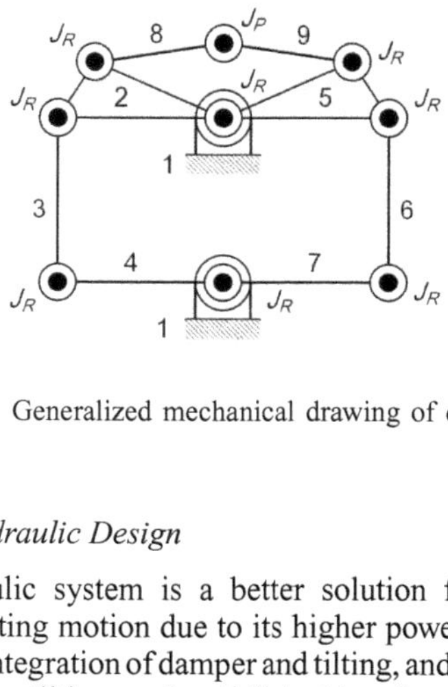

Figure 4. Generalized mechanical drawing of dual tilting wheels.

3.3 *Hydraulic Design*

A hydraulic system is a better solution for stably manipulating motion due to its higher power density, flexible integration of damper and tilting, and precisely controlling tilting angles (full locking function). For the leaning suspension design, the application of the hydraulic system can generate a sensitive response and a compact structure. Without the application of linkages, the whole weight of the leaning vehicle is reduced actually. The characteristics of the low gravity center and the superior handling are confirmed without the existence of a leaning parallelogram above or below the absorbers. Further, the hydraulic leaning system can control the front and the back suspension device simultaneously in the development of multi-wheels vehicles.

The hydraulic design of the leaning suspension system develops the lateral and the longitudinal layouts, but their principles are almost the same. The hydraulic system is pressured through the pump. The system integrated several sensors and valves to control the operation of hydraulic flow for the vehicle's motion. Moreover, the hydraulic system has to support the weight of the vehicle.

The Hydraulic Tilting System (HTS in short) is laterally arranged and uses the hydro-pneumatic technology patent (M. Moroni & R. Marabese 2020), as shown in Figure 5. The left and the right hydraulic pipes of the HTS could be regarded as shock absorbers. One end is installed on the frame, and the other is installed on the swing arm connected to the hub. In addition to the left and right hydraulic pipes, a third hydraulic pipe is installed in the middle to connect with the left and right pipes. In fact, the whole system is like a shock absorber using gas and hydraulic oil to absorb the impact and ease the action.

When the hydraulic leaning vehicle turns, the hydraulic system begins to adjust the height of the tires due to the different forces on these tires. Based on the force balance of the hydraulic system, the hydraulic oil flows from the hydraulic pipe in the high stress to the pipe in the low stress by the connecting pipe. The different heights required by the left and right wheels of the tricycle are achieved when turning. At the same time, the hydraulic pressure can absorb the force impact on tires. Furthermore, the system can fully lock the vehicle for parking by closing the ball valve of the connecting pipe. When closed, the hydraulic pipes of the left and right wheels will be independent and will not affect each other.

The longitudinal hydraulic design is a rare design. Because of the longitudinal layout, the track is narrower, but the wheel base is longer than the lateral mechanical design. The hydraulic device is a resolution for lateral stability. Nick Schotter's 4MC and Yamaha's concept vehicles-Tesseract and OR2T apply this leaning design, but they are four-wheelers. The leaning suspension system has the structure of dual-scythe. The motions of links like scythes are dependent, and the balance of the bike is controlled through the hydraulic system. For 4MC, the hydraulic system controls the front and the back leaning suspension. For Yamaha's vehicles, the back leaning suspension system is archived by the spatial linkages.

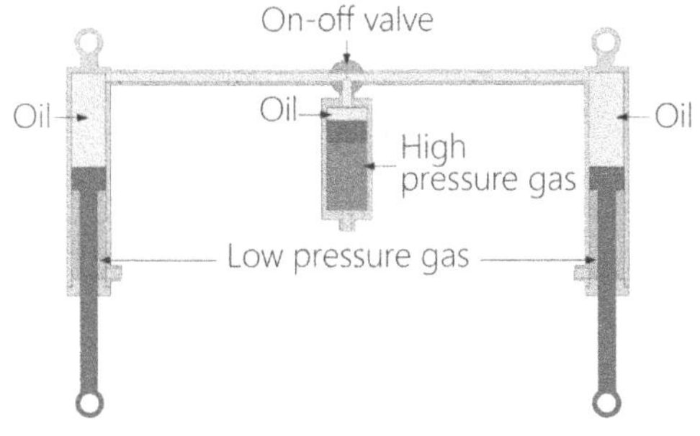

Figure 5. Hydraulic tilting system.

4 CONCLUSIONS

At the early stage of developing a tree-wheeler, the design topic of the vehicle is to transport more people and cargos, and the smooth steering is stiff. With the development of the three-wheeled vehicles, the good maneuverability, the high stability in the cornering, and the vehicle's aerodynamic characteristics for various terrains and riding conditions are gradually emphasized. This paper reviewed the development of the three-wheeler and the integrated leaning suspension system and introduced the character of the vehicle, including the narrow track, lateral stability, and the functions of leaning and crossing bumps or lumps.

The available learning systems are generally categorized. For each categorized design, the analysis of the topological structure and the operating principles are presented. The existing leaning systems included lateral mechanical design, mixed mechanical design, and hydraulic design. Moreover, the hydraulic design can be arranged laterally or longitudinally. The existing lateral mechanical designs are one parallelogram and two separate parallelograms. Based on the symmetric structure, these two designs are four-bar linkages with four revolute joints and 1 DOF. The one parallelogram is the most usual mechanical design. The mechanism above or below the absorbers affected the flexibility of maneuverability. The two separate parallelograms used the structure of double-wishbone suspension. The layout of absorbers affected their numbers applied in the design to result in the different operation processes. As to the hydraulic leaning system, the design concepts for the lateral and the longitudinal layouts are the same. The hydraulic cylinder is the one element of leaning design and the absorber. Based on the balance of pressure, the hydraulic flow is controlled to adjust the resulting height of the wheel for leaning or crossing bumps.

With the development of the leaning suspension system, the four-wheeler and the innovative multi-wheeled vehicles are generated further. Therefore, the leaning suspension system could be a landmark for the history of multi-wheel vehicles.

REFERENCES

Adhim, M. F., & Wasiwitono, U. (2017). Dynamics Analysis of Narrow Tilting Three-Wheeled Vehicle with LQG Control. *The International Journal of Mechanical Engineering and Sciences*, 1(2), 36–47.

Bartolozzi et al. (2016). *Motorcycle*, (European Patent No. 2,977,306,A1). European Patent Office.

Carabelli, S., & Tonoli, A. (2007). *A suspension tilting module for a wheeled vehicle and a wheeled vehicle equipped with said suspension tilting module* (European Patent No. 1 870 269 A1). European Patent Office.

Carabelli et al. (2008). *Suspension Tilting Module for A Wheeled Vehicle and A Wheeled Vehicle Equipped with Said Suspension Tilting Module* (U.S. patent No. 2008,0012262 A1). U.S. Patent and Trademark Office.

Doveri, M. (2004). *A laterally tilting three-wheeled vehicle* (European Patent No. 1 180 476 A1). European Patent Office.

Engineering a 3 Wheel Vehicle Chassis. (September 9, 2015), EngineerDog.com. https://engineerdog.com/2015/09/09/engineering-a-3-wheel-vehicle-chassis/

Gomes, B. (2018, August 30). *Honda files patent for a tilting three-wheeler*. iMotorbike. https://imotorbike.my/news/en/2018/08/honda-files-patent-for-a-tilting-three-wheeler

Haridas et al. (2019). Design and Fabrication of Tilting Trike. *International Research Journal of Engineering and Technology*, 6(5), 4305–4307.

Hinchliffe, M. (n.d.). *Kawasaki leaning toward three-wheelers*. Motobikewriter. https://motorbikewriter.com/kawasaki-leaning-toward-three-wheelers/

Mercier, D. (2012). Leaning Vehicle with Tilting Front Wheels and Suspension Therefor (U.S. patent No. 8,317,207 B2). U.S. Patent and Trademark Office.

Moroni, M., & Marabese, R. (2020). *Improved suspension* (European Patent No. EP 3 429 873 B1). European Patent Office.

Payne, E (2012). *The History of the 3-Wheeled Vehicle*, 3-wheelers.com. http://www.3wheelers.com/history.html.

Peugeot motorcycles (2021). *Peugeot motorcycle*, Retrieved June 10, 2021, from "https://www.peugeot-motocycles.fr/gamme/metropolis-active-2020

Rajnathn et al. (2020). Design of Tilting Reverse Trike. *International Research Journal of Engineering and Technology*, 7(5), 5041–5046.

Saeedi, M. A., & Kazemi, R. (2013). Stability of Three-Wheeled Vehicles with and without Control System. *International Journal of Automotive Engineering*, 3(1), 343–355.

Sindha, J., Chakraborty, B., & Chakravarty, D. (2017). Automatic stability control of three-wheeler vehicles – recent developments and concerns towards a sustainable technology. *Proceeding IMech E Part D: Journal of Automobile Engineering*, pp.1–17.

Suhre et al. (2007). *Leaning Suspension Mechanics* (U.S. patent No. 2007/0075517 A1). U.S. Patent and Trademark Office.

Takano, K. (2014). The Development of Tricity. *Yamaha Motor Technical Review*, 50, 2014.

Tang, C., He, L., & Khajepour, A. (2018). Design and analysis of an integrated suspension tilting mechanism for narrow urban vehicles. *Mechanism and Machine Theory*, 120, 225–238.

Terada, K., Sano, T., & Togawa, T. (2017). Investigation of the behavior of three-wheel vehicles when they pass over a low μ road surface. *SAE/JSAE 2016 Small Engine Technology Conference & Exhibition*, 30–37. (Also presented in *YAMAHA Motor Technical Review,* 110–119.)

Tonoli et al. (2007). *Vehicle with tilting suspension system* (European Patent No. 1 813 450 A1). European Patent Office.

System Innovation for a Troubled World – Tak Lam et al. (Eds)
© 2023 The Author(s), ISBN: 978-1-032-45525-9

A hybrid channel allocation scheme for dense VANETs

Jeng-Ji Huang* & Yu-Chi Kao
Department of Electrical Engineering, National Taiwan Normal University, Taipei, Taiwan

David Shiung
Department of Electronics Engineering, National Changhua University of Education, Changhua, Taiwan

ABSTRACT: Vehicular ad-hoc networks (VANETs) have recently emerged as a promising wireless networking technology to provide a wide range of services and applications for the development of intelligent transportation systems (ITS). Recently, several research works have been focusing on the assignments of data channels. However, most of them rely only on local information, which can be the usage or the quality of each data channel, and is typically gathered from a vehicle's one-hop or two-hop neighbors. Since clustering of vehicles has widely been discussed in VANETs, we thus assume in this paper that global information containing the usage of each data channel across all clusters in the system can be obtained via inter-cluster communications. Based on this assumption, we propose a hybrid channel assignment scheme in VANETs to improve the capacity by making data channels be used more compactly, as a higher capacity leads directly to a lower call blocking probability, or better quality of service (QoS) perceived by users. That is, the local information is utilized in the proposed scheme to avoid both hidden terminal and exposed terminal problems, whereas the global information is then aided in selecting the channel most heavily used in the entire system. We then investigate factors that may have significant impact on capacity improvement, including vehicle densities, transmission ranges of vehicles, offered traffic load, and the total number of vehicles in a VANET. Simulation results show that the proposed scheme can improve the capacity by up to 20%, compared to one that uses simply local information.

1 INTRODUCTION

Vehicular ad hoc networks (VANETs) have been widely considered one of the most successful applications of mobile ad hoc networks. This is because vehicles equipped with a wireless communication device can self-organize and form a network to exchange information with other vehicles or roadside units (RSUs) (S. F. Hasan et al. 2013). Therefore, VANETs can support many services and applications for intelligent transportation systems (ITS), including cooperative collision avoidance (CCA), road traffic management, and entertainment and business applications.

However, the high mobility of vehicles leads directly to the fast-changing of network topology. Therefore, some technical issues need to be carefully addressed in VANETs. For instance, the connectivity has analytically been evaluated in Y. Zhuang et al. (2011), and the routing in sparse VANETs has been discussed in J.-J. Huang (2015). The delivery of safety messages for CCA applications has attracted a lot of research attention, e.g., H. A. Omar et al. (2013). The medium access control (MAC) layer in VANETs can either be contention-based (IEEE Standard 802.11p-2010 2010), reservation-based (H. A. Omar et al. 2013), or a combination of both (L. Zhang et al. 2014).

On the other hand, scalability issues may exist in a dense VANET when a fully distributed protocol is employed. Clustering, which is half-centralized and half-distributed, is clearly suitable for a dense VANET, not to mention its advantage of enabling efficient resource allocations. Research works, such as A. Ahizoune et al. (2010), have been discussing clustering in VANETs and many of them focus on designing protocols for transmissions of non-real-time data messages, along with a contention-free broadcast protocol for delivery of safety messages.

Regarding non-real-time transmissions on service channels (SCHs), spatial reuse is a common technique for increasing bandwidth efficiency. However, care must be taken during channel allocations because simultaneous transmissions may interfere with each other. Apart from approaches such as code division multiple access (CDMA) or orthogonal frequency division multiple access (OFDMA), the interference is managed by Hafeez et al. (2011) by separating two transmissions by a sufficient distance, or in Dang et al. (2013) by the sender negotiating with the receiving end on which SCH has the best quality for transmission.

*Corresponding Author: hjj2005@ntnu.edu.tw

DOI 10.1201/9781003377399-31

However, Hafeez et al. (2011) and Dang et al. (2013) would fail to ensure the complete removal of the hidden terminal and the exposed terminal problems. Although these two problems have well been tackled in Omar et al. (2013) by designing a protocol based on ADHOC MAC, only local information is utilized by Hafeez et al. (2011), Dang et al. (2013), and Omar et al. (2013) during channel allocations, rendering them to potentially suffer from low bandwidth efficiency. Thus, a hybrid channel allocation scheme is thus proposed in this paper to address all the issues mentioned above. In the proposed scheme, valid SCHs are first determined by utilizing local information to avoid both the hidden terminal and the exposed terminal problems. Global information is then exploited to increase bandwidth efficiency by selecting the most heavily used among valid SCHs. Selecting the most heavily used can in effect make SCHs be assigned more compactly, which implies that SCHs can be reused more frequently within the service area of the system. A smaller frequency reuse factor or a higher system capacity can thus be obtained.

2 SYSTEM MODEL

According to the Dedicated Short Range Communication (DSRC) standard, there are 7 channels throughout 75 MHz bandwidth at 5.9 GHz band, which include 6 service channels (SCHs), i.e., Ch172, Ch174, Ch176, Ch180, Ch182, and Ch184, and the dedicated control channel (CCH), i.e., Ch178, for delivering safety messages and announcements. On the other hand, based on the IEEE 1609.4 standard, which describes multichannel operations for wireless access in vehicular environments (WAVE), each WAVE device is equipped with one transceiver, and all devices have to perform time synchronization with each other as long as they stay in the VANET.

Since only one transceiver is used, a time-division structure like the IEEE 1609.4 standard is defined, where a DSRC radio can alternately switch between channels to support different applications concurrently. A synchronization interval of 0.1 sec is equally divided into two parts: a CCH interval and an SCH interval. During a CCH interval, all vehicles listen to the CCH. A time division multiple access (TDMA) reservation-based ADHOC MAC is employed on the CCH, rather than the contention-based IEEE 802.11p specified in the IEEE 1609.4 standard. A reliable broadcast service can thus be guaranteed, ensuring fast delivery of emergency safety messages.

A cluster is formed when vehicles are moving in the same direction and located within the transmission range of the cluster head vehicle R. Since cluster management is not the main focus of this paper, we assume that a cluster configuration protocol similar to Su et al. (2007) is employed. We assume that a logical connection exists between any two neighboring cluster head vehicles for the inter-cluster communications. The exchanges of information between cluster head vehicles are performed on the CCH and may involve multi-hop transmissions.

3 THE PROPOSED SCHEME

In the proposed scheme, two kinds of local information are gathered by a vehicle. One is, like ADHOC MAC, the status of TDMA slots on the control channel (CCH); another is the status of those service channels (SCHs) that are being used by the vehicle's one-hop neighbors.

In order to avoid hidden terminal problems, like ADHOC MAC (F. Borgonovo et al. 2004), a virtual frame structure consisting of K slots is utilized in the proposed scheme on the CCH. A control field named frame information (FI) is included in each transmission in the time slot reserved by a vehicle. The FI is a vector with K entries specifying the status of each of the K slots preceding the transmission. Based on received FIs, each vehicle marks a slot either as RESERVED or AVAILABLE. For more details about the FI and the reservation process in ADHOC MAC, refer to Borgonovo et al. (2004).

On the other hand, a vehicle keeps track of those SCHs that are currently being used by its one-hop neighbors. In the following, we illustrate how a call is established by reserving an SCH on the CCH.

First, in order to keep track of the status of SCHs, a *local channel usage table* is maintained at each vehicle. When a vehicle intends to send data to another vehicle, the table is looked up before a request (REQ) packet embodying a list of SCHs currently unused by the sender's one-hop neighbors is sent in its own time slot on the CCH. Similarly, the receiver acknowledges the reception of the REQ by sending a reply (REP) packet embodying a list of SCHs currently unused by the receiver's one-hop neighbors.

Second, the cluster head, based on the global information, which will later be discussed, selects the most heavily used among the SCHs that are included in both the REQ's and the REP's list. The decision is then announced in the cluster head vehicle's own time slot on the CCH. The decision is then announced in the cluster head vehicle's own time slot on the CCH.

Third, the sender and the receiver, after reading the decision, send individually an acknowledgment (ACK) packet in their own time slots on the CCH, in order to notify their own one-hop neighbors of the decision. At the cluster head vehicle, after receiving an ACK sent, the field corresponding to the number of links using the SCH is increased by one in a *cluster channel usage table*. Note that a call may be blocked due to either all SCHs being busy or any of the transmission pairs being occupied by another transmission.

By reading the ACK packet broadcast by the sender or the receiver, a vehicle, being a one-hop neighbor of either side, can be notified of the imminent data transmissions, and its local channel usage table can accordingly be updated by adding a new record containing the following information: the vehicle, the SCH, and the way it will be used, i.e., for transmission (Tx) or reception (Rx). The record is immediately deleted after the call is finished.

Note that the exposed terminal problems are avoided in the proposed scheme when constructing the REQ's and the REP's list; namely, only those SCHs that are being used to receive (send) by the one-hop neighbors of the sender (receiver) are excluded from the REQ's (REP's) list.

Finally, the global information about the usage of SCHs across different clusters is exchanged through inter-cluster communications between cluster head vehicles. The information exchange is event-triggered by either a call establishment or a call finish. The inter-cluster communications are performed on the CCH through a permanent logic connection, which exists between any pair of neighboring cluster head vehicles. Although two neighboring cluster head vehicles may not be located within each other's transmission range, information exchanges can be performed through multi-hop transmissions. In case of network disconnection, vehicles on the opposite lanes may help restore the network connection for inter-cluster communications (Huang, 2015), or a standby 3G or 4G connection can take over the exchange of the global information. At each cluster head vehicle, in addition to a local channel usage table and a cluster channel usage table, a *global channel usage table* is maintained, where the global information is kept. The global channel usage table looks the same as a cluster channel usage table; nevertheless, each field with respect to the number of links in the former should clearly be no smaller than that in the latter.

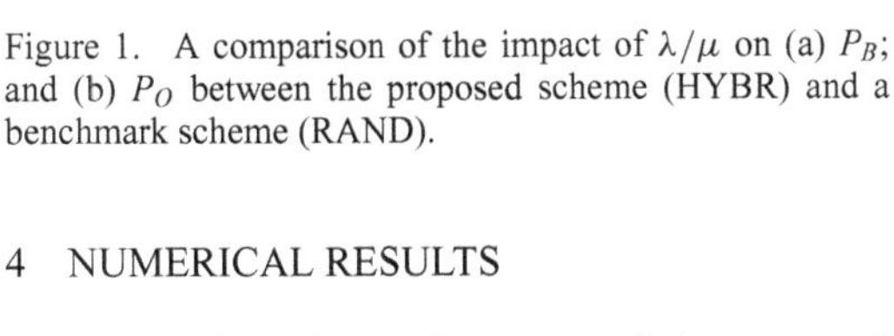

Figure 1. A comparison of the impact of λ/μ on (a) P_B; and (b) P_O between the proposed scheme (HYBR) and a benchmark scheme (RAND).

4 NUMERICAL RESULTS

In this section, the performance of the proposed scheme, termed HYBR, is evaluated via simulations written in C++. In order to demonstrate the net effect of the global information, the proposed scheme is compared with a benchmark scheme, termed RAND, using only local information. That is, because of no global information in the baseline scheme, a service channel (SCH) is randomly selected by the cluster head vehicle among valid ones, which appear in both the REQ's and the REP's list. Except for this, the two schemes are exactly the same. Each simulation result is obtained from the average of 100 independent runs, and the simulation time is 3,600 seconds in each run. In the simulations, a highway segment of length 3,000 m and a mobility model similar to Omar et al. (2013) is employed, where 100 vehicles repeatedly run on the highway segment with a Gaussian (100,20) distributed speed V km/hr. The call arrival process is Poisson with rate λ (1/second) and the call durations are *i.i.d.* exponential random variables with mean $1/\mu = 1$ second. As mentioned in Section 3, a call may be blocked due to either all SCHs being busy (a B event), or any of the transmission pairs being occupied by another transmission (an O event). In Figure 1, the occurrence rate of the B event, denoted by P_B, and the occurrence rate of the O event, denoted by P_O, are compared separately between HYBR and RAND, where the transmission range of a vehicle $R = 150, 200$, or 250 m, and the data traffic intensity λ/μ ranges between 2 and 10.

As shown in Figure 1, HYBR significantly outperforms RAND in terms of P_B, whereas the two schemes are comparable with respect to P_O. HYBR can reduce the P_B of RAND by up to 85.4%. On average, the improvement of the former over the latter on P_B is 76%. The performance gain comes mainly from a more compact use of SCHs or a better channel reuse under HYBR. The results clearly demonstrate that exploiting the global information is very beneficial for increasing capacity in VANETs.

5 CONCLUSION

In this paper, a channel allocation scheme is proposed for a dense VANET, which utilizes both local and global information. Numerical results show that our proposed scheme significantly outperforms the scheme which does not use global information.

REFERENCES

S. F. Hasan, N. H. Siddique, and S. Chakraborty, 2013. *Wireless Pers. Commun.* 69 1261–1287.

Y. Zhuang, J. Pan, Y. Luo, and L. Cai, 2011. *IEEE J. Sel. Areas. Commun.* 29 187–196.

J.-J. Huang, 2015. *IEEE Commun. Lett.* 19(7) 1193–1196.

H. A. Omar, W. Zhuang, and L. Li, 2013. *IEEE Trans. Mobile Comput.* 12 1724–1736.

IEEE Standard 802.11p-2010, 2010. *IEEE* 1–51.

L. Zhang, Z. Liu, R. Zou, J. Guo, and Y. Liu, 2014. *Wireless Pers. Commun.*74 1197–1212.

A. Ahizoune, A. Hafid, and R. B. Ali, 2010. *IEEE Conference on Local Computer Networks* 48–55.

K. A. Hafeez, L. Zhao, and B. N.-W. Ma, 2011. *EURASIP Journal on Wireless Communications and Networking* 117.

D. N. M. Dang, H. N. Dang, C. T. Do, and C. S. Hong, 2013. *IEEE Wireless Communications and Networking Conference (WCNC)* 351–355.

H. Su and X. Zhang, 2007. *IEEE Trans. Veh. Technol.* 56(6), 3309–3323.

F. Borgonovo, A. Capone, M. Cesana, and L. Fratta, 2004. *Wireless Networks* 10, 359–366.

System Innovation for a Troubled World – Tak Lam et al. (Eds)
© 2023 The Author(s), ISBN: 978-1-032-45525-9

Research on multisensory design of electric shaver from the perspectives of Kansei engineering

Jeng-Chung Woo*
School of Design•Straits Institute of Technology, Fujian University of Technology, Fuzhou, China
Design Innovation Research Center of Humanities and Social Sciences Research Base of Colleges and Universities in Fujian Province, Fuzhou, China

Zhe-Hui Lin, Feng Luo & Guo-Qing Pan
School of Design•Straits Institute of Technology, Fujian University of Technology, Fuzhou, China

ABSTRACT: The market scale of electric shaver in China has reached 26.3 billion¥ in 2021. Consumers currently place an increasing emphasis on the emotional image conveyed by products rather than just concerning with functional satisfaction. To meet consumers' expectations, the emotional message conveyed by product design is essential under multisensory channels. This reasearch first collected 220 electric shaver samples and 135 pairs of consumers' Kansei words, and then reduced them into 34 representative samples using multidimensional scale and clustering analysis, with 4 groups of representative Kansei words selected via the expert group. Moreover, consumers' Kansei images were evaluated via questionnaire using the Semantic Differential scales, with 416 valid samples acquired in total. Meanwhile, design elements of the samples (incl. Item and Category) were classified by ways of morphological analysis and audio software. At last, an associated model of the electric shaver was established between the overall design elements and user's Kansei evaluation under the multisensory channel of visual model and auditory audio taking advantage of Quantification Theory Type I. The proposed model can provide defined design indexes and references in multisensory design, facilitating designers to design in a logical and scientific manner rather than designing as per experience.

Keywords: Kansei Engineering; Quantification Theory Type I; Multisensory Design; Electric shaver.

1 INTRODUCTION

The electric shaver market has reached 26.3 billion¥ along with the development of the manufacturing industry in China (Zhang 2022). Besides, factors affecting consumer behavior involve emotional and personalized needs according to the market, rather than just product function and structure due to the social development and upgraded consumption needs. Kansei engineering is an important method for studying the emotional needs of a product to investigate the multiple linear relationships between users' emotional needs and design elements from quantitative and qualitative perspectives (Nagamachi 1995). Previously, Kansei engineering was principally applied in modeling research. For instance, Tomio studied car interiors to improve users' riding experience with the help of Kansei engineering (Jindo et al. 1997); and Shergian conducted Kansei research on the modeling of bamboo weaving clocks to increase product sales (Shergian et al. 2015). Moreover, Kansei engineering was also applied in some sound studies. For example, Keiichi developed a 3D audio walking guidance system for visually impaired people upon conducting Kansei evaluation of white noise and natural sound (Muramatsu et al. 2019); Hideyoshi studied the sound quality of vacuum cleaner in response to consumer's explicit and potential perceptual needs using Kansei engineering (Yanagisawa 2011) and emphasized that auditory message also plays an important role in the interaction between users and product, rather than just the visual message. Apparently, in the context of an emphasis on the user experience, multisensory design will provide a better way to guide design research.

Razza deconstructed and carry out Kansei evaluation on design elements of shaver modeling using Quantification Theory Type I (QTTI) (Razza 2015). Atamer disclosed that a sound study is in favor of better sales of products in a highly competitive market after the proposal of the sound of electric shaver is an important attribute that reflects varying emotional needs of men and influences, affecting the preference for purchase (Atamer 2016). However, the discussion on multisensory experience was absent in the

*Corresponding Author: wwwjc2020@outlook.com

DOI 10.1201/9781003377399-32

previous study on electric shavers from the perspective of modeling and sound.

With the perspectives of mutisensory design, this study is based on Kansei Engineering and QQTI to quantify the design elements (modeling and sound) of electric shaver and Kansei evaluation in order to analyze their association and establish a multiple linear prediction model. Based on the result, the relationship model will be applied in the optimized design of product modeling.

2 LITERATURE REVIEW

2.1 *Kansei engineering*

Kansei Engineering developed by Nagamachi is an approach for quantifying perceptual needs, which has been extensively utilized in a variety of fields. Ngip studied the design elements of sunglasses in Malaysia for selecting popular sunglasses by applying Kansei engineering in product marketing (Chuan N. et al. 2013). Kim built a Kansei evaluation system for mining the emotional expectations of recliners and proposed an approach for extracting emotional expectation (Kim W. et al. 2019). Yanagisawa carried out UX design upon constructing the user's Kansei evaluation as a cognitive process associated with 4 levels (Yanagisawa H. 2017).

2.2 *Quantification theory type I*

Quantification Theory Type I is the most widely used theory in the study of Kansei Engineering. which can predict a single dependent variable from multiple independent variables using a multiple linear equations. In this way, a quantitative and qualitative analysis can be completed (Atamer S. 2016). Nagamachi used QTTI to design cosmetic containers and Taufik used QTTI combined with and Kansei engineering to systematically analyze the appearance of detergent packaging. (Nagamachi M. 1995, Djatna T. et al. 2015).

In this research, items stand for different types of product design elements, whereas categories represent different classifications under various items. To analyze the association between design elements and Kansei evaluation through linear prediction model.

2.3 *Multisensory design*

The multisensory experience generated by people during contact is the result of the experience of products, services or spaces. The dominance of the visual sense in perceiving products and environments has been supported and emphasized in mainstream theories for years (Berendt 1988). Nowadays, the emergence of the multisensory design indicates that people's overall perception of products is defined jointly by the role of multiple senses (Dal Palù et al. 2015). In 2008, Schifferstein defined multisensory design as a design method in which a product conveys concrete impressions and message that constitutes a whole to users through multiple senses (Schifferstein et al. 2008). Moreover, Schifferstein discussed the roles played by different senses by comparatively analyzing the detailed degree of the message conveyed by a product in a single-sensory and multisensory manner. At the same time, Kuo summarized the design principle of bicycle saddles in the study of how sentiments of female road cyclists are affected by bicycle saddles via visual and tactile senses using the Semantic difference scales (SDS) (Kuo et al. 2020).

Izumi improved the design of electric shaver to facilitate enterprises to lower costs with the adoption of the TRIZ concept and information integration technology(Izumi et al 2015), T. Miyano reinforced the corrosion resistance of electric shaver blades by way of ion beam enhanced deposition(Miyano T. 1994). Unlike the previous study focusing on functions of electric shaver.

This research is conducted for studying electric shaver from the perspectives of multisensory design of modeling and sound to better satisfy the user's emotional experience. Product design and research, in recent years, have been extensively conducted using the strategy of combining multisensory design with varied research methods, contributing to the designer's scientific design activities and improving user experience.

3 MATERIALS AND METHODS

3.1 *Selection of representative samples*

First, this research deconstructed design elements of modeling and sound of electric shaver using Kansei engineering in combination with QTTI. Then, multiple linear prediction models were constructed by conducting seven-point Likert questionnaire survey on representative samples and Kansei words. The research is specifically divided into the following aspects: (i) Representative samples are filtered out using cluster analysis. (ii) Representative Kansei words are selected through the expert group. (iii) Item and category deconstruction is performed on product design elements. (iv) Association between design elements and Kansei evaluation is established through the QTTI method.

Pictures for samples collected from the e-commerce platform with complicated backgrounds were removed. Afterward, 220 electric shaver samples were acquired.

After the first round of screening, 120 samples were obtained after removing the pictures with higher similarities, which were then reduced to 80 samples upon the selection of 5 graduate students with a discipline background in design. At the same time, grey-scale processing was performed to remove the colors of the pictures in the preliminary library in order to lower the influence of subjective color preference on the study. Next, the 80 samples were classified as per the similarity and divided into 12 to 19 groups according to sample representativeness by 25 graduate students majoring in design and industrial experts with having their backgrounds in product design using the expert group method. Based on this, these samples were encoded in an 80 × 80 dissimilarity matrix. Further, analysis was performed in SPSS 23.0 software using the multivariate scaling method. the pressure coefficient of 0.04334 and RSQ of 0.97915. Finally, a clustering tree was obtained by dividing electric shaver samples into 17 groups by the Ward method using cluster analysis, as shown in Figure 1.

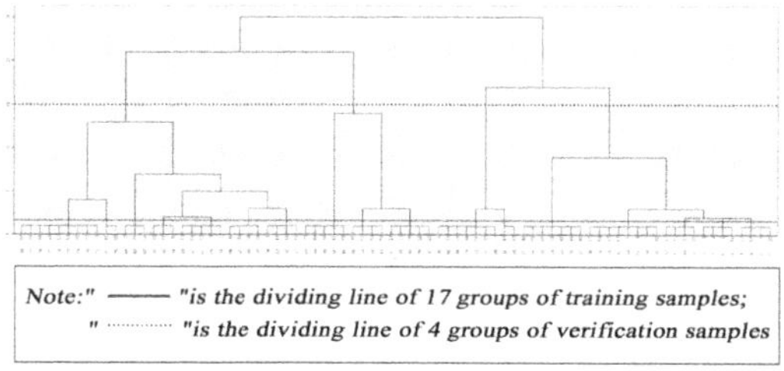

Figure 1. Cluster Analysis.

Seventeen groups were selected for clustering to consider the accuracy of the predictive model established in the later study and the load of respondents who received questionnaires. Morphological element deconstruction analysis, SDS, and QTTI multiple linear regression analysis should be adopted on electric shaver samples in the subsequent experiment. Afterward, samples in 17 groups were voted by 5 experts with a background in product design, the 2 samples with the greatest number of votes in each group were selected from each group as representative samples. More than that, 1 sample was selected from each of the 4 groups as verification samples (Lai et al. 2005). Hence, 34 representative samples can be acquired, as shown in Figure 2.

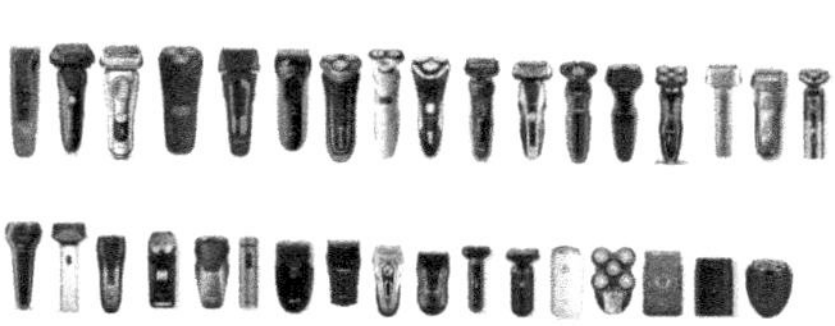

Figure 2. 34 Representative Samples.

3.2 *Selection of representative words*

A total of 135 Kansei words with respect to modeling and sound of electric shaver were first collected from channels, such as journal papers and comments of online shopping platforms, which were then reduced to 40 Kansei words after removing unclear semantics and ambiguity by 3 graduate students majoring in design. For lowering the experimental subjects' burdens in a later stage, another 30 graduate students and undergraduates majoring in design were invited to evaluate the appropriateness of Kansei words among the 40 groups provided using the five-point Likert scale. After that, Kansei words are finalized, as shown in Figure 3.

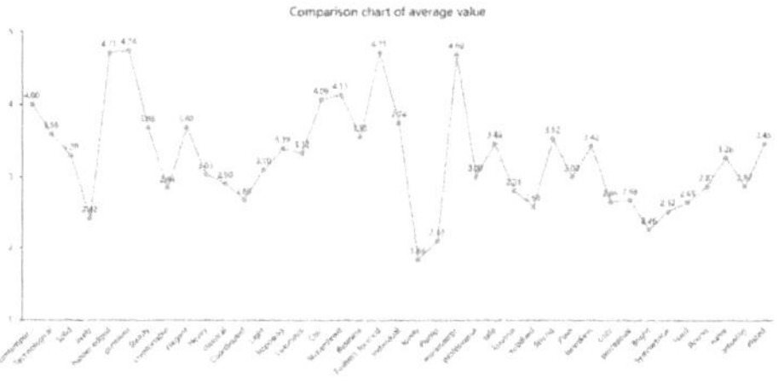

Figure 3. Comparison of Average Scores of Kansei Words.

Fashion-forward, minimalistic, harder-edged, and premium are the 4 words with the highest score reaching more than 4.5 points in the ranking of scores of Kansei words, with a standard deviation of less than 1. Then, they are paired with their antonyms, forming four groups of Kansei words, including "traditional/fashion-forward", "Soft/Harder-edged", "Cheap/Premium", and "Sophisticated/Minimalistic".

3.3 *Deconstruction of Design Elements*

Design elements of electric shaver samples were analyzed with the help of morphological analysis and Adobe Audition software. By doing so, five modeling items, including head, neck, handle, switch, and switch mode, were confirmed together with two sound items, including sound frequency and sound roughness.

Concerning modeling items, the ratio of neck shape (head-to-neck ratio/P_1) is Head width (b_1)/ Neck width (b_2), and the P_1 distribution of the sample is shown in Figure 4. The neck shape can be divided into narrow neck connection, average neck connection, and broad neck connection. When the ratio of neck length to head length is less than 3%, no neck will be classified. Streamlined handle shape (slender ratio/P_2) hasthe maximum width of the handle (b_3)/ the handle length (b_4), and the distribution of sample P_2 is shown in Figure 5. Furthermore, the streamlined handle is divided into a narrow streamline, average streamline, and broad streamline. The cylindrical handle shape (slender ratio/P_3) is the cylinder diameter/the handle length, which is merely divided into long and short cylinders for only seven shaver samples with cylindrical handles.

Measurement of sound frequency (P_4) and sound roughness was performed on 34 samples using Adobe

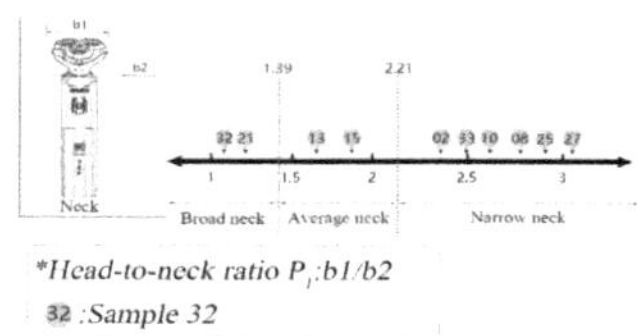

Figure 4. Distribution of Head-to-neck Ratio.

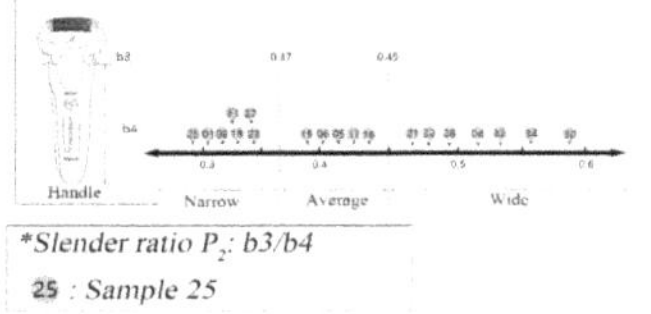

Figure 5. Distribution of Slender Ratio.

Audition in the sound project. As sound frequency is one of the attributes of sound, people may have varied sensations of sound at different frequency bands. When high-frequency bands are dominated in the sound, a sharp perception will be presented in audio sensory. When low-frequency bands dominate the sound, a low and deep sound will be perceived. When middle-frequency bands dominate the sound, a stiff sense will be perceived. Hence, the user's image perception of an electric shaver might be affected by the varying sound frequency of the product. Sound roughness (De Baene et al. 2004) is used for describing the noisy, sharp, and stable auditory perception of sound, which exerts an impact on the user's image perception of a product. This is because the greater the roughness, the noisier the auditory perception will be.

Frequencies of the 34 samples are classified into high, medium, and low categories with reference to the Scientific Pitch Notation. By analyzing sound roughness with Adobe Audition(Atamer S 2016), a greater range in frequency fluctuation indicates greater frequencies, being more complicated in the auditory perception, which is defined as "fluctuated". On the contrary, a gentle range in frequency fluctuation indicates low frequency in sound, being more stable in auditory perception, which can be defined as "stable" As can be seen in Figure 6, the complicated sound waveform on the left is defined as "fluctuated", the frequency range from 2090Hz to 2800Hz, which is defined as "high", whereas the gentle one on the right is defined as "stable", the frequency ranges from 35Hz to 65Hz is defined as "low".

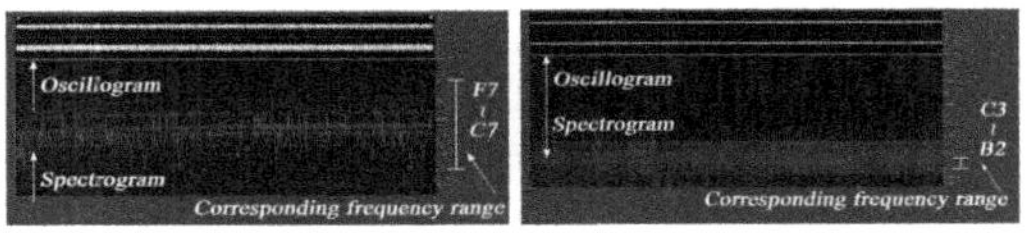

Figure 6. "Fluctuated" and "Stable" Examples/Example of Audio Frequency.

According to the analysis of product design elements on 34 electric shaver samples, design elements are divided into 7 items and 30 categories, as shown in Table 1.

Table 1. List of Design Elements.

Item	Category						
Shaver head	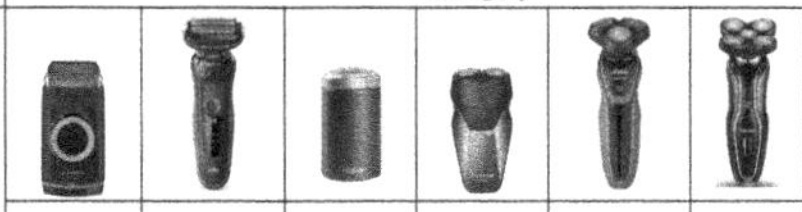						
	Flat 1-blade shaver head	Flat Multiblade shaver head	Rotary 1 shaver head	Rotary 2 shaver heads	Rotary 3 shaver heads	Rotary 5 shaver heads	
Neck	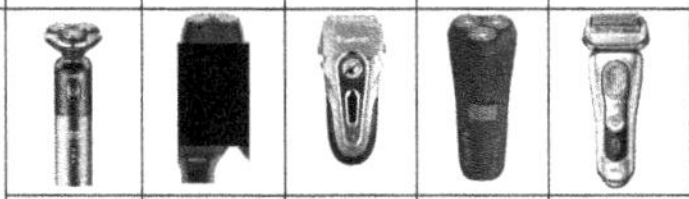						
	Narrow neck connection	Average neck connection	Broad neck connection	Smooth connection without neck	Two-segment connection without neck		
Handle	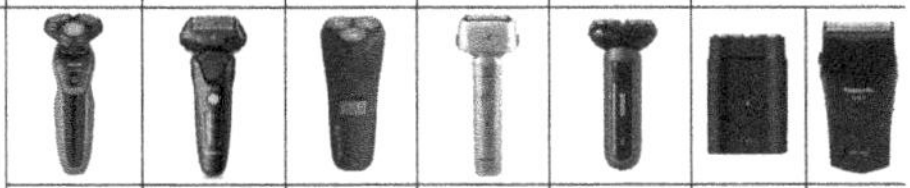						
	Narrow streamline	Average streamline	Wide streamline	Long cylinder	Short cylinder	Rounded rectangle	Angular shape
Switch (2:1)							
	Round	Rectangle	Rounded square	Oval-like	Triangle-like		
Switch mode (2:1)	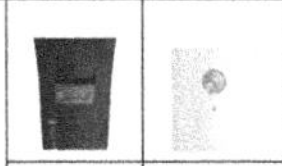						
	Sliding	Push-button					
Sound frequency	$P_4 \geq 2060Hz$		$2060Hz > P_4 > 550Hz$		$P_4 \leq 550Hz$		
	High		Average		Low		
Sound roughness	Stable				Fluctuated		

questionnaire subjects were required to evaluate the samples on SDS that can express the Kansei evaluation of consumers of varying degrees. In this study, 478 questionnaires were distributed to male participants. By excluding 62 invalid questionnaires, 416 valid questionnaires were applied for QTTI analysis, as shown in Table 2.

The above table shows that the coefficient of determination R_2 of "traditional/fashion-forward" is 0.825, which means 82.5% of the changes in the dependent variable can be explained by the multiple linear prediction equation. For the other three sets of Kansei words, their coefficients of determination are 91.1%, 87.4%, and 88.4%, respectively, presenting a favorable degree of fitting.

(i) Item impact analysis

It can be seen that the correlation between various design element items and Kansei words is represented by the partial correlation coefficient of an item, the greater the coefficient indicates the stronger the correlation with the image, and the higher the influencing capability of the image. Since handle > shaver head> neck> switch> switch mode> sound roughness> sound frequency is the design items most associated with the "Traditional/Fashion-forward", design elements with partial correlation coefficient among the top should be prioritized in the design of "Traditional/Fashion-forward" electric shavers. Besides, the item with a greater impact on the consumer's emotional image is recommended when there is a clash of modeling designs.

4 RESULTS AND DISCUSSIONS

4.1 *Construction of QTTI-based Correlation Model*

A total of 34 samples and 4 sets of Kansei words were used to form the seven-point Likert scale. Then,

Table 2. QTT I Results.

		Traditional/Fashion-forward			Cheap/Premium		
Items	Categories	CS	PO	S	CS	PO	S
Shaver head	Flat 1-blade shaver head	−0.406	0.620	2	−0.594	0.810	1
	Flat multi-blade shaver head	0.316			0.307		
	Rotary 1 shaver head	−0.513			−0.238		
	Rotary 2 shaver heads	−0.186			−0.231		
	Rotary 3 shaver heads	0.234			0.278		
	Rotary 5 shaver heads	0.212			0.467		
Neck	Narrow neck connection	0.291	0.482	3	0.292	0.512	4
	Average neck connection	0.265			0.048		
	Broad neck connection	−0.088			0.033		
	Smooth connection without neck	−0.217			−0.177		
	Two-segment connection without neck	0.263			0.206		
Handle	Narrow streamline	0.635	0.700	1	0.516	0.640	3
	Average streamline	−0.021			−0.054		
	Wide streamline	−0.117			−0.137		
	Long cylinder	−0.103			0.425		
	Short cylinder	−0.438			−0.212		
	Rounded rectangle	−0.363			−0.214		
	Angular shape	0.698			0.023		
Switch	Round	−0.028	0.454	4	−0.054	0.224	6
	Rectangle	−0.354			−0.073		
	Square-like	−0.147			0.020		
	Oval-like	0.267			0.082		
	Triangle-like	0.304			0.035		
Switch mode	Sliding	0.329	0.425	5	0.055	0.118	7
	Push-button	−0.100			−0.017		
Sound frequency	High	−0.083	0.173	7	−0.226	0.495	5
	Average	−0.035			0.121		
	Low	0.135			0.104		
Sound roughness	Stable	0.129	0.224	6	0.270	0.667	2
	Rough	−0.098			−0.207		
C			3.789			3.431	
R			0.908			0.954	
R2			0.825			0.911	
Shaver head	Flat 1-blade shaver head	−0.462	0.716	2	0.264	0.821	1
	Flat multi-blade shaver head	0.413			−0.219		
	Rotary 1 shaver head	−0.464			0.265		
	Rotary 2 shaver heads	−0.368			0.255		
	Rotary 3 shaver heads	0.306			−0.201		
	Rotary 5 shaver heads	0.337			−0.337		
Neck	Narrow neck connection	0.054	0.419	5	−0.266	0.653	3
	Average neck connection	−0.074			−0.237		
	Broad neck connection	−0.220			0.095		
	Smooth connection without neck	−0.121			0.105		
	Two-segment connection without neck	0.339			0.004		

(continued)

Table 2. Continued

		Traditional/Fashion-forward			Cheap/Premium		
Items	Categories	CS	PO	S	CS	PO	S
Handle	Narrow streamline	−0.475	0.806		−0.294	0.713	2
	Average streamline	−0.312			−0.153		
	Wide streamline	−0.309			0.102		
	Long cylinder	0.759			0.238		
	Short cylinder	0.770		1 1	0.260		
	Rounded rectangle	−0.194			0.137		
	Angular shape	−0.797			−0.251		
Switch	Round	−0.092	0.272	6	0.100	0.453	5
	Rectangle	0.312			0.030		
	Square-like	−0.021			−0.124		
	Oval-like	−0.007			−0.024		
	Triangle-like	0.064			0.016		
Switch mode	Sliding	0.085	0.133	7	−0.233	0.575	4
	Push-button	−0.026			0.071		
Sound frequency	High	−0.263	0.619	3	−0.056	0.190	7
	Average	−0.216			0.018		
	Low	0.556			0.040		
Sound roughness	Stable	0.301	0.533	4	0.072	0.316	6
	Rough	−0.230			−0.055		
C			3.641			4.228	
R			0.935			0.940	
R2			0.874			0.884	

Note: CS: category scores; PO: partial correlation coefficient; S: sort; C: constant term.

Influences of various items under the Kansei words of "Cheap/Premium" are ranked as: shave head > sound roughness > handle > neck > sound frequency > switch > switch mode.

Influences of various items under the Kansei words of "Soft/Harder-edged" are ranked as: handle > shaver head > sound frequency > sound roughness > neck > switch > switch mode.

Influences of various items under the Kansei words of "Sophisticated/Simple" are ranked as: shaver head > handle > neck > switch mode > switch > sound roughness > sound frequency.

(ii) Category impact analysis
Category score indicates the correlation between categories and Kansei words of various modeling items. Specifically, a positive value indicates that categories and Kansei words are positively correlated, being more apt to the Kansei word on the right, whereas a negative value indicates the category and Kansei words are negatively correlated, tending toward the Kansei word on the left.

Under the Kansei words of "Traditional/Fashion-forward", the "handle" item gained the highest partial coefficient of correlation; and the negative categories score belongs to the "traditional" image with more negative values indicating more "traditional" images, which can be ranked as "short cylinder", "rounded rectangle", "broad streamline", "long cylinder", and "average streamline" from large to small. Besides, the positive category score pertains to the "fashion-forward" image, which can be ranked as "angular shape" and "narrow streamline" from large to small. The importance of design elements of other Kansei words in the table can be analyzed according to Table 2.

(iii) Analysis of sound design elements In the Kansei word "Cheap/Premium", influences of sound roughness and sound frequency are ranked as 2 and 5, respectively, indicating that sound at the stable and moderate frequency can elevate the premium sense of a product, whereas the sound at the rough, too high/too low frequency, brings a sense of the cheap product. In the Kansei word "Soft/Harder-edged", influences of sound frequency and roughness are ranked as 3 and 4, respectively, indicating that the harder-edged feature of the product can be enhanced with the stable and deep sound, otherwise it brings a soft sense. The rest sound item sand categories can be analyzed according to Table 2.

4.2 *Construction of multiple linear regression equation of QTTI*

With 30 categories of electric shaver as independent variables, Kansei evaluation as dependent variables, and scores of various categories as coefficients, 4 sets of equations between design elements and Kansei words can be constructed below in accordance with the multiple linear regression equation of QTTI.

"Traditional/Fashion-forward":

$$\begin{aligned} y_k = &-0.406X_{11} + 0.316X_{12} - 0.513X_{13} - 0.186X_{14} \\ &+0.234X_{15} + 0.212X_{16} \\ &+0.291X_{21} + 0.265X_{22} - 0.088X_{23} \\ &-0.217X_{24} + 0.263X_{25} \\ &+0.635X_{31} - 0.021X_{32} - 0.1170X_{33} \\ &-0.103X_{34} - 0.438X_{35} - 0.363X_{36} + 0.698X_{37} \\ &-0.028X_{42} - 0.354X_{43} - 0.147X_{44} \\ &+0.304X_{45} \\ &+0.329X_{51} - 0.100X_{52} \\ &-0.083X_{61} - 0.035X_{62} + 0.135X_{63} \\ &+0.129X_{71} - 0.098X_{72} \\ &+3.789 \end{aligned} \quad (1)$$

"Cheap/Premium":

$$\begin{aligned} y_k = &-0.594X_{11} + 0.307X_{12} - 0.238X_{13} - 0.231X_{14} \\ &+0.278X_{15} + 0.467X_{16} \\ &+0.292X_{21} + 0.048X_{22} + 0.033X_{23} \\ &-0.177X_{24} + 0.206X_{25} \\ &+0.516X_{31} - 0.054X_{32} - 0.137X_{33} \\ &-0.425X_{34} - 0.212X_{35} - 0.214X_{36} + 0.023X_{37} \\ &-0.054X_{41} - 0.073X_{42} + 0.020X_{43} + 0.082X_{44} \\ &+0.035X_{45} \\ &+0.055X_{51} - 0.017X_{52} \\ &-0.226X_{61} + 0.121X_{62} + 0.104X_{63} \\ &+0.270X_{71} - 0.207X_{72} \\ &+3.431 \end{aligned} \quad (2)$$

"Soft/Harder-edged":

$$\begin{aligned} y_k = &-0.462X_{11} + 0.413X_{12} - 0.464X_{13} - 0.368X_{14} \\ &+0.306X_{15} + 0.337X_{16} \\ &+0.054X_{21} - 0.074X_{22} - 0.220X_{23} \\ &-0.121X_{24} + 0.339X_{25} \\ &-0.475X_{31} - 0.312X_{32} - 0.309X_{33} \\ &+0.759X_{34} + 0.770X_{35} - 0.194X_{36} + 0.797X_{37} \\ &-0.092X_{41} + 0.312X_{42} - 0.021X_{43} - 0.007X_{44} \\ &+0.064X_{45} \\ &+0.085X_{51} - 0.026X_{52} \\ &-0.263X_{61} - 0.216X_{62} + 0.556X_{63} \\ &+0.301X_{71} - 0.230X_{72} \\ &+3.641 \end{aligned} \quad (3)$$

"Sophisticated/Minimalistic":

$$\begin{aligned} y_k = &\ 0.264X_{11} + 0.219X_{12} - 0.265X_{13} - 0.255X_{14} \\ &+0.201X_{15} + 0.337X_{16} \\ &-0.266X_{21} - 0.237X_{22} - 0.095X_{23} \\ &-0.105X_{24} + 0.004X_{25} \\ &-0.294X_{31} - 0.153X_{32} - 0.102X_{33} \\ &+0.238X_{34} + 0.260X_{35} - 0.137X_{36} + 0.251X_{37} \\ &-0.100X_{41} + 0.030X_{42} - 0.124X_{43} - 0.024X_{44} \\ &+0.016X_{45} \\ &+0.233X_{51} - 0.071X_{52} \\ &-0.056X_{61} - 0.018X_{62} + 0.040X_{63} \\ &+0.072X_{71} - 0.0550X_{72} \\ &+4.22 \end{aligned} \quad (4)$$

Furthermore, the paired sample t-test was performed between the actual value of 30 training samples and the predicted value of QTTI via SPSS. The result shows that the p-values of the tested prediction models of the four groups of Kansei words are greater than 0.05, indicating no significant difference. Therefore, the reliability of the four groups of prediction models can be proven.

5 CONCLUSION

In this research, representative electric shaver samples were selected using methods such as multidimensional scale and clustering analysis. Then, Kansei words of the product were deconstructed into 7 items (including head, neck, etc.) and corresponding 30 categories using the morphological analysis. At last, correlation models of product design elements and Kansei image were constructed through QTTI multiple linear regression analysis in combination with quantified Kansei evaluation and product design elements. Based on systematic analysis methods, the methods proposed in this study are applicable for the research and application associated with the multisensory design of products. Also, this can provide designers with clear design indicators and references for traditional black-box design.

ACKNOWLEDGMENT

This work was supported by the Fujian University of Technology [grant numbers GY-S21081, 2021], and the Design Innovation Research Center of Humanities and Social Sciences Research Base of Colleges and Universities in Fujian Province.

REFERENCES

Atamer, S., & Altinsoy, M. E. 2017. DAGA.
Atamer, S. 2016. In INTER-NOISE and *NOISE-CON Congress and Conference Proceedings.Institute of Noise Control Engineering* 5185–5192.
Chuan, N. K., Sivaji, A., Shahimin, M. M., & Saad, N. 2013. *Procedia-Social and Behavioral Sciences* 97, 707–714.
Dal Palù, D., De Giorgi, C., Lerma, B., & Buiatti, E. 2018. *In Frontiers of sound in design*. Springer, Cham. pp. 31–46.
De Baene, W., Vandierendonck, A., Leman, M., Widmann, A., & Tervaniemi, M. 2004. *Biological psychology* 67(3), 319–330.
Desmet, P. M. 2012. *International Journal of Design* 6(2).
Djatna, T., & Kurniati, W. D. 2015. *Procedia Manufacturing* 4, 115–123.
Izumi, H., & Sawaguchi, M. 2015. *Procedia engineering* 131, 569–576.
Jindo, T., & Hirasago, K. 1997. *International journal of industrial ergonomics* 19(2), 105–114.
Kim, W., Ko, T., Rhiu, I., & Yun, M. H. 2019. 74, 145–153.
Kuo, J. Y., Chen, C. H., Roberts, J. R., & Chang, D. 2020. *International Journal of Industrial Ergonomics* 80, 103039.
Lai, H. H., Lin, Y. C., & Yeh, C. H. 2005. *Computers & Operations Research* 32(10), 2689–2711.
Miyano, T., & Kitamura, H. 1994. *Surface and Coatings Technology* 65(1-3), 179–183.
Muramatsu, K., Tsukada, S., Kaede, K., & Watanuki, K. 2019. *IEEE International Conference on Systems, Man and Cybernetics* (SMC). 3205–3209.
Nagamachi, M. 1995. *International Journal of industrial ergonomics* 15(1), 3–11.
Razza, B., & Paschoarelli, L. C. 2015. 3, 6228–6236.
Schifferstein, H. N., & Spence, C. 2008. Elsevier. pp. 133–161.
Shergian, A., & Immawan, T. 2015. *Agriculture and Agricultural Science Procedia* 3, 184–188.
Yanagisawa, H. 2011. Springer, London. 289–310.
Yanagisawa, H., Miyazaki, C., & Bouchard, C. 2017. In DS 87-8 *Proceedings of the 21st International Conference on Engineering Design* (ICED 17) Vol 8: Human Behaviour in Design, Vancouver, Canada, 21-25.08. 159–168.
Zhang Hongyuan. 2022. Electrical appliances, (02): 46–47.

System Innovation for a Troubled World – Tak Lam et al. (Eds)
© 2023 The Author(s), ISBN: 978-1-032-45525-9

Picosecond dual-pulse laser ablation properties of fused silica

Ping-Han Wu
Department of Mechanical Engineering, National Taiwan University, Taipei, Taiwan
Laser and Additive Manufacturing Technology Center, Industrial Technology Research Institute, Tainan, Taiwan

Hong-Tsu Young & Kuan-Ming Li*
Department of Mechanical Engineering, National Taiwan University, Taipei, Taiwan

ABSTRACT: In this research, the scribing line widths of picosecond laser dual pulses in fused silica with different time delays were investigated. This study revealed that the widest scribing line and the lowest threshold of the dual-pulse ablation were at a time delay of 100 ps. The underlying physical mechanisms were discussed, and it was found that the first sub-pulses initially excited the electrons of the fused silica and subsequently induced free-electron plasma scattering and self-trapped exciton seed generation at different time delays, which either decreased or increased the material absorption of the second sub-pulses and then varied the scribing line widths and ablation thresholds of the fused silica. The results of the picosecond dual-pulse ablation were helpful for understanding the mechanisms of the interaction of dual pulses with transparent materials.

1 INTRODUCTION

Ultrashort lasers are suitable tools for processing wide-bandgap materials owing to their high peak power and minimum thermal effect. However, a single ultrafast laser pulse can only bring limited impact on the properties of the microprocessing by varying laser parameters like wavelength, pulse duration, and pulse energy. To achieve effective microprocessing, we require multiple pulses or pulse shaping in ultrafast laser fabrication. Many experimental investigations concerning the material processes of ultrashort-laser dual pulses with wide-bandgap and transparent materials can be found in the literature [1-16]. Such processes involve laser ablation [1-4], laser drilling [5-7], laser texturing [8-10], laser internal modification [11-13], and laser micro welding [14-16]. Regarding the ablation process, Zhou et al. revealed that the feature position of the overlapping area of femtosecond (fs) laser dual pulses depends on the time delay between the dual pulses [1]. Xu et al. investigated the ablation size and recast on fused silica, and found they can be controlled by shaping femtosecond pulse trains to modify the electron dynamics in the materials [2]. Leng et al. studied the fs-laser dual-pulse train processing of fused silica and found that the ablation size caused an abrupt rise in the time delay range of 150–275 fs due to the enhancement of the photon-electron coupling efficiency [3]. Chowdhury et al. reported a pair of fs laser dual pulses ablated in fused silica, and found that the ablation depth peaked at zero time delay and decreased at about 1 ps time delay. This is caused by the plasma screening induced after the materials were exposed to the first sub-pulse [4].

Even though plentiful studies were working on the fs-laser dual-pulse processes, the damage mechanisms in fused silica treated with picosecond (ps)-laser dual pulses have not been thoroughly researched. We studied the ablation properties of 1064 nm ps-laser dual pulses in fused silica. Compared to single-pulse ablation, we first revealed a rise in the widths of the scribing lines due to ps-laser dual pulse ablation with several laser output powers and time delays. The laser ablation thresholds of the ps-laser dual pulse were also researched in this work.

2 EXPERIMENTAL SETUP

Figure 1 showed the schematic illustration of the experimental setup based on the Mach-Zehnder interferometer for line scribing the fused silica using ps-laser dual pulses. A solid stated ps laser source (PX-series, Edge-Wave, Germany) was used to generate 10-ps pulses with a wavelength of 1064 nm at a repetition rate of 100 kHz. A linear polarized single pulse was converted into p- and s-polarized (p-p, s-p) pulses by rotating the half-wave plate (1/2λ WP). These pulses were separated by a polarized beam splitter (PBS). The p-p sub-pulses were directed to Arm 1 and passed through the quarter-wave plate (1/4λ WP) twice by a 0° mirror. They were then transformed into s-p sud-pulses and reflected by PBS. Similarly, the s-p sub-pulses on Arm 2 were transformed into p-p sub-pulses and passed through

*Corresponding Author: kmli@ntu.edu.tw

DOI 10.1201/9781003377399-33

the PBS. The transformed p-p and s-p sub-pulses were combined by the PBS and became cross-polarized dual pulses. Different polarizations were used for the first and second sub-pulses to prevent their interfering with each other and to permit each process to be controlled independently. The collinearity of the two beams was ensured by transmitting them over ten meters and by checking the coincidence of the profiles on a beam profiler (Ophir). The time delay system caused the p-p sub-pulse to reach the target before the s-p sub-pulse, and the time delay varied from 0 to 200 ps. Experiments with longer delays were not carried out due to limitations of the travel length during the delay stage.

The fused silica was moved by the x-y linear stages and in a perpendicular direction to the laser. The processed fused silica was scribed, and the widths of the scribing lines were measured by an optical microscope (OM). A 10X objective lens (NA=0.24) and a translation velocity of 5 mm/s were used for the line-scribing experiments of the fused silica. The laser output powers of the single pulses and the dual pulses (p-p + s-p sub-pulses) for the experiments were always maintained identically.

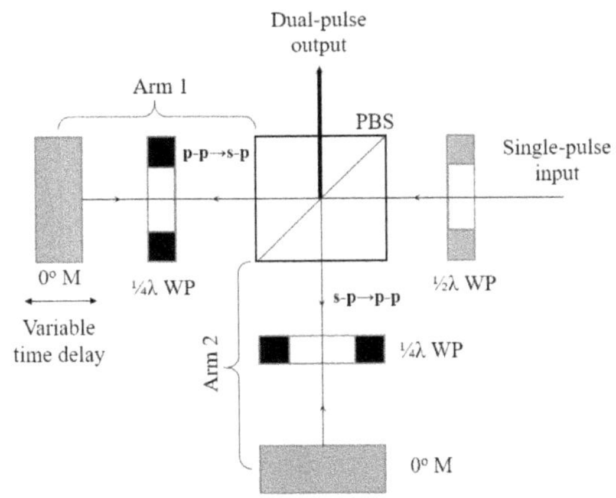

Figure 1. Experimental setup for generating the dual pulses.

3 RESULTS AND DISCUSSION

In all the experiments performed in this study, the dual pulses are composed of the same sub-pulse laser output power. The time delays were adjusted between 10 ps and 200 ps. For comparison, the experiments of the single pulses were also executed. Figure 2(a) ~ (e) depicts the relationship between the time delays and the widths of the scribing lines with different laser output powers. We also measured the laser scribing line widths which were ablated by single pulses with p and s polarizations, respectively. The laser output powers of the line scribing in these experiments varied between 3 W to 11 W at 2 W intervals. Two results in Figure 2 were interesting to discuss. First, comparing the single-pulse scribing, all data curves show the enhancement of the line widths of the dual-pulse scribing at the time delay from 50 ps to 200 ps and the widest scribing line was at the time delay of 100 ps. Second, the line widths of the scribing gradually decreased as the time delays were increased from 150 ps to 200 ps. Figure 3 shows the OM images of the scribing lines ablated in the material with the laser output power of 11 W by the dual pulses at the time delay of 100 ps and those ablated by single pulses. The scribing lines from the dual-pulse ablation also displayed the wider line widths as compared to single-pulse ablation.

When a spatial Gaussian laser beam scribes the line patterns in the material, the line widths of the scribed region W will be related to the irradiation laser fluence F (F=(Laser output power/Rep. rate)/($\pi * \omega_0^2$)) by:

$$W^2 = 2\omega_0^2 \ln\left(\frac{F}{F_{th}}\right)$$

where F_{th} is the ablation threshold of the laser scribing and ω_0 is the effective beam radius at the interaction surface. We measured the laser scribing line widths which were ablated with different fluences of the dual pulses and single pulses with p and s polarizations, respectively. The linear relation was between W^2 and the logarithm of F. We then performed an extrapolation of the linear fit to $W^2 = 0$ to find the ablation thresholds of the single pulses and dual pulses in variable time delays, as shown in Figure 4. A significant decrease in the ablation thresholds was also observed. The minimum ablation threshold was obtained at the time delay of 100 ps. The ablation threshold gradually increased as the time delay was increased from 150 ps to 200 ps. Except for the time delay of 10 ps, the ablation thresholds of the dual pulses with different time delays from 50 ps to 200 ps were still smaller than the thresholds of the single pulses with p- and s-polarizations, respectively. The time delays required for dual-pulse scribing to reduce the material ablation threshold were exactly the same as the delay time required to increase the scribing line width.

To consider these unexpected results of dual-pulse scribing, three underlying physical mechanisms of laser ablation in the fused silica by dual-pulse irradiation have been made.

1. The first sub-pulses of the dual pulses excited electrons in the fused silica from the valance band (VB) to the conduction band (CB) by multiphoton absorption. Further, the free-electron absorption on the trailing edge of the first sub-pulse developed into avalanche absorption [17, 18]. Then, the generated free electrons relaxed to the localized trapping states such as self-trapped excitons (STEs) which were created to localize the electronic excitation energy in the materials. At the time delays of 50 ps and 100 ps, it was less than the relaxation time of STEs generated by the first sub-pulse irradiation [14], the energy of the second sub-pulse could be coupled into the material by STEs. Therefore, these STEs driven the avalanche absorption in the material with absorbing the energy of the second sub-pulse. The effect of the STE-seeded avalanche absorption of the second sub-pulses at the

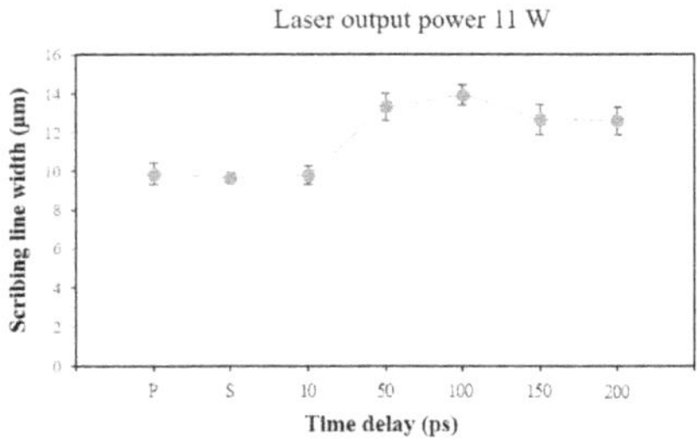

(a)

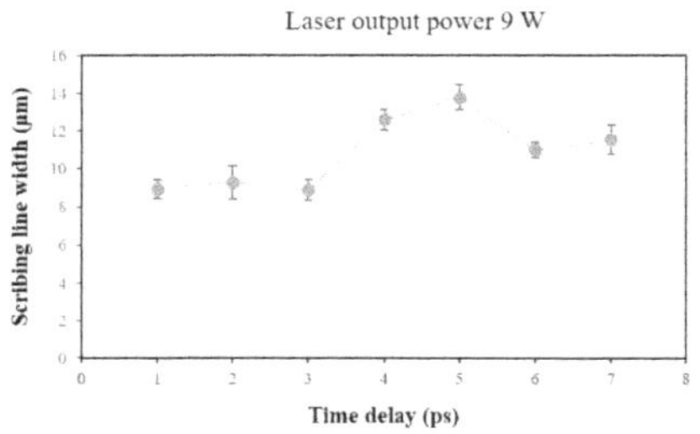

(b)

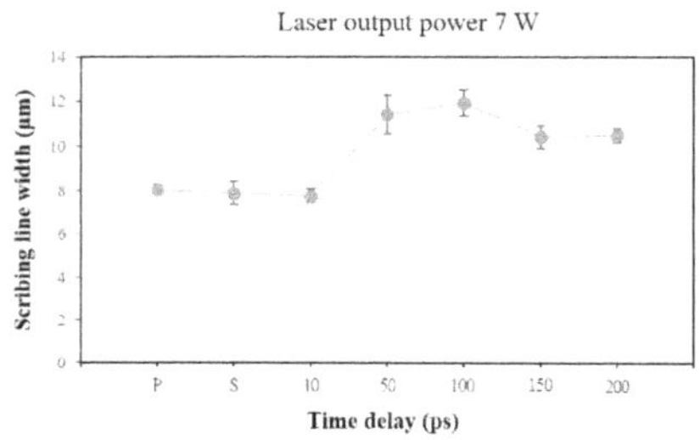

(c)

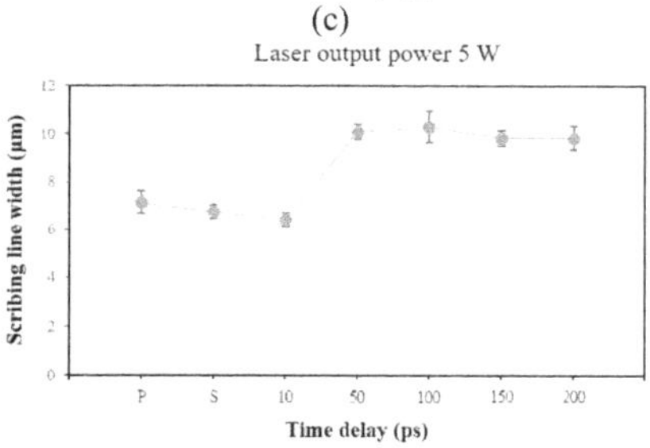

(d)

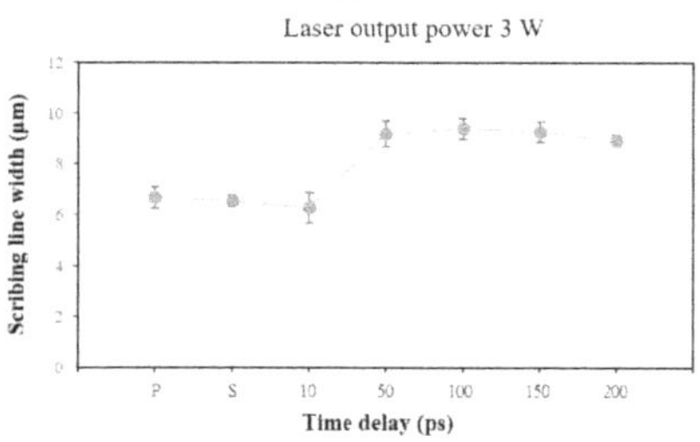

(e)

Figure 2. (a)–(e) Results for the relationship between the time delays and the widths of the scribing lines with different laser output powers. For comparison, the experiments of the single pulses were also executed.

100-ps time delay was maximized [19]. As a consequence, the widest scribing line and the lowest damage threshold were observed at this time delay.

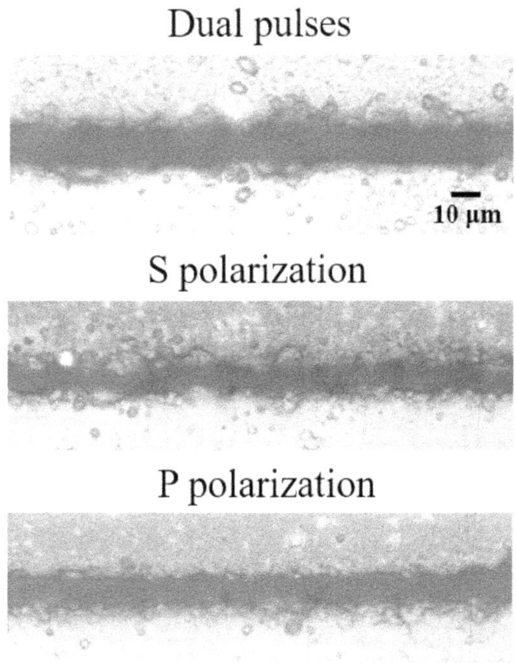

Figure 3. OM image comparison of the scribing lines ablated in the material with the laser output power of 11 W by the dual pulses at the time delay of 100 ps and single pulses (p- and s-polarizations).

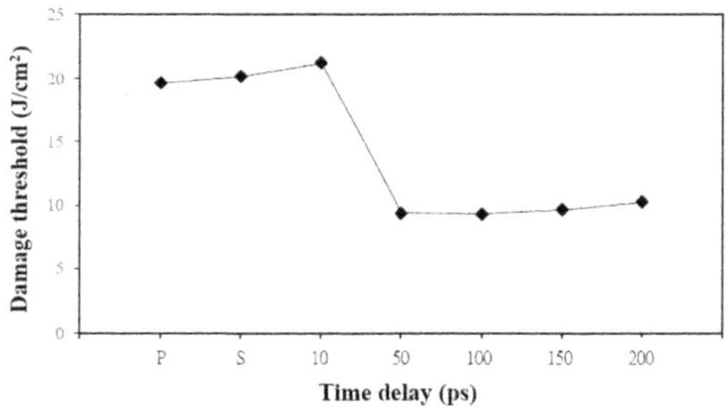

Figure 4. Results for Dependence of ablation thresholds on time delays. The ablation thresholds of the single pulses were also shown in the figure.

2. At the time delays of 150 ps and 200 ps, these time delays were longer than the relaxation time from the electrons in the CB to STEs after the first sub-pulse irradiation [14], so some STEs from the first sub-pulse began to relax. The localized states generated in the relaxation process also provided seed electrons for increasing the absorption of the second sub-pulses, so the scribing line widths and ablation thresholds at these time delays were still wider and smaller than those of the single pulses, respectively. However, the energy levels of the localized states were lower than the levels of the STEs. Then, the energy of the second sub-pulse required for electrons to move from the localized states to the conduction band was higher than the absorption case in the 100-ps time delay [15].

Therefore, the curves of the scribing line widths and ablation thresholds at these time delays in Figures 2 and 3 had a slightly downward and an upward trend, respectively.

3. At the time delay of 10 ps, which was less than the relaxation time of the free electrons after the first sub-pulse irradiation [14], when the second sub-pulses irradiated in the materials, the free electron density of the conduction band was increased due to the excitation of the electrons by the first sub-pulses. The free electrons at this moment presented metallic properties in their light responses, such as higher reflectivity and limited penetration. These optical properties shielded or reflected the second sub-pulses, which should have been absorbed by the material and subsequently led to less ablation [4, 17]. This free-electron plasma scattering [4] of the dual-pulse ablation at a 10 ps time delay led to an ablation threshold higher than that of the single-pulse ablation.

4 CONCLUSION

Ps laser irradiation using dual pulses was used to ablate the fused silica. The scribing line widths were significantly wider than those for irradiation by single pulses and these wider scribing lines resulted in the smaller ablation thresholds. The underlying physical mechanism could be explained by the STE or the seeds of the localized state generated by the first sub-pulses increasing the absorption of the second sub-pulses. This research on picosecond dual-pulse ablation could provide an effective way for fused silica micromachining and be helpful for understanding the interaction mechanisms of dual pulses in transparent materials.

REFERENCES

[1] B. Zhou, et al., *Optics letters*, 2020. **45**(7): pp. 1994–1997.

[2] C. Xu, et al., *Chinese Optics Letters*, 2013. **11**(4): p. 041403.

[3] N. Leng, et al., *Applied Physics A*, 2012. **109**(3): pp. 679–684.

[4] I. H. Chowdhury, X. Xu, and A. M. Weiner, *Applied Physics Letters*, 2005. **86**(15): p. 151110.

[5] L. Jiang, et al., *Applied optics*, 2014. **53**(31): pp. 7290–7295.

[6] L. Jiang, et al., *Optics letters*, 2012. **37**(14): pp. 2781–2783.

[7] J. del Hoyo, et al., *Nanophotonics*, 2021. **10**(3): pp. 1089–1097.

[8] S. Höhm, et al., *Applied surface science*, 2013. **278**: pp. 7–12.

[9] S. Richter, et al., *Optics letters*, 2012. **37**(4): pp. 482–484.

[10] L. Jiang, et al., *Optics express*, 2012. **20**(19): pp. 21505–21511.

[11] T. Nagata, M. Kamata, and M. Obara, *Applied Physics Letters*, 2005. **86**(25): p. 251103.

[12] T. Nagata, M. Kamata, and M. Obara. *Optical waveguide fabrication with tailored femtosecond laser pulses*. in *High-Power Laser Ablation V*. 2004. International Society for Optics and Photonics.

[13] D. Wortmann, M. Ramme, and J. Gottmann, *Optics Express*, 2007. **15**(16): pp. 10149–10153.

[14] S. Wu, et al., *Optics express*, 2013. **21**(20): pp. 24049–24059.

[15] S. Wu, et al., *Optics express*, 2012. **20**(27): pp. 28893–28905.

[16] K. Sugioka, et al., *Optics letters*, 2011. **36**(14): pp. 2734–2736.

[17] P. Balling and J. Schou, *Rep Prog Phys*, 2013. **76**(3): p. 036502.

[18] S. S. Mao, et al., *Applied Physics A*, 2004. **79**(7): pp. 1695–1709.

[19] A. Tamura, et al. *Double pulse explosion drilling of transparent materials*. in *Laser-based Micro-and Nanoprocessing XII*. 2018. International Society for Optics and Photonics.

System Innovation for a Troubled World – Tak Lam et al. (Eds)
© 2023 The Author(s), ISBN: 978-1-032-45525-9

Evolution toward coordinated multipoint architecture in small cell enhancement system operation scenarios for LTE-A technologies

Chia-Lun Wu , Tsung-Tao Lu, Bau-Lin Chen, Jwo-Shiun Sun, Hsin-Piao Lin & Yu-Sian Huang
Department of Electronic Engineering, National Taipei University of Technology, Taipei, Taiwan

Chin-Tan Lee*
Department of Electronic Engineering, National Quemoy University, Kinmen, Taiwan

ABSTRACT: This paper explores the applications of coordinated multipoint operation (CoMP) of enhanced node B (eNB) in future wireless networks for device-to-device (D2D) communication architecture, such as metro, high-speed railway, and subway systems. Based on theory and engineering principles in practice, this paper proposes guidelines and formulas for link budget design to help designers fully control and analyze the uplink and downlink singles of output power among fiber repeaters, mobile stations, and base stations. Finally, the results of this leakage cable system are successfully applied to enhance coverage design for mass rapid transit (MRT) system which is easily installed cellular over fiber solutions for small cell enhancement (SCE) system architecture for LTE-A technologies access are becoming ubiquitous self-organizing network (SON) to the internet of everything real case hierarchy of telecommunications.

Keywords: Coordinated MultiPoint Operation (CoMP), Enhanced Node B (eNB), Device to Device (D2D), Small Cell Enhancement (SCE), Self-Organizing Network (SON).

1 INTRODUCTION

This paper investigates the small cell enhancement (SCE) [1] of fiber repeater (FR) systems that applied radio frequency (RF) over LTE-A technologies of coordinated multipoint operation (CoMP) [2]–[3] and high order mutliuser multi-input multi-output (MU-MIMO) in device to device (D2D) architecture for self-organizing network (SON) operation scenarios [4]–[5] that play an important role in the operation of multi-vendor heterogeneous networks (HetNets). The use cases based on CoMP are addressed in detail, as are various architecture options and tradeoffs related to the implementation of HetNets [1].

In Telecommunication Certification Body (TCB) workshop Santa Clara, CA, Steve Martin office of engineering and technology equipment authorization Branch Federal Communications Commission (FCC) laboratory in which SCE rules are applicable to Radio over Fiber (RoF) [6]–[7] and can access devices operating from 700MHz to 2300MHz [8]–[9].

The definition of SCE includes the use of the transmission line between the substation and the customer terminal device. SCEs are a part of systems that transmits RF energy by conduction or radiation [10]–[11].

*Corresponding Author: ktlee@nqu.edu.tw

This paper is organized as follows. The design rules of CoMP for LTE-A [12] are illustrated in Section II. Then studies of the key factors and solution of the fiber repeater over the CSE system of D2D architecture for SON will be discussed in Sections III and IV respectively. Finally, the conclusion is made in Section V.

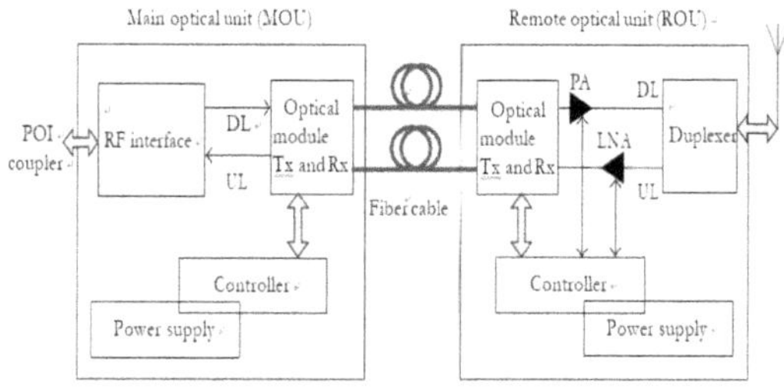

Figure 1. SCE block diagram.

2 DESIGN RULES OF SMALL CELL ENHANCEMENT SYSTEMS

As the CoMP implies, it shall allow the optimization of transmission and reception from multiple distribution points (MDP) which could be either multiple cells or remote radio heads (RRH), in a coordinated way.

DOI 10.1201/9781003377399-34

CoMP will enable joint transmission and/or reception to mobile devices.

Based on CoMP scenarios, the RF over SCE concepts of the FR operation principle is shown in Figure 1. FR is basically divided into two parts. The first part, main distributing point (MDP), is the main optical unit (MOU). MOU converts RF signals in Down Link (DL) to optical signals, and converts optical signals to RF signals in Up Link (UL). The second part RRH is the remote optical unit (ROU), which detects DL signals and modulates UL signals.

The MDP [13] system block diagram is illustrated in Figure 2. Because the RF signals coming from the eNB in DL are approximately +13dBm [14], the signal level would be attenuated for E/O module receiving. The optical signals are separated into multichannel signals by ROUs. Therefore, the DL RF module is composed of an adjustable digital attenuator and splitter.

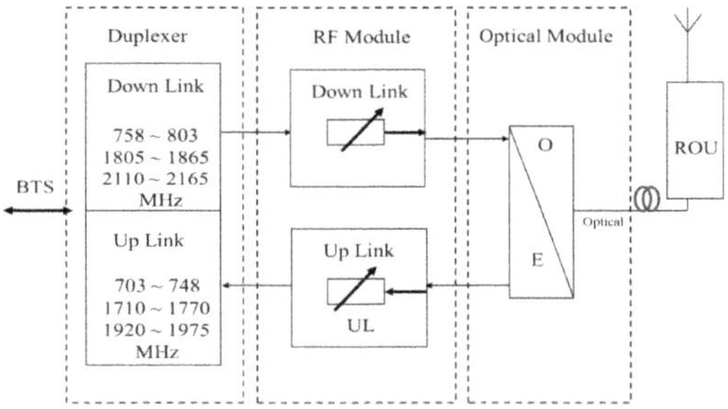

Figure 2. MDP block diagram.

When MOU receives optical signals from RRH through O/E optical modules, the RF signal conversions are processed in UL RF modules. Therefore, The RF module is composed of an adjustable digital attenuator and combiner. It distributes the UL signals power which is an attenuated RF signal level in eNB Noise Floor $\geqq -121$ dBm. Because eNB services a lot of mobile stations, input noise signal, which is above -121dBm, will block communication of weak signal cell phones and drop calls rating.

The RRH [15] block diagram is shown in Figure 3, the remote optical unit in a wireless terminal serves the mobile station (MS).

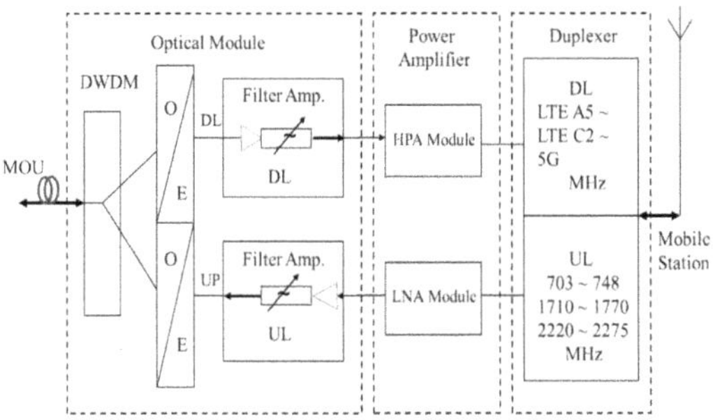

Figure 3. RRH block diagram.

In the part of DL, the O/E module of RRH receives optical signal from the MDP and converts the optical signal into an RF signal. Then the RF signal is amplified by power amplifier and transmitted through the antenna. Therefore, the basic DL architecture of RRH is the combination of an O/E module, IF filtering high-frequency conversion module, and power amplifier module.

As the same principle, in the part of UL, mobile station signal is received by the RRH antenna, and then modulated to optical signal via another independent fiber back to the MDP, and then the MOU functions of UL of the optical signal from O/E module converts into RF signal back to the base station. Thus, the basic UL architecture of RRH is the combination of the low noise amplifier module, IF filtering, frequency conversion module, and O/E module.

3 STUDIES OF THE KEY FACTORS

In the cellular communication system between eNB is BTS and MS, the thermal noise signal, produced by added FR, drops the receiving sensitivity of BTS directly.

A. *Calculation gain of fiber repeater*

The use of FR in a wireless communication network causes leveling on BTS from which the FR receives the RF signal. In UL, high interfering signal power at the BTS receiver will limit communication service in MS. The transmission power in MS would guarantee the required Eb/N0, where Eb is energy bit and N0 is spectral noise density. In DL, this effect will cause the same problem in the MS receiver. The thermal noise power of BTS, P_{th}, can be expressed as (1).

$$P_{th} = 10\log[KTB] \tag{1}$$

where

K: Boltzmamnn's constant (1.38×10^{-23} J/K)

T: Absolute temperature (290°K)

B: Signal bandwidth (200 KHz/Channel)

The noise power of BTS receiver, P_{bts}, can be expressed as (2).

$$P_{bts} = 10\log[KTB] + NF_{bts} \tag{2}$$

where

NF_{bts}: Noise Figure of the BTS

If the NF_{bts} is replaced by 2 dB, the value of P_{bts} is -119 dBm.

When the FR injects and links between power sources and BTS, the noise power of the receiver is the sum of the noise power in BTS and FR.

B. *Injected by fiber repeater of BTS deviation*

In UL, large interfering signal power at the BTS receiver will limit communication service in MSs. The transmission power in MS would guarantee the required Eb/N0. In particular, the FR which is an active component will amplify the signals coming from transmission interfering signals due to thermal noise power. Therefore, the output noise power of FR, P_{rep} can be expressed as (3)

$$P_{rep} = 10\log[KTB] + P_{bts} + N_{Frep} + G_{rep} \quad (3)$$

where

NF_{rep}: Noise Figure of the fiber repeater

G_{rep}: Gain of the fiber repeater

All of the path loss from BTS to the FR is L_p. Therefore, the noise power signal injected by the FR in BTS receiver, P'_{rep} [16] can be written as (4)

$$\begin{aligned} P'_{rep} &= P_{rep} - L_p \\ &= 10\log[KTB] + P_{th,bts} + NF_{rep} + G_{rep} - L_p \\ &= -119dBm + P_{th,bts} + NF_{rep} + G_{rep} - L_p \quad (4) \end{aligned}$$

When FR is injected in the BTS receiver, the total noise power, $P_{btstotal}$ [17], can be presented as (5)

$$\begin{aligned} P_{bts,total} &= P_{bts} - P'_{rep} \\ &= 10\log[10P_{bts} + 10P'_{rep}] \\ &= P_{bts} + 10log\left[1 + 10^{\frac{NFrep+NF_{bts}+Grep-L_p}{10}}\right] \quad (5) \\ &= Pbts + \Delta Nbts \end{aligned}$$

Where

$$\Delta N_{bts} = 10log\left[1 + 10^{\frac{NFrep+NF_{bts}+Grep-L_p}{10}}\right]$$

The above analysis can demonstrate that the injected FR in BTS receiver increases the noise power more than one without FR. Furthermore, many factors affect the noise floor in BTS receivers.

The related parameters include the noise figure of BTS, the noise figure of the FR, the gain of FR, and the path loss from BTS to FR. The equation can be written as (5).

Case 1: When NF_{rep}, $NF_{bts} + G_{rep}$, $L_p = 0$, the noise floor has increased by ΔN_{bts}, which equals 3 dB, in receiver port of BTS.

Case 2: When NF_{rep}, $NF_{bts} + G_{rep}$, $L_p = -6$, then the noise floor has increased by ΔN_{bts}, which equals 0.97 dB, in receiver port of BTS. Therefore, the invaded FR is not an influence on BTS receiver.

In general, the noise figure of BTS, NF_{bts} is 2 dB. Following equation (3), the BTS noise power injected by the FR P_{rep} is -125dBm.

In the engineering practice, the noise figures of BTS and FR are constant, the noise floor increase of BTS receives port, and ΔN_{bts} has an influence on the path loss of the BTS transmitter to FR and FR G_{rep}.

The noise figure of BTS, NF_{bts}, is 2 dB. The noise figure of FR, NF_{rep}, is 5 dB as usual. The gain of FR, G_{rep}, should be lower than 8 dB path loss. By limiting G_{rep}, the ΔN_{bts} would be controlled under 1 dB.

In the network design, if the range of the objective coverage is relatively great, it needs to connect in parallel to a lot of FRs. In this case, the noise floor of BTS receiver, $P_{btstotal}$, is the noise figure and the thermal noise power signal injected by the FR due to BTS receive port, P'_{rep} shown to be $P_{reptotal} = 10Log[10P_{bts} + 10(P'_{rep})]$. The quantity of FR is n. In order to control the $P_{reptotal}$, the ΔN_{bts} must be under 1dB. The BTS noise floor has increased by FR as shown in Table 1.

Table 1. The BTS noise floor increased by FR.

$10Log[1+n10^{(NF_{rep}-NF_{bts}-G_{rep}-L_p)/10}]dB$

n	**ΔN_{bts}:(n) $NF_{rep} - NF_{bts} + G_{rep} - L_p$(dB)**	**P_{bts}: The thermal noise power of BTS receive(dB)**
0	0	3
1	−6	1.77
2	−9	0.98

The ΔN_{bts} signal increment by FR due to the BTS receiver is equal. For the case of n FRs injected in BTS receiver, the noise power P'_{rep}, contributed by one of FRs, should be followed as $P'_{rep} < -125 - 10\text{Log } n$.

If every one of the FRs has the same path loss L_p, the FR gain in n FRs, is smaller than one in one FR. In n FRs, the dropped value for FR gain is about 10log *n*. Generally, UL gain of FR is set in the BTS receiver by some considerations:

(1) Path loss L_p from BTS transmitter to FR,

(2) Numbers of FR in parallel to BTS receiver.

In general, the FR directly couples to the signal, which comes from BTS, and the path loss of FR is L_p for the coupling of the coupler. The same principle, UL of FR gains, needs more than a small coupling loss of about 8 dB.

Two parallel fiber repeaters would be needed in practice. In general, it could be chosen a high coupling ratio of coupler, when the input optic fiber signal of FR is 0 dBm.

Therefore, the path loss L_p coming from BTS to FR is 39 dB. The FR of UL gain is 27 dB, so FR invades the BTS receive port whose sensitivity is not influenced.

The calculation of the UL budget FR is shown in Table 2.

4 SOLUTION OF THE MDP AND RRH

A. System structure and link budget

Real case 1: MRT R3–R8 implement and measurement practices.

1. Indoor coverage requirements and inspection methods.

Measurements were made in three different areas. In MRT underground environment, including multiservices stations and tunnels of quality [18], the reliability of indoor signal objective coverage should be above 95% for requirements and acceptances. That can be listed in Table 3.

2. System structure and link budget

The SCE system of SON network CoMP structure needs tunnel environment factor and signal objective

Table 2. The calculation of the link budget FR.

The calculation of the link budget fiber repeater	
UL Grep (dB)	**UL Grep (dB) = (Lp) + (NF_{bts}) – (NF_{rep}) + Δ*Nbts*** = 39+2-5+(-9) = 27
n	**Δ*Nbts*: number of fiber repeater (n)**
0	0
1	−6
2	−9
NF_{bts}(dB)	2
NF_{rep} (dB)	5
Path Loss (Lp)(dB)	**Path Loss (Lp) = Coupling Loss + Cable Loss + Combiner Loss 39**
Coupling Loss(dB)	30
Cable Loss(dB)	2
Combiner Loss(dB)	7

coverage plan to establish an automated neighbor relation function (ANRF) and configuration of physical cell identity (PCI) for optimization of mobility and data efficiency using multiple radio access and layer technology (MultiRAT-Layer) [12],[19] as shown in Figure 4.

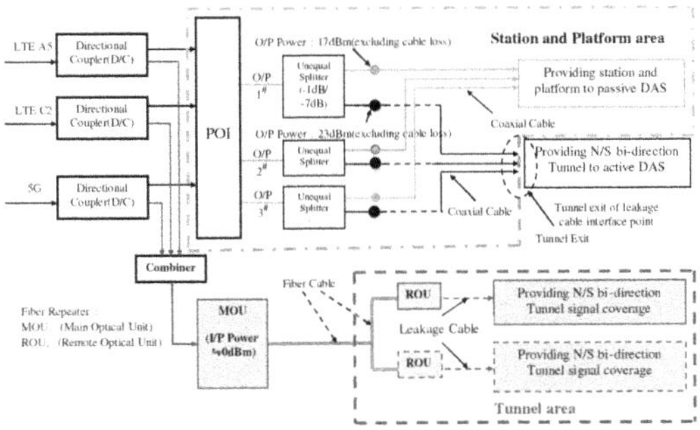

Figure 4. The MRT in-building coverage of SCE system network architecture for Optimization MultiRAT and Multi-Layer.

3. Power spectrum density of calculations

R4 to R4A leakage cable is installed at well number three of south and north tunnels is one respectively, leakage cable connected to ROU output port: first, to mobility load balancing (MLB) then improve overall cell throughput second, for mobility robustness optimization (MRO) improves handover successful ratio [12],[20]–[21] For example, R4 to R4A was shown in Figure 5.

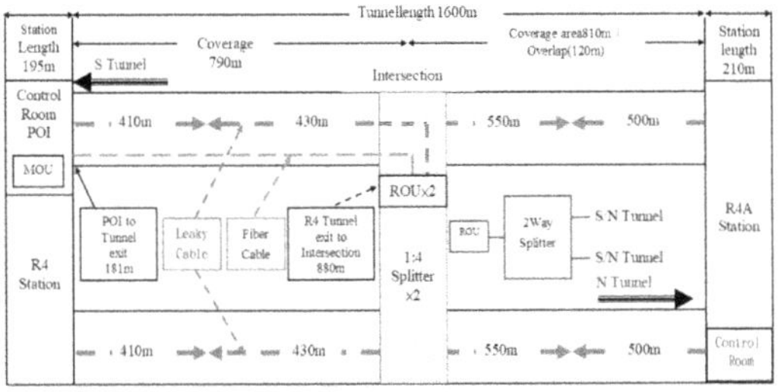

Figure 5. The SCE system structure to optimize MLB and MRO.

Real case 2: R4 Station [R4 to R4A] Control room of POI + MOU × 1, N/S tunnel intersection exit ROU × 2. Control room, intersection exit and system network plan calculation of link budget of D2D architecture for optimization coverage and capacity optimization (CCO) improve system coverage and throughput to solve coverage hole, weak coverage and overshooting problems [12]–[22], as shown in Figure 6.

1. Station R4 to 4A of Power spectrum density

(1) R4 Station: 195 m, R4 Station BTS provided R4 to R4A in tunnel signal coverage. Control room to

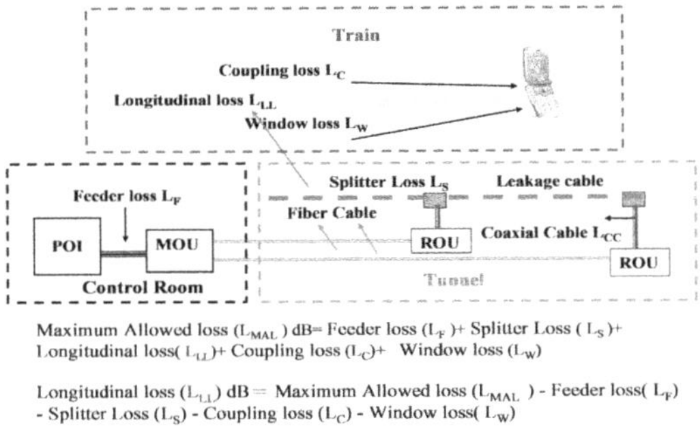

Figure 6. System network plan and calculation of link budget of Device to Device (D2D) architecture for optimization CCO.

Table 3. The quality plan in indoor coverage requirements and acceptances.

System	Area A	SSreq dBm	Area B	SSreq dBm	Area C	SSreq dBm	Coverage (%)
LTE-A	Tunnel, Station and Platform	≧−80	Parking area Hall, Shopping mall, Information service, Sell ticket area, Electric escalator	≧−85	ElevatorControl room, METRO machine room, Toilet, office, Exit on tunnel	≧−85	95%
5G		≧−93		≧−92		≧−98	95%
LTE-A 5G		≧−85	Tunnel and LaBa area			≧−93	97%
							90%

tunnel exit is 181 m, 1-5/8~, coaxial cable loss is 8 dB, and tunnel length is 1,600 m.

(2) Intersection: In the south, two ROU pieces of equipment are installed; in the north, one. Point of interface (POI) output port connects to MOU. MOU output ports connect fiber cable to the intersection ROU input port.

(3) R4 to R4A leakage cable is installed at well number three of the south tunnel and well number three of the north tunnel.The leakage cable connects to the ROU output port.

2. R4 to 4A in the tunnel of signal coverage canalization

(1) R4 tunnel exit to tunnel interior; R4A tunnel exit to tunnel interior; POI+1-5/8; leakage cable:

The practical analysis of the signal coverage area from the POI to the N/S tunnel intersection exit connected to the 1-5/8" leakage cable of the equation can be written as (6), (7), and (8).

$$P_{o(ROU)} = P_{t(bts)} - L_{POI} - L_{CC} \quad (6)$$

Where

$P_{o(ROU)}$ is the output power of ROU to leakage cable.

$P_{t(bts)}$ is the output power of BTS.

L_{POI} is insertion loss of POI.

L_{CC} is coaxial cable loss (about 181m).

$P_{o(ROU)} = 33\text{dBm}-9\text{dB}-8\text{dB}=16\text{dBm}$

$$L_{TOT} = P_{o(ROU)} - SS_{req} - S_{req} \quad (7)$$

Where:

L_{TOT} is the total maximum allowed loss.

SS_{req} is signal strength requirement (−85 dB)

$L_{TOT(-85dB)} = 16 - (-85dB) = 101dB$

$$L_{LL} = \boldsymbol{L_{TOT}} - L_F - L_S - L_C - L_W \quad (8)$$

Where

L_{LL} is the longitudinal loss.

L_F is the Feeder loss. (5 dB)

L_S is the Splitter Loss. (3.5 dB)

L_C is the coupling loss. (66 dB)

L_W is the Window loss. (12 dB)

$L_{LL} = 101 - 5 - 3.5 - 62 - 12 = 22$ (dB)

(2) Intersection Exit to R4 inside of tunnel, Intersection exit to R4A inside of tunnel:

The practical analysis of signal coverage area from intersection exit ROU connected to 1-5/8″ leakage cable.

Following equation (8).

Intersection Exit to R4 inside of tunnel:

$Po_{ROU} = 27$dBm, Splitter × 2 (1:4).

1 5/8″ leakage cable is 480 m.

$L_{TOT(-85dB)} = 27, (-85 \text{ dB}) = 112$ dB

$L_{LL} = 112 - 4 - 7 - 66 - 12 = 23$ (dB).

Intersection Exit to R4A inside of tunnel:

1 5/8″ leakage cable is 600 m.

$L_{LL} = 112 - 4 - 1 - 66 - 12 = 29$ (dB).

B. Measurement in practice

1. System Structure and Cell Coverage Plan

The system network structure needs a tunnel environment factor and signal objective coverage plan, as shown in Figure 4.

2. Measurement Tools

The measurement even was made walking in the TRA R3 station and moving on a train along the TRA red line southern tunnel from R3 station through four stations R4–R6 to R7 station. The measurement was conducted using Ericsson's test mobile system (ETMS). It consists of an HP portable computer, HTC/Samsung Apple mobile unit, and the ETMS investigation 10.1 data collection software. A detailed description of ETMS can be found in reference [23]. Equipped with ETMS, one can measure the performance of any LTE/5G mobile communication network [24]. ETMS mobile guarantees accuracy, mentioned in the 3GPP TS specifications [25]–[26]. According to that, it can measure with absolute accuracy of ±5dB from −105dBm to −40dBm under normal conditions. The dynamic range for measurement of the received level is −110dBm to −35dBm. However, in dedicated mode, ETMS can measure a maximum Rx Level of −13dBm.

3. Measurement Results

Typical measurements include coverage of a cell is estimated by CPICH and received signal code power (RSCP) for the primary common pilot channel. The standard deviation of RSCP caused by free space attenuation together with slow and fast fading indicates the smoothness of RSCP in the network. The large variations in RSCP lead to greater slow-fading margins. The quality of the coverage is indicated by CPICH-*Ec/No*, the received energy per chip on CPICH divided by the power density in the band. The CPICH- *Ec/No* is defined as the ratio between RSCP and received signal strength indicator (RSSI) of measured quality indicators for the SCE network system.

(1) The part of station and platform

CPICH-RSCP is used to evaluate the signal strength of the SCE system. However, it should be known that a high RSCP recorded may well be due to the presence of high co-channel interference. The signal strength could be especially high if the interference component is in phase with the carrier. Assuming minimal effect from interference, RSCP should be -75 dBm ~ -100 dBm for an acceptable quality. CPICH-*Ec/No* is used to evaluate the transmission bit error rate of the cellular system. Table 4 shows the R4A station of the relation between CPICH-RSCP and CPICH-*Ec/No*.

Figure 7 shows RSCP on the secondlevel platform of the R4A station. The platform is covered by an antenna, which is placed downward in the middle of the platform ceiling. Mobile unit of RSCP and *Ec/No* measured walking around the platform for optimization of mobility and data efficiency. Furthermore, multiple radio access and layer technology (Multi-RAT-Layer) is in development [27].

(2) The part of the tunnel

The tester seat, near the wall on the train, was locatedaway from the leakage cable and, traveling through the tunnel as shown in Figure 8 that shows N tunnel measure position.

Table 4. R4A station of relation between CPICH-RSCP and CPICH-Ec/No.

RSCP Signal Level	Numbers	Percent	Cumulative	Percent to Cumulative	Ec/No Quality Index	Numbers	Percent	Cumulative	Percent to Cumulative
-35<=x	0	0.0	0	0.0	0	0	0.0	0	0.0
-45<=x<=-35	127	13.5	127	13.5	1-3	114	12.1	114	12.1
-55<=x<-45	153	29.9	280	43.4	4-6	140	27.1	254	39.2
-65<=x<-55	261	11.5	541	54.9	7-9	248	14.4	502	53.6
-75<=x<-65	353	33.8	894	95.6	10-12	390	41.8	892	95.4
-85<=x<-75	33	2.3	927	99.0	13-15	33	3.5	925	98.9
-95<=x<-85	8	0.8	935	100.0	16-18	10	1.1	935	100.0
-110<=x<-95	0	0	935	0.0	19-21	0	0	935	0.0

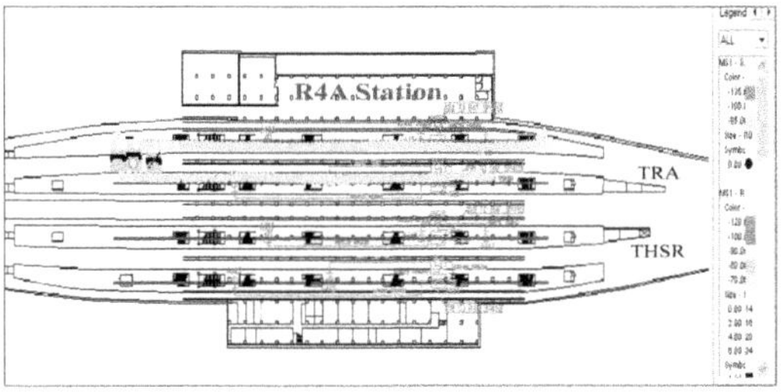

Figure 7. RSCP on the second level platform of the R4A station.

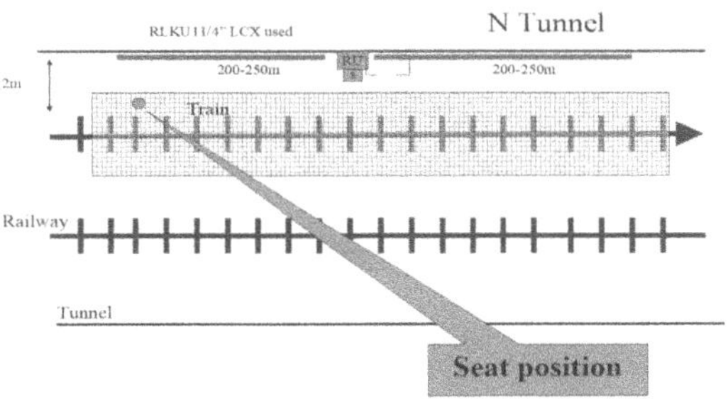

Figure 8. N tunnel measure position.

CPICH-Ec/No is used to evaluate the transmission bit error rate of the SCE system. Table 5 shows the R3 to R4A N tunnel relation between CPICH-RSCP and CPICH-Ec/No.

Figure 9 shows the RSCP on the second level N tunnel of the R3 to R4A station. The N tunnel is covered by leakage cable, which Andrew radiated mode leakage cable was installed 25 cm from the southern-sided wall of the tunnel and 3.5 m high from the floor of the tunnel.

Following tunnel acceptances of the quality plan, it includes: Train tracking tests of the bi-direction of MLB and MRO functions for each system twice. In the tunnel, signal handoff is smooth. If the signal intensity is high when cross-connection is made in accordance with standard regulation, then optimization is not required. Coverage and capacity optimization

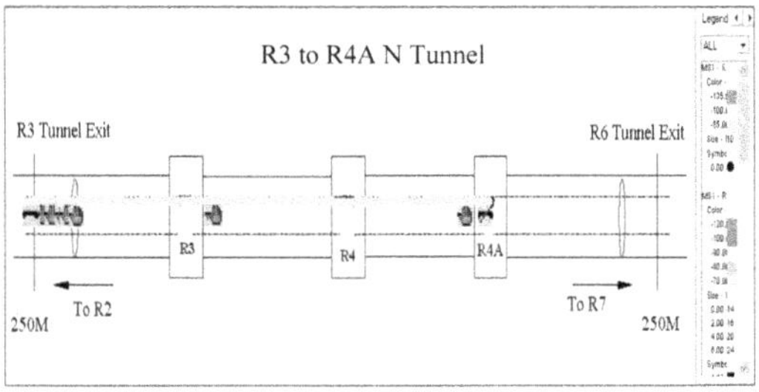

Figure 9. RSCP on the second level N Tunnel of the R3 to R4A station.

(CCO) improves system coverage, and the throughput performance function is excellent, as shown in Figure 10.

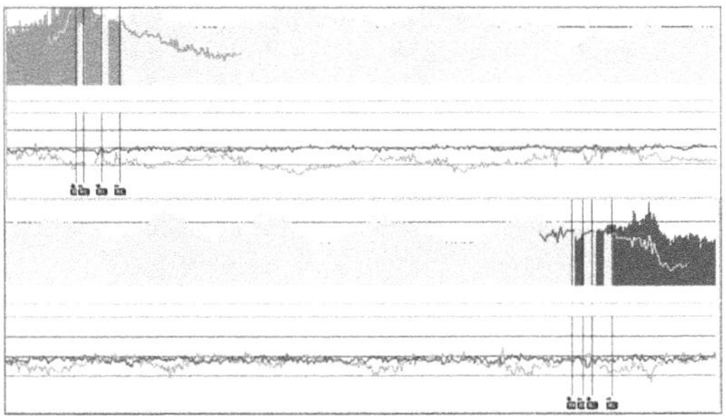

Figure 10. RSCP on the second level N Tunnel of the R3 to R4A station.

5 CONCLUSION

This paper expressed the design principles of FR in detail, which includes:

(1) Calculation of the gain to solve the MDP and RRH implementation issues.
(2) Power balance calculation between UL and DL.
(3) Establishment of the requirement for indoor coverage and measurement for acceptance procedure.
(4) Prove of Multi-RAT, Multi-Layer, MLB, MRO, and CCO system operation scenarios can solve issues with SON.

Table 5. R3 TO R4A N tunnel relation between CPICH-RSCP and CPICH-EC/NO.

RSCP Signal Level	Numbers	Percent	Cumulative	Percent to Cumulative	Ec/No Quality Index	Numbers	Percent	Cumulative	Percent to Cumulative
-35<=x	0	0.0	0	0.0	0	0	0.0	0	0.0
-45<=x<=-35	92	2.5	92	2.5	1-3	70	2.2	70	2.2
-55<=x<-45	321	10	413	12.9	4-6	606	16.7	676	18.9
-65<=x<-55	730	22.8	1143	35.7	7-9	359	13.4	1035	32.3
-75<=x<-65	1736	54.2	2879	95.9	10-12	2060	64.3	3095	96.6
-85<=x<-75	322	4	3201	99.9	13-15	105	3.3	3200	99.6
-95<=x<-85	3	0.1	3204	100.0	16-18	4	0.1	3204	100.0
-110<=x<-95	0	0	3204	0.0	19-21	0	0	3204	0.0

An achievement of tunnel engineering in indoor coverage toward CoMP operation scenarios in SCE system architecture helps design engineers and maintenance engineers properly perform planning analysis and maintenance operations.

Based on the above-mentioned points, MDP and RRH equipment can solve the transmission problem of open areas for indoor coverage of wireless communications.

REFERENCES

[1] NGMN Alliance Pushes for New Work Program", Press (Release 16), pp. 03–120, Mar. 2020.

[2] 3GPP TR 36.742 "Study on Self-Organizing Networks (SON) for enhanced Coordinated MultiPoint (eCoMP)," (Release 14), pp. 03–45, Aug. 2017.

[3] 3GPP TR 36.746 "Study on further enhancements to LTE Device to Device (D2D), UE to network relays for Internet of Things (IoT) and wearables," (Release 14), pp. 03–13, Aug. 2016.

[4] 3GPP TR 36.897, "Study on Elevation Beamforming/ Full-Dimension (FD) MIMO for LTE." (Release 13), pp. 18–30. Jun. 2015.

[5] 3GPP TS 36.176 "Enhanced CRS and 4Rx SU-MIMO Interference Mitigation Performance Requirements for LTE," (Release 15), pp. 18–30, Nov. 2016.

[6] B. J. Koshy, and P. M. Shankar, "Spread-Spectrum Techniques for Fiber-Fed Micro-cellular Networks," *IEEE Trans*. on Vehicular Technology, vol. VT-48, No. 3, May 1999, pp. 847–857.

[7] M. V. S. N. Prasad and R. Singh, "Terrestrial mobile communications train measurements in Western India," *IEEE Trans. Veh. Technol.*, vol. 52, no. 3, pp. 671–682, May 2003.

[8] Y. P. Zhang, "Practical performance of digital cellular system in mass rapid transit environments," *Int. J. Commun. Syst.*, vol. 18, pp. 143–157, Dec. 2004.

[9] H. A. Raweshidy and S. Komaki, *Radio over Fiber Technologies for Mobile Communications Networks*, Artech House, Norwood, MA, USA, pp. 183–214, 2002.

[10] S. T. Hwang, B. G. Kim, S. K. Lee, J. H. Lee, H. L. Kim, Y. G. Lee, and G. W. Kim, "ROF technologies for in-building wireless systems," *IEICE Transactions on Electronics*, vol. E90-C, no. 2, pp. 345–350 Feb. 2012.

[11] Matthew Baker, Chairman 3GPP TSG RAN WG1 "From Macro to Small Cells: Enhancements for Small Cells in 3GPP" Small Cells Summit 2013.

[12] 3GPP TR 36.819, "Coordinated multipoint operation for LTE physical layer aspects."

[13] Andrew Company, "ION-M85/18/21 System Overview," Worldwide Headquarters 3 Westbrook Corporate Center Suite 900/1800/2100/2300 Westchester, Illinois, USA, 2016.

[14] 3GPP TS 25.143, V6.4.0 "Radio Access Network UTRA Repeater Conformance Testing" (Release 6), pp. 27–28, Mar, 2006.

[15] Andrew Company, "ION-M85/18/21 Design Principle," Worldwide Headquarters 3 Westbrook Corporate Center Suite 900/1800/2100/2300 Westchester, Illinois, USA, 2017.

[16] Theodore S.Rappaport, *Wireless Communication Principles & Practice*, Prentice-Hall, NJ, pp. 300–320, 1996.

[17] K. A. Persson, A. Alping ,and D. Wake "WCDMA radio-over-fiber transmission experiment using electro-absorption transceiver," *IEE Electronics Letters*, vol. 41, no. 13, pp. 512–514, Jun. 2005.

[18] D. Didascalou, J. Maüer, and W. Wiesbeck, "Subway tunnel guided electromagnetic wave propagation at mobile communications frequencies," *IEEE Trans. Antennas Propag.*, vol. 49, no. 11, pp. 1590–1596, Nov. 2001.

[19] M. Lienard and P. Degauque, "Propagation in wide tunnels at 2 GHz: A statistical analysis," *IEEE Trans. Veh. Technol.*, vol. 47, no. 4, pp. 1322–1328, Nov. 1998.

[20] P. Mariage, M. Lienard, and P. Degauque, "Theoretical and experimental approach of the propagation of high frequency waves in road tunnels," *IEEE Trans. Antennas Propag.*, vol. 42, no. 1, pp. 75–81, Jan. 1994.

[21] C. L. Holloway, D. A. Hill, R. A. Dalke, and G. A. Hufford, "Radio wave propagation characteristics in lossy circular waveguides such as tunnels, mine shafts and boreholes," *IEEE Trans. Antennas Propag.*, vol. 48, no. 9, pp. 1354–1366, Sep. 2000.

[22] 3GPP RP-122009 "Study on LTE Device to Device Proximity," December 2012.

[23] C. H. Cox, *Analog Optical Links Theory and Practice*, Cambridge University Press, pp. 159–200, Dec, 2000.

[24] Ericsson Review, "Signal Strength measurement in LTE,LTE-A and 5G TEMS Mobiles," 284 23–3026 Uen Rev AB, Taiwan, pp. 1–45. 2020.

[25] 3GPP TS 25.106, V6.4.0 "Radio Access Network UTRA repeater radio transmission and reception," (Release 15), pp. 20–23, Mar. 2016.

[26] 3GPP TR 36.842, "Study on Small Cell enhancements for E-UTRA and E-UTRAN – Higher layer aspects."

[27] T. Isotalo, J. Niemela, and J. Lempiainen, "Measurements for Distributed Antennas in WCDMA Indoor Network" pp. 1–5, Jan. 2020.

System Innovation for a Troubled World – Tak Lam et al. (Eds)
© 2023 The Editor(s), ISBN: 978-1-032-45525-9

Author index

Smart Science, Design and Technology

The main goal of this series is to publish research papers in the application of "Smart Science, Design & Technology". The ultimate aim is to discover new scientific knowledge relevant to IT-based intelligent mechatronic systems, engineering and design innovations. We would like to invite investigators who are interested in mechatronics and information technology to contribute their original research articles to these books.

Mechatronic and information technology, in their broadest sense, are both academic and practical engineering fields that involve mechanical, electrical and computer engineering through the use of scientific principles and information technology. Technological innovation includes IT-based intelligent mechanical systems, mechanics and systems design, which implant intelligence to machine systems, giving rise to the new areas of machine learning and artificial intelligence.

ISSN : 2640-5504
eISSN : 2640-5512

1. Engineering Innovation and Design: Proceedings of the 7th International Conference on Innovation, Communication and Engineering (ICICE 2018), November 9–14, 2018, Hangzhou, China

   Edited by Artde Donald Kin-Tak Lam, Stephen D. Prior, Siu-Tsen Shen, Sheng-Joue Young & Liang-Wen Ji

   ISBN: 978-0-367-02959-3 (Hbk + multimedia device)
   ISBN: 978-0-429-01977-7 (eBook)
   DOI: https://doi.org/10.1201/9780429019777

2. Smart Science, Design & Technology: Proceedings of the 5th International Conference on Applied System Innovation (ICASI 2019), April 12–18, 2019, Fukuoka, Japan

   Edited by Artde Donald Kin-Tak Lam, Stephen D. Prior, Siu-Tsen Shen, Sheng-Joue Young & Liang-Wen Ji

   ISBN: 978-0-367-17867-3 (Hbk)
   ISBN: 978-0-429-05812-7 (eBook)
   DOI: https://doi.org/10.1201/9780429058127

3. Innovation in Design, Communication and Engineering: Proceedings of the 8th Asian Conference on Innovation, Communication and Engineering (ACICE 2019), October 25–30, 2019, Zhengzhou, P.R. China

   Edited by Artde Donald Kin-Tak Lam, Stephen D. Prior, Siu-Tsen Shen, Sheng-Joue Young & Liang-Wen Ji

   ISBN: 978-0-367-17777-5 (Hbk)
   ISBN: 978-0-429-05766-3 (eBook)
   DOI: https://doi.org/10.1201/9780429057663

4. System Innovation in a Post-Pandemic World: Proceedings of the IEEE 7th International Conference on Applied System Innovation (ICASI 2021), September 24–25, 2021, Alishan, Taiwan

   Edited by Artde Kin-Tak Lam, Stephen D Prior, Sheng-Joue Young, Siu-Tsen Shen, Liang-Wen Ji

   ISBN: 978-1-032-24392-4 (Hbk)
   ISBN: 978-1-003-27847-4 (eBook)
   DOI: https://doi.org/10.1201/9781003278474

5. System Innovation for a Troubled World: Applied System Innovation VIII. Proceedings of the IEEE 8th International Conference on Applied System Innovation (ICASI 2022), April 21–23, 2022, Sun Moon Lake, Nantou, Taiwan

   Edited by Artde Donald Kin-Tak Lam, Stephen D. Prior, Siu-Tsen Shen, Sheng-Joue Young & Liang-Wen Ji

   ISBN: 978-1-032-45525-9 (Hbk)
   ISBN: 978-1-003-37739-9 (eBook)
   DOI: https://doi.org/10.1201/9781003377399

For Product Safety Concerns and Information please contact our EU
representative GPSR@taylorandfrancis.com
Taylor & Francis Verlag GmbH, Kaufingerstraße 24, 80331 München, Germany

www.ingramcontent.com/pod-product-compliance
Ingram Content Group UK Ltd.
Pitfield, Milton Keynes, MK11 3LW, UK
UKHW052211190926
475702UK00006B/972
* 9 7 8 1 0 3 2 4 5 5 2 5 9 *